CISM Courses and Lectures

The series presents lecture notes, monographs, edited works and proceedings in the field of Mechanics, Engineering, Computer Science and Applied Mathematics.
Purpose of the series is to make known in the international scientific and technical community results obtained in some of the activities organized by CISM, the International Centre for Mechanical Sciences.

International Centre for Mechanical Sciences

Courses and Lectures Vol. 560

For further volumes:
www.springer.com/series/76

Holm Altenbach · Tomasz Sadowski
Editors

Failure and Damage Analysis of Advanced Materials

Springer

Editors

Holm Altenbach
Otto von Guericke University Magdeburg, Magdeburg, Germany

Tomasz Sadowski
Lublin University of Technology, Lublin, Poland

ISSN 0254-1971
ISBN 978-3-7091-1834-4 ISBN 978-3-7091-1835-1 (eBook)
DOI 10.1007/978-3-7091-1835-1
Springer Wien Heidelberg New York Dordrecht London

All contributions have been typeset by the authors
Printed in Italy

Printed on acid-free paper

Springer is part of Springer Science+Business Media (www.springer.com)

PREFACE

Failure as a limit state of the material behavior is well known from engineering practice. Different types of failure can be identified: transition from the elastic to plastic state, loss of stiffness, loss of fracture resistance at different scale levels, ultimate strength, and fatigue. In addition, failure can be accompanied by various types of damage. The course was discussed basic concepts and new developments in failure and damage analysis with focus on advanced materials such as composites, laminates, sandwiches and foams, and also new metallic materials. Starting from some mathematical foundations (limit surfaces, symmetry considerations, invariants) new experimental results and their analysis will be presented. Finally, new concepts for failure prediction and analysis were introduced and discussed.

The classical strength criteria developed intensively in the 19th and 20th century are mostly based on the comparison of the stress state (usually three-dimensional) with some scalar-valued properties estimated in tests. Such a phenomenological approach can be easily extended to other types of limit states of a material (for example, plastic behavior, and damage or fracture toughness). But even in the case of classical, but anisotropic structural materials, predictions are not always satisfactory and the effort required for their experimental confirmation can increase dramatically. Furthermore, in the case of advanced materials additional effects such as load dependent material response should be taken into account. These effects can induce mechanisms leading to different behavior in tension and compression.

Considering advanced metallic and non-metallic materials new methods of failure and damage prediction were discussed. Based on experimental results the traditional methods will be revised. In some cases it is enough to extend the classical approaches (for example, for metallic sheet material). In other situations (foams, composites) this is not satisfying since the different mechanisms cannot be adequately presented.

The lecture notes contains 5 parts. Part 1 (Classical and Non-Classical Failure Criteria) was prepared by Holm Altenbach & Vladimir Kolupaev. The following items are discussed: examples of failure behavior, theory of invariants and symmetry, classical isotropic models, compressibility and incompressibility, non-classical , and anisotropic *models. Part 2 (Constitutive Description of Isotropic and Anisotropic Plasticity for Metals) is written by Frédéric Barlat & Myoung-Gyu Lee and contains: modeling of advanced metallic materials, plasticity in metallic materials, isotropic and anisotropic yield criteria, state variable evolution and hardening, influence of constitutive description on failure prediction. Liviu Marsavina presented in his Part 3 (Failure and Damage in Cellular Materials): behavior of cellular materials in compression and tensile, fracture toughness of cellular materials under static and dynamic loading, effect of density, forming direction, loading speed and size effect, predicting properties of cellular materials using micromechanical models, comparison between polymer and metallic foams behavior. Neil McCartney (Part 4: Analytical Methods of Predicting Performance of Composite Materials) presents: predicting properties of undamaged lamina, predicting properties of undamaged laminates, principles controlling fracture processes in composites, prediction of ply cracking in general symmetric laminates, prediction of ply cracking in laminates subject to loading that includes bend deformation, some other important issues. Ramesh Talreja (Part 5: Analysis of Failure in Composite Structures) discusses the following problems: clarification of strength, fracture and damage in heterogeneous solids, role of constraint in lamina failure, homogenization and representative volume element concepts, continuum damage and internal variables, damage modes, thermodynamics framework for composite response with damage, damage evolution, synergistic damage mechanics. During the course were presented 6 lectures by Tomasz Sadowski on damage and failure criteria for micromechanical modeling of multiphase polycrystalline composites and joints of different materials, multiscale approach in material modeling, deformation damage theory defects initiation and propagation, experimental verification of damage and failure criteria in complex materials, modeling of hybrid joints of structural parts degradation with application of cohesive zone model. The lectures were not published by health reasons. People interested in these lectures can contact directly Tomasz Sadowski (`sadowski.t@gmail.com`).*

Last but not least we have to thank Mrs. Dr.-Ing. Anna Girchenko. She unified all manuscripts, which were finally submitted as LaTeXfiles.

Holm Altenbach and Tomasz Sadowski

CONTENTS

Classical and Non-Classical Failure Criteria

Holm Altenbach* and Vladimir A. Kolupaev**

* Lehrstuhl für Technische Mechanik, Fakultät für Maschinenbau, Otto-von-Guericke-Universität, Magdeburg, Germany

** Fraunhofer Institute for Structural Durability and System Reliability LBF, Darmstadt, Germany

Abstract In material science or structural mechanics, failure is generally the loss of load carrying capacity of a material unit or structural element. This definition introduces the fact that failure can be examined in different scales (microscopic, mesoscopic, macroscopic). In addition, one has to distinguish among brittle, ductile, and intermediate material behavior. In structural mechanics, if the structural response is beyond the initiation of nonlinear material behavior, failure is related to the determination of the integrity of the structure.

In principle, failure criteria correspond to phenomenological material behavior modeling. They describe the occurrence of failure at different loading conditions. Although there are no physical principles on which failure criteria can be based on, there are still a lot of suggestions available in the literature. Similarly due to the lack of generally accepted failure criteria, the formulation is up to now under research.

The criteria based on the introduction of some empirical assumptions for critical values defined by the stress or strain state are denoted as the engineering one. In addition, characteristics of the stored strain energy or power can also be used. Based on some of these hypotheses and their consequences failure criteria will be discussed here.

1 Examples of Failure Behavior

As mentioned earlier, regarding failure behavior, one has to distinguish among absolute brittle, ideal ductile, and intermediate material behavior. The first one is related to fracture, while the second one to yield. The intermediate behavior includes the combined occurrence of the brittle and ductile failure and is related to the majority of materials. In addition to above failures, the variety of other types of failure will be briefly discussed.

H. Altenbach, T. Sadowski (Eds.), *Failure and Damage Analysis of Advanced Materials*,
CISM International Centre for Mechanical Sciences
DOI 10.1007/978-3-7091-1835-1_1 © CISM Udine 2015

1.1 Failure

Failure is related to the material and to the structure. In the first case the observation scale plays an important role hence various failure definitions exist and we have various evidences. The microscopic material failure is related to crack initiation, growth and propagation. As usually this approach can be applied to the fracturing of specimens and simple structures affected by well defined global loadings.

The most popular failure models are micro-mechanical models, which combine continuum mechanics and classical fracture mechanics (Besson et al., 2003). These models are based on the assumption that during inelastic deformation one should observe:

- microvoid nucleation and growth until local plastic neck or fracture of the intervoid matrix occurs, and
- coalescence of neighboring voids.

Finally, the macroscopic fracture results when macrocracks occurs. It is known that the first model of this type was proposed by Gurson (1977) and extended by Tvergaard and Needleman (Tvergaard, 1981, 1982; Needleman and Tvergaard, 1984; Tvergaard and Needleman, 1984; Needleman and Tvergaard, 1987). Another approach is based on continuum damage mechanics (CDM) and thermodynamics and was proposed by Rousselier (1981, 2001a,b).

Both models can be characterized as a modification of the von Mises yield potential (von Mises, 1913). The modification is based on the inclusion the damage behavior. The damage is represented by void volume fraction of cavities (porosity f). In this sense this concept is a combination of the phenomenological classical approach with some micromechanical elements.

Macroscopic material failure is defined in terms of critical load, strain or energy storage. Li (2001) presented the following classification of macroscopic failure:

- stress or strain failure,
- energy type failure,
- damage failure, and
- empirical described failure.

With respect to this classification different failure criteria can be formulated.

Regarding material behavior models as usual five observation scales are considered Li (2001):

- the structural element scale,
- the macroscopic scale where engineering stresses and strains are defined,
- the mesoscale which is represented by a typical void, small crack or inclusion,

- the microscale (scale of crystallites or grains), and
- the atomic scale.

In modern theories the material behavior at one level is considered as a collective of its behavior at a sublevel which corresponds to the Curie-Neumann principle (Neumann, 1885; Paufler, 1986; Voigt, 1910). An efficient deformation and failure model should be consistent at every level. Below the attention will be paid only on phenomenological criteria on the macroscopic or structural level because they reflect a lot of effects of the material behavior in a relatively simple way in engineering applications.

Different types of *"failure"* can be identified in the engineering practice:

- transition from the elastic to plastic state,
- loss of stiffness,
- loss of fracture resistance at different scale levels,
- ultimate strength,
- fatigue, etc.

In this sense failure means that the material approaches a certain limit state.

It is not so easy to find a suitable definition of failure since the its formulation depends, for example, on the application field. WIKIPEDIA offers the following explanation[1]:

Definition 1.1 (Failure - General statement)**.** Failure is the state or condition of not meeting a desirable or intended objective, and may be viewed as the opposite of success.

The same source gives another explanation for engineering applications.

Definition 1.2 (Failure - Engineering statement)**.** A engineering failure analysis is focussed on the questions how a component or product fails in service or if failure occurs in manufacturing or during production processing.

Last but not least let us introduce a specific statement.

Definition 1.3 (Failure in the Sense of the Course)**.** Failure is a limit state of the material behavior and/or loss of carrying capacity of structural element or the whole structure.

The last statement corresponds to the engineering practice. It means that the structure or elements of the structure are unable to fulfil all prescribed functions for some time. The limit state is defined with respect to the application case.

Such a statement can be related to the stress-strain diagram (Fig. 1). For example, if a structure can be exploited only in the elastic range the

[1] `http://en.wikipedia.org/wiki/Failure` (August 18th, 2014)

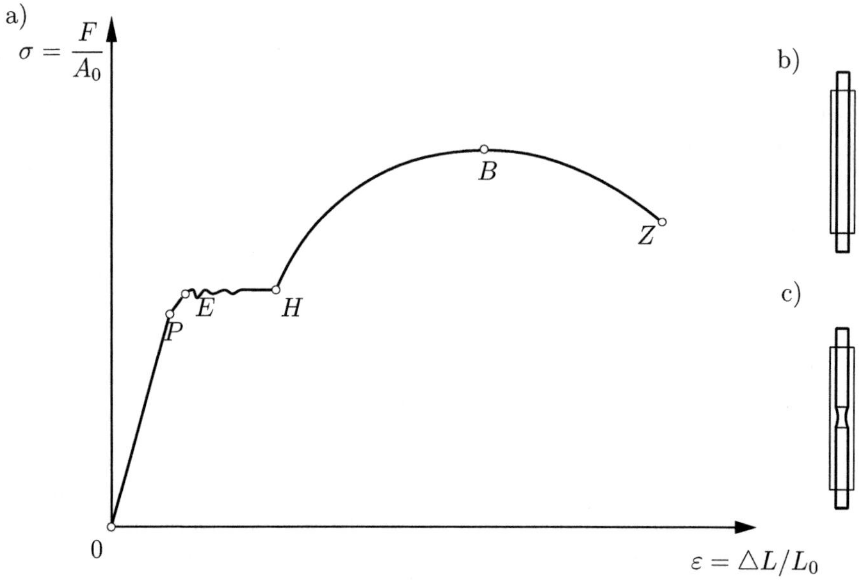

Figure 1. Stress-strain diagram for a ductile material: a) Engineering stresses σ vs. strains ε (P - proportional limit, E - elastic limit, H - beginning of hardening, B - ultimate strength, Z - rupture strength), b) Proportional elongation, c) Necking.

point P in the stress-strain diagram is the limit state. Other limit states are the transition from the elastic to the plastic range (point E), the beginning of necking (point B), the fracture (point Z), etc. Note that all these limit cases are related to the diagram which is experimentally estimated in an one-dimensional tension experiment. But this is an exceptional loading case in mechanical or civil engineering.

As usual we have multi-axial loading cases resulting various values of the stress tensor. The limit state should be independent from the values of the stress tensor components. That means we need invariant limit estimates instead of the limit values for each tensorial component which vary with the change of the coordinate system. In addition we have to notice, that for different materials we obtain different experimental stress-strain curves (Fig. 2). In Fig. 2 the following symbols are used: σ_{m} is the ultimative stress (strength) and σ_{y} is the yield stress. x denotes fracture at the fracture stress σ_{b}.

In the classical theory the material behavior at tension and compression is assumed to be the same (different signs, but the absolute values of the

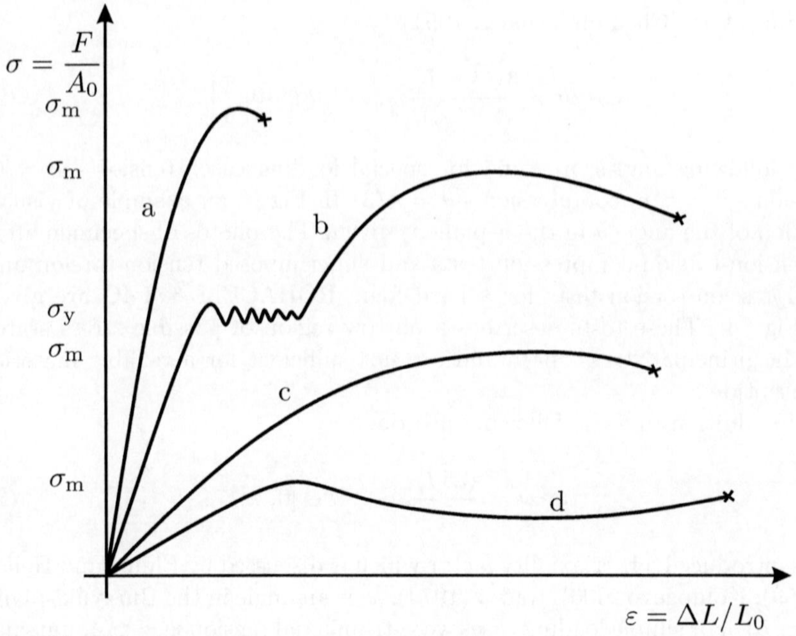

Figure 2. Various types of stress-strain diagrams: a) Brittle material (for example ceramics) with mostly elastic behavior, b) Ductile material with dominant hardening (for example mild steel), c) Ductile material without significant yield point (for example non-ferrous metals and alloys), d) Ductile material with dominant softening (for example plastics).

failure stresses are the same)

$$\sigma_+ \approx |\sigma_-|.$$

For torsion behavior it is accepted that the equivalent stress-strain diagram in the von Mises sense is the same like at tension or compression. Note that in this case

$$\tau_* = \frac{\sigma_+}{\sqrt{3}}. \tag{1}$$

Examples showing that such behavior cannot be observed for some construction materials and those are given in Altenbach et al. (1995a); Yu (2004).

For the realistic description of material behavior at complex loadings multi-axial tests are necessary. Let us introduce the measure of the complex loading state (Chen and Zhang, 1991)

$$\cos 3\theta = \frac{3\sqrt{3}}{2} \frac{I_3'}{(I_2')^{3/2}}, \qquad \theta \in \left[0, \frac{\pi}{3}\right]. \tag{2}$$

The following angles are valid for special loading case: tension - $\theta = 0$, torsion - $\theta = \pi/6$, compression - $\theta = \pi/3$. In Fig. 3 an example of visualization of the angle θ in the π-plane is given. The photos of specimen after a tension-torsion-compression tests and superimposed tension-torsion and compression-torsion tests for a hard foam ROHACELL® 71 IG are given in Fig. 4. These tests describes a narrow region of possible stress states in the principal stress space and are not sufficient for a reliable material description.

In addition to θ the following invariant

$$\tan \psi = \frac{\sqrt{3\, I_2'}}{I_1}, \qquad \psi \in [0, \pi] \tag{3}$$

was introduced, cf. triaxiality factor which is discussed by Finnie and Heller (1959); Kolupaev (2006); Radaj (1974). ψ is an angle in the Burzyńki-plane (Fig. 5). For simple loading cases we get: uniaxial tension $\psi = \pi/4$, uniaxial compression $\psi = -\pi/4$, torsion $\psi = \pi/2$, thinwalled tubular specimen with closed ends under inner pressure $\psi = \pi/6$, balanced biaxial tension $\psi = \arctan(1/2)$ among others. For the measurements in Fig. 4 it follows $\psi \in [-\pi/4, \pi/4]$.

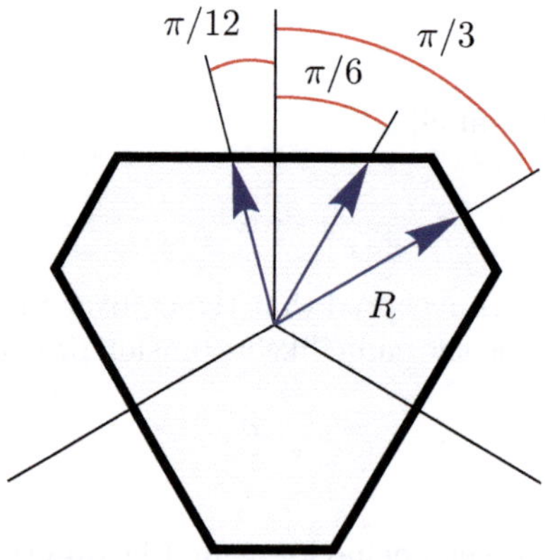

Figure 3. Model for incompressible material behavior of trigonal symmetry in the π-plane in polar coordinates $R(\theta)$.

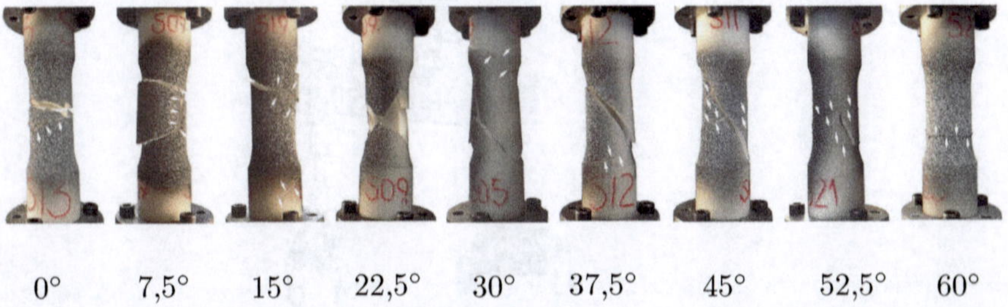

Figure 4. Tension-torsion-compression tests on tubular specimen with $d_{\text{inner}} = 30$ mm, $D_{\text{outer}} = 48$ mm in the test region (the loading state is characterized by angle θ), material: polymethacrylimide (PMI) hard foam ROHACELL® 71 IG, Evonik Röhm GmbH, Darmstadt.

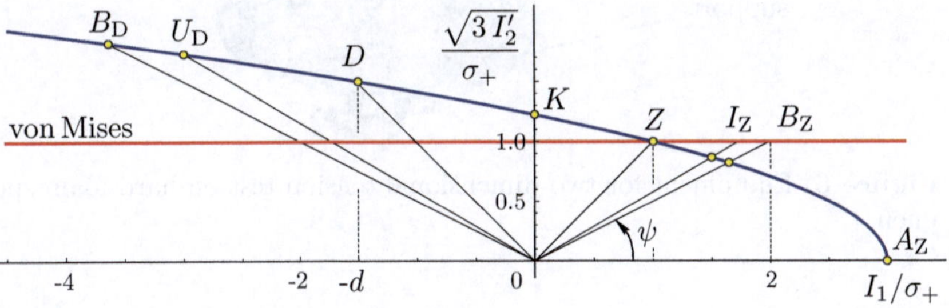

Figure 5. Rotationally symmetric models in the Burzyński-plane normalized with respect to σ_+: cylinder of von Mises (1913) and paraboloid of Balandin (1937), I_1/σ_+ - hydrostatic axis.

The stresses in the Burzyński-plane are normalized with respect to the failure stress at tension σ_+, so that the equivalent stress $\sigma_{\text{eq}} = 1$ results. The horizontal line describe the von Mises model. The whole surface is obtained by the rotation of this line about the axis I_1 (Życzkowski, 1981). It is obvious that this model does not restrict the hydrostatic stresses. The second line describes the rotationally symmetric paraboloid. This line contain the point A_Z, which limits the hydrostatic tension stress (balanced triaxial loading $\sigma_I = \sigma_{II} = \sigma_{III}$). Further points belong to the plane stress state: Z - tension, B_Z - balanced biaxial tension, K - torsion, D - compression, B_D - balanced biaxial compression. The points I_Z and U_D describe two tests on thin-walled tube specimens with closed ends under inner and outer pressure.

For formulation of improved failure criteria multi-axial tests in the entire region $\theta \in [0, \pi/3]$, $\psi \in [0, \pi]$ should be realized. The settings $\psi = 0$

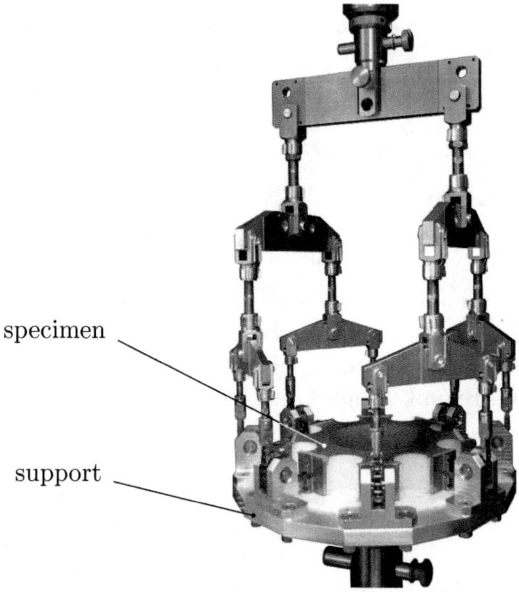

Figure 6. Equipment for two-dimensional tension test on hard foam specimen.

and $\psi = \pi$ describe hydrostatic tension and compression, respectively. Unfortunately, a hydrostatic tension test can be hardly performed (Kolupaev et al., 2013a) and further multi-axial tests are expensive. A simple two-dimensional test can be realized with the help of the equipment presented in Fig. 6. The loading cases $\sigma_{\rm I} = \sigma_{\rm II}$, $\sigma_{\rm III} = 0$ and $\sigma_{\rm I} = 2\,\sigma_{\rm II}$, $\sigma_{\rm III} = 0$ are suitable tests (see Fig. 7).

1.2 Need of Criteria

From the discussion in the previous subsection follows that the basic experimental observations are coming from one-dimensional tests (basic tests of material testing, see Blumenauer, 1996). In general the stress or strain states in a structure are multi-axial states. With the help of criteria one compares the multi-axial material behavior with a single selected property of the material, which is specific for the given material and influenced by the loading, temperature, etc. in each application. At the same time all other material properties (for example the Young modulus) are ignored. Sometime it is necessary to extend this assumption for a more accurate material

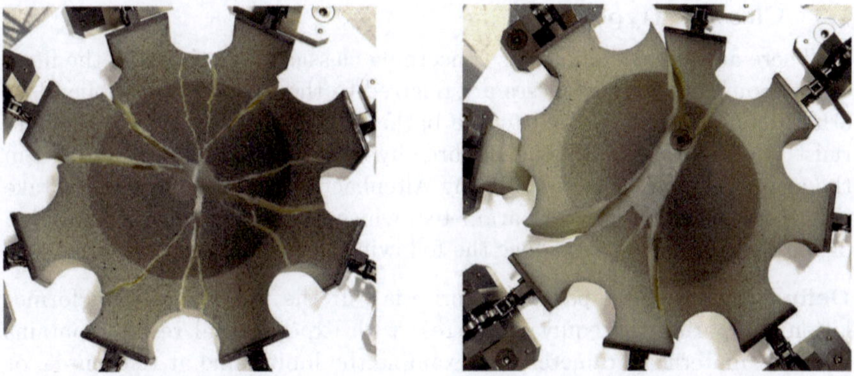

Figure 7. Biaxial tension loadings on hard foam. left: $\theta = \pi/3, \psi = \arctan(1/2)$ (material ROHACELL® 110 IG), right: $\theta = \psi = \pi/6$ (material ROHACELL® 71 IG).

modeling - possible ways of changing are shown below.

Failure criteria can be formulated as functions of stresses and strains. In Christensen (2013) are presented some arguments concerning the best choice. Let us assume that we have a stress-based criterion. In this case the equivalent stress plays the main role.

Definition 1.4 (Equivalent stress). If comparison of multi-axial and one-dimensional states is based on a stress criterion a scalar variable - the equivalent stress σ_{eq} should be introduced. This variable can be explicitly estimated for any arbitrary stress state and compared with a uniaxial state, for example tension test with the limit stress σ_+ (the value of σ_+ can be chosen with respect to Figs. 1 or 2). In this case the criterion can be formulated as follows

$$\sigma_{\text{eq}} = \sigma_+. \tag{4}$$

Instead of the equivalence expression (4) one can introduce formulations with other limit values on the right-hand side of Eq. (4). Such values are the limit value at compression σ_- or at torsion τ_*.

It is obvious that the equivalent stress expression should be not influenced by the loading direction if the material behavior is assumed to be isotropic. In this case we have to construct from the stress measure invariants of the stress tensor, which is a symmetric 2nd rank tensor. This results in formulations which depends on the maximum of three linear-independent invariants. In the case of anisotropic materials the formulation of equivalent stress expressions is more complicated.

1.3 Classical Hypotheses

There are several definitions concerning classical hypotheses in the literature. Some of the hypotheses are referred to the year of first publication, other definitions are related to the bachelor course in "Strength of Materials" (there are offered as usual three hypotheses which are often used in the engineering practice). Following Altenbach et al. (1995a) here we take into account the number of parameters which are included the formulation of the criteria. So we introduce the following definition.

Definition 1.5 (One-parameter criteria). If the failure criterion formulation comparing the equivalent stress with experimental results contains only one material parameter (for example the limit value at tension σ_+ or at torsion τ_*) the criterion is named one-parameter criterion.

This definition is often used for "classical criterion". In this sense this formal definition allows to distinguish classical and non-classical criteria by the number of parameters. Mostly as the material parameter the limit stress value at tension is used σ_+.

As usual three one-parameter failure hypotheses are presented in the bachelor course of "Strength of Materials". They result in the formulation of three criteria:

- the maximum of normal stress criterion (Rankine, 1876),
- the maximum of shear stress criterion (Tresca, 1868), and
- the maximum of distorsion energy criterion (von Mises, 1913).

Note that there are more one-parameter criteria in the literature, but their application is limited.

Any strength criteria should reflect the three-dimensional material behavior in a proper manner. The comparison of arbitrary stress states with material properties from one-dimensional test results is not reliable in many cases, if one has only one test from which one threshold can be established.

1.4 First Improvements

After establishing the above mentioned criteria the further development was characterized by the formulation of phenomenological criteria including, for instance, instead of one material parameter obtained for the limit state additional parameter or parameters have been adopted for better adjustment. Such criteria, for example, are the

- maximum strain criterion for brittle solids (Barré de Saint-Venant, 1871; Bach, 1902; Grashof, 1878; Filonenko-Borodich, 1960; Résal, 1898; Sauter and Wingerter, 1990; Timoshenko and Goodier, 1987),
- Mohr-Coulomb failure criterion for cohesive-frictional solids (Coulomb, 1776; Mohr, 1900a,b), and

- Drucker-Prager criterion for pressure-dependent solids (Drucker and Prager, 1952; Mirolyubov, 1953).

These criteria belong to the group of so-called two-parameter criteria. Following this classification in Altenbach et al. (1995a), criteria containing up to six parameters are introduced. This variation of formulation regarding strength criteria will be discussed in the Subsect. 4.1. In Altenbach et al. (2014) another formalism for constructing equivalent stress expressions has been suggested. Even in this case the criteria can be distinguished by the number of introduced material parameters. But there can be introduced much more parameters which consequently would result in identification difficulties.

All criteria approximate the given experimental data and extrapolate the experiences in the whole range of validity. An "exact" formulation of a limit criterion cannot be presented in general since one has the problem with the insufficient input information and scattering. In addition, there are no physical principles (for example, like balance equations in Continuum Mechanics) on which such formulation can be based.

A growing interest for constructing new criteria and enhancement of the experimental data can be observed in the recent years. A statistics concerning the increasing number of publications in this field is presented in Yu (2004) based on a literature survey. For a detailed discussion, analysis and systematization of various criteria we should present some mathematical tools. The main tools are the theory of invariants and the theory of symmetry with respect to the stress space. Some basics are given in the next section.

2 Invariants and Symmetries of the Stress Tensor

Invariants allows to formulate failure criteria not influenced by the choice of the coordinate system. In addition, the analysis of the capabilities of the criteria can be performed in a very simple manner. Study of the symmetries of the stress tensor allows to take into account the invariance of material properties under different types of transformations like rotation, reflexion and inversion. This chapter is based statements given in Betten (1987); Betten and Borrmann (1988); Betten (2001, 2008); Itskov (2007); Lebedev et al. (2010); Naumenko and Altenbach (2007); Schade and Neemann (2006); Zhilin (2012)[2].

[2]The chapter was reviewed by Alexander Bolchoun, who make a lot of suggestions for improvements. The authors thank so for his suggestions.

2.1 Invariants

In this subsection we pay attention on invariants which are necessary for failure criteria.

Definition 2.1 (Invariants). An invariant is a property of a class of mathematical objects that remains unchanged when transformations of a certain type are applied to the objects.

It will be shown later that the coefficients in the characteristic equation of a tensor are invariants. This can be easily checked comparing the coefficient's values for an arbitrary coordinate system and for the principal coordinate system. A counter-example in Mechanics can be given as follows: the stress tensor is represented as σ_{ij}. This representation is related to a coordinate system. Changing the coordinate system by a certain transformation all numbers σ_{ij} are changing. Assuming isotropic material behavior the invariants of the stress tensor play an important role in the formulation of the equivalent stress expression. Here we define several sets of invariants mostly used in practical situations (Altenbach et al., 1995a; Życzkowski, 1981).

Principal Invariants. The first example are the so-called principal invariants which are obtained from the solution of the eigenvalue problem for symmetric second-rank tensors. Let us postulate that the stress state is defined by the stress tensor $\boldsymbol{\sigma}$. In classical continuum mechanics this tensor is a symmetric second-rank tensor. The principal invariants follow from

$$(\boldsymbol{\sigma} - \lambda \boldsymbol{I}) \cdot \boldsymbol{n} = \boldsymbol{0}, \qquad \boldsymbol{n} \neq \boldsymbol{0}. \tag{5}$$

$\boldsymbol{I}$ is the unit second rank tensor, $\boldsymbol{n}$ denotes the principal direction (eigendirection) and λ is the principal stress (eigenvalue) of the problem. In what follows the eigenvalues are named principal stresses. For the symmetrical stress tensor the principal stresses are real-valued. Three different cases should be distinguished:

- three different solutions,
- one single and one double solution, and
- one triple solution.

A value λ is a solution of the problem (5) if and only if:

$$|\boldsymbol{\sigma} - \lambda \boldsymbol{I}| = \det(\boldsymbol{\sigma} - \lambda \boldsymbol{I}) = 0. \tag{6}$$

Equation (6) is of importance since the principal stresses can be computed with the help of this condition. This equation can be represented in the form (characteristic polynomial):

$$\lambda^3 - I_1(\boldsymbol{\sigma})\lambda^2 + I_2(\boldsymbol{\sigma})\lambda - I_3(\boldsymbol{\sigma}) = 0, \tag{7}$$

where $I_i(\boldsymbol{\sigma}), i = 1, 2, 3)$ are the principal invariants of the stress tensor (Itskov, 2007; Lebedev et al., 2010)

$$\begin{aligned}
I_1(\boldsymbol{\sigma}) &= \boldsymbol{I} \cdot\cdot\, \boldsymbol{\sigma}, \\
I_2(\boldsymbol{\sigma}) &= \frac{1}{2}\left[I_1^2(\boldsymbol{\sigma}) - I_1\left(\boldsymbol{\sigma}^2\right)\right], \\
I_3(\boldsymbol{\sigma}) &= |\boldsymbol{\sigma}| = \det \boldsymbol{\sigma} = \frac{1}{3}\left[I_1\left(\boldsymbol{\sigma}^3\right) + 3I_1(\boldsymbol{\sigma})I_2(\boldsymbol{\sigma}) - I_1^3(\boldsymbol{\sigma})\right].
\end{aligned} \tag{8}$$

Further the principal values (principal stresses) $\lambda_i, i = 1, 2, 3$ of the stress tensor are denoted by $\sigma_{\mathrm{I}}, \sigma_{\mathrm{II}}, \sigma_{\mathrm{III}}$. The following order is assumed

$$\sigma_{\mathrm{I}} \geq \sigma_{\mathrm{II}} \geq \sigma_{\mathrm{III}}. \tag{9}$$

Because $\boldsymbol{\sigma}$ is symmetric, the exists the spectral decomposition

$$\boldsymbol{\sigma} = \sigma_{\mathrm{I}}\boldsymbol{n}_{\mathrm{I}}\boldsymbol{n}_{\mathrm{I}} + \sigma_{\mathrm{II}}\boldsymbol{n}_{\mathrm{II}}\boldsymbol{n}_{\mathrm{II}} + \sigma_{\mathrm{III}}\boldsymbol{n}_{\mathrm{III}}\boldsymbol{n}_{\mathrm{III}}, \tag{10}$$

where $\boldsymbol{n}_{\mathrm{I}}, \boldsymbol{n}_{\mathrm{II}}, \boldsymbol{n}_{\mathrm{III}}$ are the eigendirections (principal directions), which can be obtained from the solution of

$$(\boldsymbol{\sigma} - \sigma_i \boldsymbol{I}) \cdot \boldsymbol{n}_i = \boldsymbol{0}, \qquad ||\boldsymbol{n}_i|| = 1. \tag{11}$$

The eigendirections are pairwise orthogonal that is $\boldsymbol{n}_i \cdot \boldsymbol{n}_j = \delta_{ij}$, where δ_{ij} is the Kronecker symbol. When the three principal stresses are distinct the eigendirections can be defined in a unique manner. For the case of multiple principal stresses the eigendirections are not unique, but the eigenvectors $\boldsymbol{n}_i$ are orthogonal.

Another set of invariants are the irreducible invariants (Schade and Neemann, 2006).

Definition 2.2 (Irreducible invariants)**.** If an invariant can be represented as a polynomial of other invariants this invariant is named reducible, otherwise irreducible.

The stress tensor has three irreducible invariants. The following set was suggested in Betten (1987):

- the linear invariant

$$J_1(\boldsymbol{\sigma}) = I_1(\boldsymbol{\sigma}) = \boldsymbol{I} \cdot\cdot\, \boldsymbol{\sigma},$$

- the quadratic invariant

$$J_2(\boldsymbol{\sigma}) = \boldsymbol{\sigma} \cdot\cdot\, \boldsymbol{\sigma},$$

- the cubic invariant

$$J_3(\boldsymbol{\sigma}) = (\boldsymbol{\sigma} \cdot \boldsymbol{\sigma}) \cdot\cdot\, \boldsymbol{\sigma}.$$

It is obvious that that the principal invariants can be expressed by these invariants as follows

$$\begin{aligned} I_1 &= J_1, \\ I_2 &= \frac{1}{2}(J_2 - J_1^2), \\ I_3 &= \frac{1}{6}(2J_3 - 3J_2J_1 + J_1^3). \end{aligned}$$

Alternately the following representation is also possible

- the linear invariant

$$J_1(\boldsymbol{\sigma}) = I_1(\boldsymbol{\sigma}) = \boldsymbol{I} \cdot\cdot\, \boldsymbol{\sigma},$$

- the quadratic invariant

$$J_2(\boldsymbol{\sigma}) = \frac{1}{2}\boldsymbol{\sigma} \cdot\cdot\, \boldsymbol{\sigma},$$

- the cubic invariant

$$J_3(\boldsymbol{\sigma}) = \frac{1}{3}(\boldsymbol{\sigma} \cdot \boldsymbol{\sigma}) \cdot\cdot\, \boldsymbol{\sigma}.$$

This set is a modification of the set of invariants suggested by Betten (1987).

In applications, for example in the theory of plasticity, the axiatoric-deviatoric invariants are often used. The axiatoric-deviatoric invariants are based on the principal invariants of the stress deviator, which can be computed from the eigenvalue problem for the deviator

$$\boldsymbol{s} = \boldsymbol{\sigma} - \frac{1}{3}\boldsymbol{\sigma} \cdot\cdot\, \boldsymbol{I} = \boldsymbol{\sigma} - \frac{1}{3}I_1(\boldsymbol{\sigma})\boldsymbol{I} \tag{12}$$

The principal deviatoric stresses result from

$$|\boldsymbol{s} - \lambda\boldsymbol{I}| = \det(\boldsymbol{s} - \lambda\boldsymbol{I}) = 0 \tag{13}$$

or

$$\lambda^3 - I_1(\boldsymbol{s})\lambda^2 + I_2(\boldsymbol{s})\lambda - I_3(\boldsymbol{s}) = 0, \tag{14}$$

which can be simplified since $I_1(\boldsymbol{s}) = 0$:

$$\lambda^3 + I_2(\boldsymbol{s})\lambda - I_3(\boldsymbol{s}) = 0. \tag{15}$$

The second and the third invariants are equal to

$$\begin{aligned} I_2(\boldsymbol{s}) &= -\frac{1}{2} J_2\left(\boldsymbol{s}^2\right), \\ I_3(\boldsymbol{s}) &= |\boldsymbol{s}| = \det \boldsymbol{s} = \frac{1}{3} J_3\left(\boldsymbol{s}^3\right). \end{aligned} \tag{16}$$

For a better separation of incompressible and compressible material behavior in models the second, the third deviatoric and the axiatoric invariant

$$I_1(\boldsymbol{\sigma}) = \boldsymbol{\sigma} \cdot\cdot \boldsymbol{I} \tag{17}$$

will be used. In the case of incompressible behavior $I_1(\boldsymbol{\sigma}) = 0$. It is easy to show that $I_2(\boldsymbol{s})$ and $I_3(\boldsymbol{s})$ are reducible invariants.

The next example of a set of invariants are the cylindrical invariants. They are introduced in Novozhilov (1951a,b) and defined as follows

- the axiatoric invariant (17)

$$I_1(\boldsymbol{\sigma}) = \boldsymbol{\sigma} \cdot\cdot \boldsymbol{I},$$

- the second invariant of the stress deviator (15)

$$I_2(\boldsymbol{s}) = -\frac{1}{2} J_2(\boldsymbol{s}),$$

- and the stress angle (2)

$$\cos 3\theta = \frac{3\sqrt{3}}{2} \frac{\det \boldsymbol{s}}{(I_2'(\boldsymbol{s}))^{3/2}}, \qquad \theta \in \left[0, \frac{\pi}{3}\right], \tag{18}$$

These invariants are widely discussed in Chen and Zhang (1991); Ottosen and Ristinmaa (2005); Życzkowski (1981) among others. This belongs also to the reducible invariants.

Remark 2.3. There are other sets of invariants. The so-called spherical invariants are presented, for example, in Tamuzh and Lagzdyn'sh (1968); Lagzdin and Tamuzh (1971); Lagzdiņš et al. (1992); Lagzdiņš and Tamužs (1996). Applications for these invariants are given in the phenomenological theory of fracture or plasticity. The advantage of these invariants that they can be used for isotropic and anisotropic materials.

2.2 Orthogonal Transformations

Orthogonal transformations in three-dimensional Euclidean space are stiff rotations, reflections, or combinations of a rotation and a reflection[3]. Reflections are transformations that exchange left and right, similar to mirror images. The determinant of tensors corresponding to proper rotations (without reflection) have the value +1. In the case of reflection the determinant have the value -1. An application is to find a basic set of scalar invariants for a given group of symmetry transformations, such that each invariant relative to the same group is expressible as a single-valued function of the basic set.

Definition 2.4 (Functional basis)**.** The functional basis is a set of scalar functions of the stress tensor $\boldsymbol{\sigma}$: $\{f_1(\boldsymbol{\sigma}), f_2(\boldsymbol{\sigma}), \ldots, f_n(\boldsymbol{\sigma})\}$, which are not changing under given transformation and which are functionally independent and lose their functional independency if extended by other functions.

Many applications are related to the modeling of the material behavior:

- Classical continua
  Such a problem arises in the formulation of constitutive equations for a given group of material symmetries. For example, the strain energy density of an elastic non-polar material is a scalar valued function of the deformation gradient tensor (Bower, 2010). In the special case of small strains instead of the deformation gradient tensor, the second rank symmetric strain tensor is used. For an isotropic material the strain energy density function depends only on the invariants, which follows from the principle of material frame indifference.
- Cosserat continuum
  In the theory of Cosserat continuum two strain measures are introduced, where the first strain measure is a polar tensor while the second one is an axial tensor (Eremeyev et al., 2013).
- Shell theory
  The strain energy density of a thin elastic shell is a function of two second rank tensors and one vector (Eremeyev and Pietraszkiewicz, 2006).

In all cases the problem is to find a minimum set of functionally independent invariants for the considered tensorial arguments. Details of the theory of tensor functions are discussed, for example, in Boehler (1987), reviews on representations of tensor functions are given in Rychlewski and Zhang (1991); Zheng (1994).

[3]Similar discussions can be given for the two-dimensional Euclidean space. Note that in this case so-called oriented surfaces play an important role, which result in more complicated formulaes (Zhilin, 2003, 2012).

Definition 2.5 (Orthogonal transformation). An orthogonal transformation of a scalar α, a vector $\boldsymbol{a}$ and a second rank tensor $\boldsymbol{A}$ is defined by

$$\alpha' \equiv (\det \boldsymbol{Q})^{\zeta}\alpha, \quad \boldsymbol{a}' \equiv (\det \boldsymbol{Q})^{\zeta}\boldsymbol{Q}\cdot\boldsymbol{a}, \quad \boldsymbol{A}' \equiv (\det \boldsymbol{Q})^{\zeta}\boldsymbol{Q}\cdot\boldsymbol{A}\cdot\boldsymbol{Q}^{\mathrm{T}}, \tag{19}$$

where $\boldsymbol{Q}$ is an orthogonal tensor, i.e. $\boldsymbol{Q}\cdot\boldsymbol{Q}^{\mathrm{T}} = \boldsymbol{I}$, $\det \boldsymbol{Q} = \pm 1$, $\boldsymbol{I}$ is the second rank unit tensor, $\zeta = 0$ for absolute (polar) scalars, vectors and tensors and $\zeta = 1$ for axial ones.

More details concerning polar and scalar objects one can find in Schade and Neemann (2006); Zhilin (2012) among others. We have the following examples concerning these two notions:

1. An example of the axial scalar is the mixed product of three polar vectors, i.e. $\alpha = \boldsymbol{a}\cdot(\boldsymbol{b}\times\boldsymbol{c})$.
2. A typical example of the axial vector is the cross product of two polar vectors, i.e. $\boldsymbol{c} = \boldsymbol{a}\times\boldsymbol{b}$.
3. An example of the second rank axial tensor is the skew-symmetric tensor $\boldsymbol{W} = \boldsymbol{a}\times\boldsymbol{I}$, where $\boldsymbol{a}$ is a polar vector.

Definition 2.6 (Orthogonal invariant of a second rank tensor). Consider a group of orthogonal transformations S (e.g., the material symmetry transformations) characterized by a set of orthogonal tensors $\boldsymbol{Q}$ closes with respect to multiplication. A scalar-valued function of a second rank tensor $f = f(\boldsymbol{A})$ is called to be an orthogonal invariant under the group S if

$$\forall \boldsymbol{Q} \in S: \quad f(\boldsymbol{A}') = (\det \boldsymbol{Q})^{\eta} f(\boldsymbol{A}), \tag{20}$$

where $\eta = 0$ if values of f are absolute scalars and $\eta = 1$ if values of f are axial scalars.

Any second rank tensor $\boldsymbol{B}$ can be decomposed into a symmetric and a skew-symmetric part, i.e.

$$\boldsymbol{B} = \frac{1}{2}(\boldsymbol{B}+\boldsymbol{B}^{\mathrm{T}}) + \frac{1}{2}(\boldsymbol{B}-\boldsymbol{B}^{\mathrm{T}}) = \boldsymbol{A} + \boldsymbol{a}\times\boldsymbol{I},$$

where $\boldsymbol{A}$ is a symmetric tensor and $\boldsymbol{a}$ is an associated vector. Therefore $f(\boldsymbol{B}) = f(\boldsymbol{A},\boldsymbol{a})$. For a set of symmetric second rank tensors and vectors the definition of an orthogonal invariant (20) can be generalized as follows $\forall \boldsymbol{Q} \in S$

$$f(\boldsymbol{A}'_1,\boldsymbol{A}'_2,\ldots,\boldsymbol{A}'_n,\boldsymbol{a}'_1,\boldsymbol{a}'_2,\ldots,\boldsymbol{a}'_k) = (\det \boldsymbol{Q})^{\eta} f(\boldsymbol{A}_1,\boldsymbol{A}_2,\ldots\boldsymbol{A}_n,\boldsymbol{a}_1,\boldsymbol{a}_2,\ldots,\boldsymbol{a}_k). \tag{21}$$

2.3 Invariants for the Full Orthogonal Group

Orthotropy is a type of the symmetry transformation with several applications in constitutive modeling of welds, metal forming processes (metal sheet rolling), etc. Zhilin (2003) presents orthogonal invariants for different sets of second rank tensors and vectors with respect to the full orthogonal group. It is shown that orthogonal invariants are integrals of a generic partial differential equation (basic equations for invariants). Examples of orthogonal invariants are:

- for a symmetric second rank tensor $\boldsymbol{A}$

$$J_k = \mathrm{tr}\boldsymbol{A}^k, \quad k = 1, 2, 3$$

  Note that instead of J_k it is possible to use also the principal values I_k.
- for a symmetric second rank tensor $\boldsymbol{A}$ and a vector $\boldsymbol{a}$

$$\begin{array}{rclrcl} J_k &=& \mathrm{tr}\boldsymbol{A}^k, & k &=& 1,2,3, \\ J_4 &=& \boldsymbol{a}\cdot\boldsymbol{a}, & J_5 &=& \boldsymbol{a}\cdot\boldsymbol{A}\cdot\boldsymbol{a}, \\ J_6 &=& \boldsymbol{a}\cdot\boldsymbol{A}^2\cdot\boldsymbol{a}, & J_7 &=& \boldsymbol{a}\cdot\boldsymbol{A}^2\cdot(\boldsymbol{a}\times\boldsymbol{A}\cdot\boldsymbol{a}). \end{array} \tag{22}$$

  In this set of invariants only 6 are functionally independent. The relation between the invariants can be formulated as follows

$$J_7^2 = \begin{vmatrix} J_4 & J_5 & J_6 \\ J_5 & J_6 & \boldsymbol{a}\cdot\boldsymbol{A}^3\cdot\boldsymbol{a} \\ J_6 & \boldsymbol{a}\cdot\boldsymbol{A}^3\cdot\boldsymbol{a} & \boldsymbol{a}\cdot\boldsymbol{A}^4\cdot\boldsymbol{a} \end{vmatrix}, \tag{23}$$

  where $\boldsymbol{a}\cdot\boldsymbol{A}^3\cdot\boldsymbol{a}$ and $\boldsymbol{a}\cdot\boldsymbol{A}^4\cdot\boldsymbol{a}$ can be expressed by J_l, $l = 1, \ldots 6$ applying the Cayley-Hamilton theorem.

2.4 Invariants for the Transverse Isotropy Group

Transverse isotropy is an important type of the symmetry transformation due to a variety of applications. Transverse isotropy is usually assumed in constitutive modeling of fiber reinforced materials, fiber suspensions, directionally solidified alloys, deep drawing sheets and piezoelectric materials. The invariants and generating sets for tensor-valued functions with respect to transverse isotropy are discussed, for example, in Bruhns et al. (1999). Here the idea proposed in Zhilin (2003, 2012) for the invariants with respect to the full orthogonal group is applied to the case of transverse isotropy. The invariants will be found as integrals of the generic partial differential equations. Although a functional basis formed by these invariants does not include any redundant element, functional relations between them may exist.

Consider a proper orthogonal tensor which represents a rotation about a fixed axis, i.e.

$$\boldsymbol{Q}(\varphi\boldsymbol{m}) = \boldsymbol{m}\boldsymbol{m} + \cos\varphi(\boldsymbol{I} - \boldsymbol{m}\otimes\boldsymbol{m}) + \sin\varphi\boldsymbol{m}\times\boldsymbol{I}, \quad \det\boldsymbol{Q}(\varphi\boldsymbol{m}) = 1. \tag{24}$$

$\boldsymbol{m}$ is a constant unit vector (axis of rotation), φ is the angle of rotation about $\boldsymbol{m}$. The symmetry transformation defined by this tensor corresponds to the transverse isotropy, whereby five different cases are possible (Spencer, 1987). Let us find scalar-valued functions of a second rank symmetric tensor $\boldsymbol{A}$ satisfying the condition

$$f\left[\boldsymbol{A}'(\varphi)\right] = f\left[\boldsymbol{Q}(\varphi\boldsymbol{m})\cdot\boldsymbol{A}\cdot\boldsymbol{Q}^{\mathrm{T}}(\varphi\boldsymbol{m})\right] = f(\boldsymbol{A}). \tag{25}$$

In Eq. (25) only the left-hand side depends on φ. Therefore

$$\frac{\mathrm{d}f}{\mathrm{d}\varphi} = \frac{\mathrm{d}\boldsymbol{A}'}{\mathrm{d}\varphi}\left(\frac{\partial f}{\partial\boldsymbol{A}'}\right)^{\mathrm{T}} = 0. \tag{26}$$

The derivative of $\boldsymbol{A}'$ with respect to φ can be calculated by the following rules

$$\begin{aligned}\mathrm{d}\boldsymbol{A}'(\varphi) &= \mathrm{d}\boldsymbol{Q}(\varphi\boldsymbol{m})\cdot\boldsymbol{A}\cdot\boldsymbol{Q}^{\mathrm{T}}(\varphi\boldsymbol{m}) + \boldsymbol{Q}(\varphi\boldsymbol{m})\cdot\boldsymbol{A}\cdot\mathrm{d}\boldsymbol{Q}^{\mathrm{T}}(\varphi\boldsymbol{m}),\\ \mathrm{d}\boldsymbol{Q}(\varphi\boldsymbol{m}) &= \boldsymbol{m}\times\boldsymbol{Q}(\varphi\boldsymbol{m})\mathrm{d}\varphi \Rightarrow \mathrm{d}\boldsymbol{Q}^{\mathrm{T}}(\varphi\boldsymbol{m}) = -\boldsymbol{Q}^{\mathrm{T}}(\varphi\boldsymbol{m})\times\boldsymbol{m}\,\mathrm{d}\varphi.\end{aligned} \tag{27}$$

By inserting these equations into Eq. (26) we obtain

$$(\boldsymbol{m}\times\boldsymbol{A} - \boldsymbol{A}\times\boldsymbol{m})\cdot\cdot\left(\frac{\partial f}{\partial\boldsymbol{A}}\right)^{\mathrm{T}} = 0. \tag{28}$$

Equation (28) is a linear homogeneous first order partial differential equation which characteristic system is

$$\frac{\mathrm{d}\boldsymbol{A}}{\mathrm{d}s} = (\boldsymbol{m}\times\boldsymbol{A} - \boldsymbol{A}\times\boldsymbol{m}). \tag{29}$$

Any system of n linear ordinary differential equations has not more than $n-1$ functionally independent integrals (Courant, 1989). In our case of a second rank tensor which has 6 independent components we can obtain 5 integrals. With the basis $\boldsymbol{e}_i$ the second rank tensor $\boldsymbol{A}$ can be represented in the form $\boldsymbol{A} = A^{ij}\boldsymbol{e}_i\boldsymbol{e}_j$. Equation (29) is a system of six ordinary differential equations with respect to the coordinates A^{ij}. The five integrals of (29) may be written down as follows

$$g_i(\boldsymbol{A}) = c_i, \quad i = 1, \ldots, 5,$$

where c_i are integration constants. Any linear combination of these five integrals is also a solution of the partial differential equation (28). Therefore the five integrals g_i represent the invariants of the symmetric tensor $\boldsymbol{A}$ with respect to the symmetry transformation (24). The solutions of (29) are

$$\boldsymbol{A}^k(s) = \boldsymbol{Q}(s\boldsymbol{m}) \cdot \boldsymbol{A}_0^k \cdot \boldsymbol{Q}^{\mathrm{T}}(s\boldsymbol{m}), \quad k = 1, 2, 3, \tag{30}$$

where $\boldsymbol{A}_0$ plays the role of the initial condition. In order to find the integrals, the variable s must be eliminated from Eq. (30). With

$$\begin{aligned} &\mathrm{tr}(\boldsymbol{Q} \cdot \boldsymbol{A}^k \cdot \boldsymbol{Q}^{\mathrm{T}}) = \mathrm{tr}(\boldsymbol{Q}^{\mathrm{T}} \cdot \boldsymbol{Q} \cdot \boldsymbol{A}^k) = \mathrm{tr}\boldsymbol{A}^k, \quad \boldsymbol{m} \cdot \boldsymbol{Q}(s\boldsymbol{m}) = \boldsymbol{m}, \\ &(\boldsymbol{Q} \cdot \boldsymbol{a}) \times (\boldsymbol{Q} \cdot \boldsymbol{b}) = (\det \boldsymbol{Q})\boldsymbol{Q} \cdot (\boldsymbol{a} \times \boldsymbol{b}) \end{aligned} \tag{31}$$

and using the notation $\boldsymbol{Q}_m \equiv \boldsymbol{Q}(s\boldsymbol{m})$ the integrals can be found as follows (Naumenko and Altenbach, 2007)

$$\begin{aligned} \mathrm{tr}(\boldsymbol{A}^k) &= \mathrm{tr}(\boldsymbol{A}_0^k), \quad k = 1, 2, 3, \\ \boldsymbol{m} \cdot \boldsymbol{A}^l \cdot \boldsymbol{m} &= \boldsymbol{m} \cdot \boldsymbol{Q}_m \cdot \boldsymbol{A}_0^l \cdot \boldsymbol{Q}_m^{\mathrm{T}} \cdot \boldsymbol{m} \\ &= \boldsymbol{m} \cdot \boldsymbol{A}_0^l \cdot \boldsymbol{m}, \quad l = 1, 2, \\ \boldsymbol{m} \cdot \boldsymbol{A}^2 \cdot (\boldsymbol{m} \times \boldsymbol{A} \cdot \boldsymbol{m}) &= \boldsymbol{m} \cdot \boldsymbol{Q}_m \cdot \boldsymbol{A}_0^2 \cdot \boldsymbol{Q}_m^{\mathrm{T}} \cdot (\boldsymbol{m} \times \boldsymbol{Q}_m \cdot \boldsymbol{A}_0 \cdot \boldsymbol{Q}_m^{\mathrm{T}} \cdot \boldsymbol{m}) \\ &= \boldsymbol{m} \cdot \boldsymbol{A}_0^2 \cdot \boldsymbol{Q}_m^{\mathrm{T}} \cdot \left[(\boldsymbol{Q}_m \cdot \boldsymbol{m}) \times (\boldsymbol{Q}_m \cdot \boldsymbol{A}_0 \cdot \boldsymbol{m})\right] \\ &= \boldsymbol{m} \cdot \boldsymbol{A}_0^2 \cdot (\boldsymbol{m} \times \boldsymbol{A}_0 \cdot \boldsymbol{m}) \end{aligned} \tag{32}$$

After some manipulations the six invariants of the tensor $\boldsymbol{A}$ with respect to the symmetry transformation (24) can be established

$$\begin{aligned} &J_k = \mathrm{tr}\boldsymbol{A}^k, \quad k = 1, 2, 3, \quad J_4 = \boldsymbol{m} \cdot \boldsymbol{A} \cdot \boldsymbol{m}, \\ &J_5 = \boldsymbol{m} \cdot \boldsymbol{A}^2 \cdot \boldsymbol{m}, \quad J_6 = \boldsymbol{m} \cdot \boldsymbol{A}^2 \cdot (\boldsymbol{m} \times \boldsymbol{A} \cdot \boldsymbol{m}) \end{aligned} \tag{33}$$

In Bruhns et al. (1999) for the case of the transverse isotropy six invariants are derived by the use of another approach. Note that only five invariants in Eqs. (33) are functionally independent. Taking into account that I_6 is the mixed product of vectors $\boldsymbol{m}$, $\boldsymbol{A} \cdot \boldsymbol{m}$ and $\boldsymbol{A}^2 \cdot \boldsymbol{m}$ the relation between the invariants can be written down as follows

$$J_6^2 = \det \begin{bmatrix} 1 & J_4 & J_5 \\ J_4 & J_5 & \boldsymbol{m} \cdot \boldsymbol{A}^3 \cdot \boldsymbol{m} \\ J_5 & \boldsymbol{m} \cdot \boldsymbol{A}^3 \cdot \boldsymbol{m} & \boldsymbol{m} \cdot \boldsymbol{A}^4 \cdot \boldsymbol{m} \end{bmatrix}. \tag{34}$$

One can verify that $\boldsymbol{m} \cdot \boldsymbol{A}^3 \cdot \boldsymbol{m}$ and $\boldsymbol{m} \cdot \boldsymbol{A}^4 \cdot \boldsymbol{m}$ are also transversely isotropic invariants. However, applying the the Cayley-Hamilton theorem they can be uniquely expressed by $I_1, I_2, \ldots I_5$ in the following way

$$\begin{aligned} \boldsymbol{m} \cdot \boldsymbol{A}^3 \cdot \boldsymbol{m} &= I_1 J_5 + I_2 J_4 + I_3, \\ \boldsymbol{m} \cdot \boldsymbol{A}^4 \cdot \boldsymbol{m} &= (I_1^2 + I_2) J_5 + (I_1 I_2 + I_3) J_4 + I_1 I_3, \end{aligned}$$

where I_1, I_2 and I_3 are the principal invariants of $\boldsymbol{A}$. Let us note that the invariant J_6 cannot be dropped (Naumenko and Altenbach, 2007).

To describe yielding and failure of oriented solids a dyad $\boldsymbol{M} = \boldsymbol{vv}$ has been used (Betten, 1985), where the vector $\boldsymbol{v}$ specifies a privileged direction. A potential is assumed to be an isotropic function of the symmetric stress tensor and the tensor generator $\boldsymbol{M}$. Applying the representation of isotropic functions the integrity basis (Betten, 2008) including ten invariants was found. In the special case $\boldsymbol{v} = \boldsymbol{m}$ the number of invariants reduces to the five $J_1, J_2, \ldots, J_5$ defined by Eqs. (33). Further details of this approach and applications in continuum mechanics are given by Boehler (1987). However, the problem is to find an integrity basis of a symmetric tensor $\boldsymbol{A}$ and a dyad $\boldsymbol{M}$, i.e. to find scalar valued functions $f(\boldsymbol{A}, \boldsymbol{M})$ satisfying the condition

$$\begin{gathered} f(\boldsymbol{Q} \cdot \boldsymbol{A} \cdot \boldsymbol{Q}^{\mathrm{T}}, \boldsymbol{Q} \cdot \boldsymbol{M} \cdot \boldsymbol{Q}^{\mathrm{T}}) = (\det \boldsymbol{Q})^{\eta} f(\boldsymbol{A}, \boldsymbol{M}), \\ \forall \boldsymbol{Q}, \quad \boldsymbol{Q} \cdot \boldsymbol{Q}^{\mathrm{T}} = \boldsymbol{I}, \quad \det \boldsymbol{Q} = \pm 1 \end{gathered} \tag{35}$$

essentially differs from the problem (25). In order to show this we take into account that the symmetry group of a dyad $\boldsymbol{M}$, i.e. the set of orthogonal solutions of the equation $\boldsymbol{Q} \cdot \boldsymbol{M} \cdot \boldsymbol{Q}^{\mathrm{T}} = \boldsymbol{M}$ includes the following elements

$$\begin{aligned} \boldsymbol{Q}_{1,2} &= \pm \boldsymbol{I}, \\ \boldsymbol{Q}_3 &= \boldsymbol{Q}(\varphi \boldsymbol{m}), \quad \boldsymbol{m} = \frac{\boldsymbol{v}}{|\boldsymbol{v}|}, \\ \boldsymbol{Q}_4 &= \boldsymbol{Q}(\pi \boldsymbol{n}) = 2\boldsymbol{nn} - \boldsymbol{I}, \quad \boldsymbol{n} \cdot \boldsymbol{n} = 1, \quad \boldsymbol{n} \cdot \boldsymbol{v} = 0, \end{aligned} \tag{36}$$

where $\boldsymbol{Q}(\varphi \boldsymbol{m})$ is defined by Eq. (24). The solutions of the problem (35) are at the same time the solutions of the following problem

$$f(\boldsymbol{Q}_i \cdot \boldsymbol{A} \cdot \boldsymbol{Q}_i^{\mathrm{T}}, \boldsymbol{M}) = (\det \boldsymbol{Q}_i)^{\eta} f(\boldsymbol{A}, \boldsymbol{M}), \quad i = 1, 2, 3, 4,$$

i.e. the problem to find the invariants of $\boldsymbol{A}$ relative to the symmetry group (36). However, Eqs. (36) includes much more symmetry elements if compared to the problem (25).

An alternative set of transversely isotropic invariants can be formulated by the use of the following decomposition (Bischoff-Beiermann and Bruhns, 1994; Bruhns et al., 1999)

$$\boldsymbol{A} = \alpha \boldsymbol{mm} + \beta(\boldsymbol{I} - \boldsymbol{mm}) + \boldsymbol{A}_{\mathrm{pD}} + \boldsymbol{tm} + \boldsymbol{mt}, \tag{37}$$

where α, β, $\boldsymbol{A}_{\mathrm{pD}}$ and $\boldsymbol{t}$ are projections of $\boldsymbol{A}$. With the projectors $\boldsymbol{P}_1 = \boldsymbol{mm}$ and $\boldsymbol{P}_2 = \boldsymbol{I} - \boldsymbol{mm}$ we may write

$$\begin{aligned}
\alpha &= \boldsymbol{m}\cdot\boldsymbol{A}\cdot\boldsymbol{m} = \mathrm{tr}(\boldsymbol{A}\cdot\boldsymbol{P}_1),\\
\beta &= \frac{1}{2}(\mathrm{tr}\boldsymbol{A} - \boldsymbol{m}\cdot\boldsymbol{A}\cdot\boldsymbol{m}) = \frac{1}{2}\mathrm{tr}(\boldsymbol{A}\cdot\boldsymbol{P}_2),\\
\boldsymbol{A}_{\mathrm{pD}} &= \boldsymbol{P}_2\cdot\boldsymbol{A}\cdot\boldsymbol{P}_2 - \beta\boldsymbol{P}_2,\\
\boldsymbol{t} &= \boldsymbol{m}\cdot\boldsymbol{A}\cdot\boldsymbol{P}_2
\end{aligned}\qquad(38)$$

The decomposition (37) is the analogue to the following representation of a vector $\boldsymbol{a}$

$$\boldsymbol{a} = \boldsymbol{I}\cdot\boldsymbol{a} = \boldsymbol{mm}\cdot\boldsymbol{a} + (\boldsymbol{I}-\boldsymbol{mm})\cdot\boldsymbol{a} = \psi\boldsymbol{m} + \boldsymbol{\tau}, \quad \psi = \boldsymbol{a}\cdot\boldsymbol{m}, \quad \boldsymbol{\tau} = \boldsymbol{P}_2\cdot\boldsymbol{a}. \qquad (39)$$

The projections introduced in Eqs. (38) have the following properties

$$\mathrm{tr}(\boldsymbol{A}_{\mathrm{pD}}) = 0, \quad \boldsymbol{A}_{\mathrm{pD}}\cdot\boldsymbol{m} = \boldsymbol{m}\cdot\boldsymbol{A}_{\mathrm{pD}} = \boldsymbol{0}, \quad \boldsymbol{t}\cdot\boldsymbol{m} = 0. \qquad (40)$$

With Eqs. (37) and (40) the tensor equation (29) can be transformed to the following system of equations

$$\begin{cases}
\dfrac{\mathrm{d}\alpha}{\mathrm{d}s} = 0,\\
\dfrac{\mathrm{d}\beta}{\mathrm{d}s} = 0,\\
\dfrac{\mathrm{d}\boldsymbol{A}_{\mathrm{pD}}}{\mathrm{d}s} = \boldsymbol{m}\times\boldsymbol{A}_{\mathrm{pD}} - \boldsymbol{A}_{pD}\times\boldsymbol{m},\\
\dfrac{\mathrm{d}\boldsymbol{t}}{\mathrm{d}s} = \boldsymbol{m}\times\boldsymbol{t}.
\end{cases}\qquad(41)$$

From the first two equations we observe that α and β are transversely isotropic invariants. The third equation can be transformed to one scalar and one vector equation as follows

$$\frac{\mathrm{d}\boldsymbol{A}_{\mathrm{pD}}}{\mathrm{d}s}\cdot\cdot\,\boldsymbol{A}_{\mathrm{pD}} = 0 \qquad\Rightarrow\qquad \frac{\mathrm{d}(\boldsymbol{A}_{\mathrm{pD}}\cdot\cdot\,\boldsymbol{A}_{\mathrm{pD}})}{\mathrm{d}s} = 0, \qquad \frac{\mathrm{d}\boldsymbol{b}}{\mathrm{d}s} = \boldsymbol{m}\times\boldsymbol{b}$$

with $\boldsymbol{b} \equiv \boldsymbol{A}_{\mathrm{pD}}\cdot\boldsymbol{t}$. We observe that $\mathrm{tr}(\boldsymbol{A}_{\mathrm{pD}}^2) = \boldsymbol{A}_{\mathrm{pD}}\cdot\boldsymbol{A}_{\mathrm{pD}}$ is a transversely isotropic invariant, too.

Finally, we have to find the integrals of the following system

$$\begin{cases}
\dfrac{\mathrm{d}\boldsymbol{t}}{\mathrm{d}s} = \boldsymbol{t}\times\boldsymbol{m},\\
\dfrac{\mathrm{d}\boldsymbol{b}}{\mathrm{d}s} = \boldsymbol{b}\times\boldsymbol{m}
\end{cases}\qquad(42)$$

The solutions of Eqs. (42) are

$$\boldsymbol{t}(s) = \boldsymbol{Q}(s\boldsymbol{m}) \cdot \boldsymbol{t}_0, \quad \boldsymbol{b}(s) = \boldsymbol{Q}(s\boldsymbol{m}) \cdot \boldsymbol{b}_0,$$

where $\boldsymbol{t}_0$ and $\boldsymbol{b}_0$ play the role of initial conditions. The vectors $\boldsymbol{t}$ and $\boldsymbol{b}$ belong to the plane of isotropy, i.e. $\boldsymbol{t} \cdot \boldsymbol{m} = 0$ and $\boldsymbol{b} \cdot \boldsymbol{m} = 0$. Therefore, one can verify the following integrals

$$\boldsymbol{t} \cdot \boldsymbol{t} = \boldsymbol{t}_0 \cdot \boldsymbol{t}_0, \quad \boldsymbol{b} \cdot \boldsymbol{b} = \boldsymbol{b}_0 \cdot \boldsymbol{b}_0, \quad \boldsymbol{t} \cdot \boldsymbol{b} = \boldsymbol{t}_0 \cdot \boldsymbol{b}_0, \quad (\boldsymbol{t} \times \boldsymbol{b}) \cdot \boldsymbol{m} = (\boldsymbol{t}_0 \times \boldsymbol{b}_0) \cdot \boldsymbol{m}. \tag{43}$$

We found seven integrals, but only five of them are functionally independent. In order to formulate the relation between the integrals we compute

$$\boldsymbol{b} \cdot \boldsymbol{b} = \boldsymbol{t} \cdot \boldsymbol{A}_{\mathrm{pD}}^2 \cdot \boldsymbol{t}, \qquad \boldsymbol{t} \cdot \boldsymbol{b} = \boldsymbol{t} \cdot \boldsymbol{A}_{\mathrm{pD}} \cdot \boldsymbol{t}.$$

For any plane tensor $\boldsymbol{A}_{\mathrm{p}}$ satisfying the equations $\boldsymbol{A}_{\mathrm{p}} \cdot \boldsymbol{m} = \boldsymbol{m} \cdot \boldsymbol{A}_{\mathrm{p}} = \boldsymbol{0}$ the Cayley-Hamilton theorem can be formulated as follows

$$\boldsymbol{A}_{\mathrm{p}}^2 - (\mathrm{tr}\boldsymbol{A}_{\mathrm{p}})\boldsymbol{A}_{\mathrm{p}} + \frac{1}{2}\left[(\mathrm{tr}\boldsymbol{A}_{\mathrm{p}})^2 - \mathrm{tr}(\boldsymbol{A}_{\mathrm{p}}^2)\right](\boldsymbol{I} - \boldsymbol{m}\boldsymbol{m}) = \boldsymbol{0}.$$

Since $\mathrm{tr}\boldsymbol{A}_{\mathrm{pD}} = 0$ we have

$$2\boldsymbol{A}_{\mathrm{pD}}^2 = \mathrm{tr}(\boldsymbol{A}_{\mathrm{pD}}^2)(\boldsymbol{I} - \boldsymbol{m}\boldsymbol{m}), \qquad \boldsymbol{t} \cdot \boldsymbol{A}_{\mathrm{pD}}^2 \cdot \boldsymbol{t} = \frac{1}{2}\mathrm{tr}(\boldsymbol{A}_{\mathrm{pD}}^2)(\boldsymbol{t} \cdot \boldsymbol{t})$$

Because $\mathrm{tr}(\boldsymbol{A}_{\mathrm{pD}}^2)$ and $\boldsymbol{t} \cdot \boldsymbol{t}$ are already defined, the invariant $\boldsymbol{b} \cdot \boldsymbol{b}$ can be omitted. The vector $\boldsymbol{t} \times \boldsymbol{b}$ is spanned on the axis $\boldsymbol{m}$. Therefore

$$\boldsymbol{t} \times \boldsymbol{b} = \gamma \boldsymbol{m}, \quad \gamma = (\boldsymbol{t} \times \boldsymbol{b}) \cdot \boldsymbol{m},$$
$$\gamma^2 = (\boldsymbol{t} \times \boldsymbol{b}) \cdot (\boldsymbol{t} \times \boldsymbol{b}) = (\boldsymbol{t} \cdot \boldsymbol{t})(\boldsymbol{b} \cdot \boldsymbol{b}) - (\boldsymbol{t} \cdot \boldsymbol{b})^2.$$

Now we can summarize six invariants and one relation between them as follows

$$\begin{aligned} \bar{J}_1 &= \alpha, \quad \bar{J}_2 = \beta, \quad \bar{J}_3 = \frac{1}{2}\mathrm{tr}(\boldsymbol{A}_{\mathrm{pD}}^2), \quad \bar{J}_4 = \boldsymbol{t} \cdot \boldsymbol{t} = \boldsymbol{t} \cdot \boldsymbol{A} \cdot \boldsymbol{m}, \\ \bar{J}_5 &= \boldsymbol{t} \cdot \boldsymbol{A}_{\mathrm{pD}} \cdot \boldsymbol{t}, \quad \bar{J}_6 = (\boldsymbol{t} \times \boldsymbol{A}_{\mathrm{pD}} \cdot \boldsymbol{t}) \cdot \boldsymbol{m}, \end{aligned} \tag{44}$$

and

$$\bar{J}_6^2 = \bar{J}_4^2 \bar{J}_3 - \bar{J}_5^2. \tag{45}$$

Let us assume that the symmetry transformation $\boldsymbol{Q}_n \equiv \boldsymbol{Q}(\pi \boldsymbol{n})$ belongs to the symmetry group of the transverse isotropy. In this case

$$f(\boldsymbol{A}') = f(\boldsymbol{Q}_n \cdot \boldsymbol{A} \cdot \boldsymbol{Q}_n^{\mathrm{T}}) = f(\boldsymbol{A})$$

must be valid. With $\boldsymbol{Q}_n \cdot \boldsymbol{m} = -\boldsymbol{m}$ we can write

$$\alpha' = \alpha, \qquad \beta' = \beta, \qquad \boldsymbol{A}'_{\mathrm{pD}} = \boldsymbol{A}_{\mathrm{pD}}, \qquad \boldsymbol{t}' = -\boldsymbol{Q}_n \cdot \boldsymbol{t}.$$

Therefore in Eqs. (44) $\bar{J}'_k = \bar{J}_k$, $k = 1, 2, \ldots, 5$ and

$$\begin{aligned}\bar{J}'_6 &= (\boldsymbol{t}' \times \boldsymbol{A}'_{\mathrm{pD}} \cdot \boldsymbol{t}') \cdot \boldsymbol{m} = ((\boldsymbol{Q}_n \cdot \boldsymbol{t}) \times \boldsymbol{Q}_n \cdot \boldsymbol{A}_{\mathrm{pD}} \cdot \boldsymbol{t}) \cdot \boldsymbol{m} \\ &= (\boldsymbol{t} \times \boldsymbol{A}_{\mathrm{pD}} \cdot \boldsymbol{t}) \cdot \boldsymbol{Q}_n \cdot \boldsymbol{m} = -(\boldsymbol{t} \times \boldsymbol{A}_{\mathrm{pD}} \cdot \boldsymbol{t}) \cdot \boldsymbol{m} = -\bar{I}_6\end{aligned}$$

Consequently

$$\begin{aligned}&f(\boldsymbol{A}') = f(\bar{J}'_1, \bar{J}'_2, \ldots, \bar{J}'_5, \bar{J}'_6) = f(\bar{J}_1, \bar{J}_2, \ldots, \bar{J}_5, -\bar{J}_6) \\ &\Rightarrow \quad f(\boldsymbol{A}) = f(\bar{J}_1, \bar{J}_2, \ldots, \bar{J}_5, \bar{J}_6^2)\end{aligned}$$

and $\bar{J}_6^2$ can be omitted due to the last relation in Eqs. (44).

2.5 Invariants for the Orthotropic Symmetry Group

The orthotropic symmetry is an important symmetry group as it widely applied in constitutive models for composite materials. Furthermore, the analysis of the material behavior under the consideration of induced anisotropy. The common approach is to introduce internal state variables, usually second rank tensors to characterize kinematic hardening and damage. The orthogonal tensors

$$\boldsymbol{Q}_1 = 2\boldsymbol{n}_1\boldsymbol{n}_1 - \boldsymbol{I}, \qquad \boldsymbol{Q}_2 \equiv \boldsymbol{n}_2\boldsymbol{n}_2 - \boldsymbol{I}, \qquad \det \boldsymbol{Q}_1 = \det \boldsymbol{Q}_2 = 1$$

represent the rotations the angle π about the axes $\boldsymbol{n}_1$ and $\boldsymbol{n}_2$. These tensors are the symmetry elements of the orthotropic (orthorhombic) symmetry group. Let us find the scalar-valued functions of a symmetric tensor $\boldsymbol{A}$ satisfying the following conditions

$$f(\boldsymbol{Q}_1 \cdot \boldsymbol{A} \cdot \boldsymbol{Q}_1^{\mathrm{T}}) = f(\boldsymbol{Q}_2 \cdot \boldsymbol{A} \cdot \boldsymbol{Q}_2^{\mathrm{T}}) = f(\boldsymbol{A}). \tag{46}$$

Replacing the tensor $\boldsymbol{A}$ by the tensor $\boldsymbol{Q}_2 \cdot \boldsymbol{A} \cdot \boldsymbol{Q}_2^{\mathrm{T}}$ we find that

$$f(\boldsymbol{Q}_1 \cdot \boldsymbol{Q}_2 \cdot \boldsymbol{A} \cdot \boldsymbol{Q}_2^{\mathrm{T}} \cdot \boldsymbol{Q}_1^{\mathrm{T}}) = f(\boldsymbol{Q}_2 \cdot \boldsymbol{A} \cdot \boldsymbol{Q}_2^{\mathrm{T}}) = f(\boldsymbol{A}). \tag{47}$$

Consequently the tensor $\boldsymbol{Q}_3 = \boldsymbol{Q}_1 \cdot \boldsymbol{Q}_2 = 2\boldsymbol{n}_3\boldsymbol{n}_3 - \boldsymbol{I} = \boldsymbol{Q}(\pi\boldsymbol{n}_3)$ belongs to the symmetry group, where the unit vector $\boldsymbol{n}_3$ is orthogonal to $\boldsymbol{n}_1$ and $\boldsymbol{n}_2$. Consider three tensors $\boldsymbol{A}'_i$ formed from the tensor $\boldsymbol{A}$ by three symmetry transformations i.e., $\boldsymbol{A}'_i \equiv \boldsymbol{Q}_i \cdot \boldsymbol{A} \cdot \boldsymbol{Q}_i^{\mathrm{T}}$. Taking into account that $\boldsymbol{Q}_i \cdot \boldsymbol{n}_i = \boldsymbol{n}_i$ (no summation over i) and $\boldsymbol{Q}_i \cdot \boldsymbol{n}_j = -\boldsymbol{n}_j$, $i \neq j$ we can write

$$\begin{aligned}&\operatorname{tr}(\boldsymbol{A}_i'^k) = \operatorname{tr}(\boldsymbol{A}^k), \quad k = 1, 2, 3, \quad i = 1, 2, 3, \\ &\boldsymbol{n}_i \cdot \boldsymbol{A}'_i \cdot \boldsymbol{n}_i = \boldsymbol{n}_i \cdot \boldsymbol{Q}_i \cdot \boldsymbol{A} \cdot \boldsymbol{Q}_i^{\mathrm{T}} \cdot \boldsymbol{n}_i = \boldsymbol{n}_i \cdot \boldsymbol{A} \cdot \boldsymbol{n}_i, \quad i = 1, 2, 3, \\ &\boldsymbol{n}_i \cdot \boldsymbol{A}_i'^2 \cdot \boldsymbol{n}_i = \boldsymbol{n}_i \cdot \boldsymbol{Q}_i \cdot \boldsymbol{A}^2 \cdot \boldsymbol{Q}_i^{\mathrm{T}} \cdot \boldsymbol{n}_i = \boldsymbol{n}_i \cdot \boldsymbol{A}^2 \cdot \boldsymbol{n}_i, \quad i = 1, 2, 3.\end{aligned} \tag{48}$$

The above set includes 9 scalars. The number can be reduced to 7 due to the obvious relations

$$\mathrm{tr}(\boldsymbol{A}^k) = \boldsymbol{n}_1 \cdot \boldsymbol{A}^k \cdot \boldsymbol{n}_1 + \boldsymbol{n}_2 \cdot \boldsymbol{A}^k \cdot \boldsymbol{n}_2 + \boldsymbol{n}_3 \cdot \boldsymbol{A}^k \cdot \boldsymbol{n}_3, \quad k = 1, 2.$$

Therefore the orthotropic scalar-valued function of the symmetric second rank tensor can be represented as a function of the following seven arguments

$$\begin{aligned} J_1 &= \boldsymbol{n}_1 \cdot \boldsymbol{A} \cdot \boldsymbol{n}_1, \quad J_2 = \boldsymbol{n}_2 \cdot \boldsymbol{A} \cdot \boldsymbol{n}_2, \quad J_3 = \boldsymbol{n}_3 \cdot \boldsymbol{A} \cdot \boldsymbol{n}_3, \\ J_4 &= \boldsymbol{n}_1 \cdot \boldsymbol{A}^2 \cdot \boldsymbol{n}_1, \quad J_5 = \boldsymbol{n}_2 \cdot \boldsymbol{A}^2 \cdot \boldsymbol{n}_2, \quad J_6 = \boldsymbol{n}_3 \cdot \boldsymbol{A}^2 \cdot \boldsymbol{n}_3, \quad J_7 = \mathrm{tr}\boldsymbol{A}^3. \end{aligned} \tag{49}$$

Instead of J_4, J_5, J_6 and J_7 in (49) one may use the arguments

$$\begin{aligned} \tilde{J}_1 &= (\boldsymbol{n}_1 \cdot \boldsymbol{A} \cdot \boldsymbol{n}_2)^2, \quad \tilde{J}_2 = (\boldsymbol{n}_2 \cdot \boldsymbol{A} \cdot \boldsymbol{n}_3)^2, \quad \tilde{J}_3 = (\boldsymbol{n}_1 \cdot \boldsymbol{A} \cdot \boldsymbol{n}_3)^2, \\ \tilde{J}_4 &= (\boldsymbol{n}_1 \cdot \boldsymbol{A} \cdot \boldsymbol{n}_2)(\boldsymbol{n}_1 \cdot \boldsymbol{A} \cdot \boldsymbol{n}_3)(\boldsymbol{n}_2 \cdot \boldsymbol{A} \cdot \boldsymbol{n}_3). \end{aligned} \tag{50}$$

The invariants $\tilde{J}_1, \tilde{J}_2, \tilde{J}_3, \tilde{J}_4$ can be uniquely expressed through $J_1, \ldots, J_7$ by use of the following relations

$$\begin{aligned} J_4 &= J_1^2 + \tilde{J}_1 + \tilde{J}_3, \quad J_5 = J_2^2 + \tilde{J}_1 + \tilde{J}_2, \quad J_6 = J_3^2 + \tilde{J}_2 + \tilde{J}_3, \\ J_7 &= 2J_1(J_4 - J_1^2) + 2J_2(J_5 - J_2^2) + 2J_3(J_6 - J_3^2) + \tilde{J}_4. \end{aligned} \tag{51}$$

Let us note that if $\boldsymbol{A}$ is a polar tensor, then the invariants (49) and (50) are also applicable to the class of the orthotropic symmetry characterized by the following eight symmetry elements

$$\boldsymbol{Q} = \pm \boldsymbol{n}_1\boldsymbol{n}_1 \pm \boldsymbol{n}_2\boldsymbol{n}_2 \pm \boldsymbol{n}_3\boldsymbol{n}_3. \tag{52}$$

We derived the generic partial differential equation for the case of the transverse isotropy. Applying this approach one may find a set of functionally independent invariants among all possible invariants. Let us formulate the generic partial differential equation for the case of orthotropic symmetry. To this end let us find the scalar valued arguments of the tensor $\boldsymbol{A}$ from the following condition

$$f(\boldsymbol{A}', \boldsymbol{n}_1'\boldsymbol{n}_1', \boldsymbol{n}_2'\boldsymbol{n}_2', \boldsymbol{n}_3'\boldsymbol{n}_3') = f(\boldsymbol{A}, \boldsymbol{n}_1\boldsymbol{n}_1, \boldsymbol{n}_2\boldsymbol{n}_2, \boldsymbol{n}_3\boldsymbol{n}_3), \tag{53}$$

where

$$\boldsymbol{A}' = \boldsymbol{Q} \cdot \boldsymbol{A} \cdot \boldsymbol{Q}^{\mathrm{T}}, \qquad \boldsymbol{n}_i' = \boldsymbol{Q} \cdot \boldsymbol{n}_i, \qquad \forall \boldsymbol{Q}, \quad \det \boldsymbol{Q} = 1.$$

The symmetry group of a single dyad is given by Eqs. (36). It can be shown that the symmetry group of three dyads $\boldsymbol{n}_i\boldsymbol{n}_i$ includes eight elements

(52). Among all rotation tensors $\boldsymbol{Q}$ the three rotations $\boldsymbol{Q}_1$, $\boldsymbol{Q}_2$ and $\boldsymbol{Q}_3$ belong to the symmetry group of $\boldsymbol{n}_i\boldsymbol{n}_i$. Therefore Eq. (53) is equivalent to the following three equations

$$\begin{aligned}
f(\boldsymbol{Q}_1\cdot\boldsymbol{A}\cdot\boldsymbol{Q}_1^{\mathrm{T}},\boldsymbol{n}_1\boldsymbol{n}_1,\boldsymbol{n}_2\boldsymbol{n}_2,\boldsymbol{n}_3\boldsymbol{n}_3) &= f(\boldsymbol{A},\boldsymbol{n}_1\boldsymbol{n}_1,\boldsymbol{n}_2\boldsymbol{n}_2,\boldsymbol{n}_3\boldsymbol{n}_3),\\
f(\boldsymbol{Q}_2\cdot\boldsymbol{A}\cdot\boldsymbol{Q}_2^{\mathrm{T}},\boldsymbol{n}_1\boldsymbol{n}_1,\boldsymbol{n}_2\boldsymbol{n}_2,\boldsymbol{n}_3\boldsymbol{n}_3) &= f(\boldsymbol{A},\boldsymbol{n}_1\boldsymbol{n}_1,\boldsymbol{n}_2\boldsymbol{n}_2,\boldsymbol{n}_3\boldsymbol{n}_3),\\
f(\boldsymbol{Q}_3\cdot\boldsymbol{A}\cdot\boldsymbol{Q}_3^{\mathrm{T}},\boldsymbol{n}_1\boldsymbol{n}_1,\boldsymbol{n}_2\boldsymbol{n}_2,\boldsymbol{n}_3\boldsymbol{n}_3) &= f(\boldsymbol{A},\boldsymbol{n}_1\boldsymbol{n}_1,\boldsymbol{n}_2\boldsymbol{n}_2,\boldsymbol{n}_3\boldsymbol{n}_3).
\end{aligned}$$

Consequently, the scalar-valued arguments of $\boldsymbol{A}$ found from (53) satisfy three Eqs (46) and (47). To derive the generic partial differential equation for invariants we follow the approach of Zhilin (2003). Let $\boldsymbol{Q}(\tau)$ be a continuous set of rotations depending on the real parameter τ. In this case

$$\frac{\mathrm{d}}{\mathrm{d}\tau}\boldsymbol{Q}(\tau)=\boldsymbol{\omega}(\tau)\times\boldsymbol{Q}(\tau)\quad\Rightarrow\quad\frac{\mathrm{d}}{\mathrm{d}\tau}\boldsymbol{Q}^{\mathrm{T}}(\tau)=-\boldsymbol{Q}^{\mathrm{T}}(\tau)\times\boldsymbol{\omega}(\tau),$$

with the conditions $\boldsymbol{Q}(0)=\boldsymbol{I},\boldsymbol{\omega}(0)=\boldsymbol{\omega}_0$,where the axial vector $\boldsymbol{\omega}$ has the sense of the angular velocity of rotation. Taking the derivative of Eq. (53) with respect to τ we obtain the following partial differential equation

$$(\boldsymbol{\omega}\times\boldsymbol{A}'-\boldsymbol{A}'\times\boldsymbol{\omega})\cdot\left(\frac{\partial f}{\partial\boldsymbol{A}'}\right)^{\mathrm{T}}+\sum_{i=1}^{3}(\boldsymbol{\omega}\times\boldsymbol{n}_i'\boldsymbol{n}_i'-\boldsymbol{n}_i'\boldsymbol{n}_i'\times\boldsymbol{\omega})\cdot\left(\frac{\partial f}{\partial\boldsymbol{n}_i'\boldsymbol{n}_i'}\right)^{\mathrm{T}}=0,\quad(54)$$

where $\boldsymbol{A}'(\tau)=\boldsymbol{Q}(\tau)\cdot\boldsymbol{A}\cdot\boldsymbol{Q}^{\mathrm{T}}(\tau),\boldsymbol{n}_i'(\tau)=\boldsymbol{Q}(\tau)\cdot\boldsymbol{n}_i$. For $\tau=0$ Eq. (54) takes the form

$$(\boldsymbol{\omega}_0\times\boldsymbol{A}-\boldsymbol{A}\times\boldsymbol{\omega}_0)\cdot\left(\frac{\partial f}{\partial\boldsymbol{A}}\right)^{\mathrm{T}}+\sum_{i=1}^{3}(\boldsymbol{\omega}_0\times\boldsymbol{n}_i\boldsymbol{n}_i-\boldsymbol{n}_i\boldsymbol{n}_i\times\boldsymbol{\omega}_0)\cdot\left(\frac{\partial f}{\partial\boldsymbol{n}_i\boldsymbol{n}_i}\right)^{\mathrm{T}}=0.\quad(55)$$

Taking into account the following identities

$$(\boldsymbol{a}\times\boldsymbol{A})\cdot\boldsymbol{B}=\boldsymbol{a}\cdot(\boldsymbol{A}\cdot\boldsymbol{B})_\times,\quad\boldsymbol{B}\cdot(\boldsymbol{A}\times\boldsymbol{a})=\boldsymbol{a}\cdot(\boldsymbol{B}\cdot\boldsymbol{A})_\times,$$

valid for any vector $\boldsymbol{a}$ and any second rank tensors $\boldsymbol{A}$ and $\boldsymbol{B}$, Eq. (55) can be transformed to

$$\begin{aligned}
\boldsymbol{\omega}_0\cdot\Bigg[\boldsymbol{A}\cdot\left(\frac{\partial f}{\partial\boldsymbol{A}}\right)^{\mathrm{T}}-\left(\frac{\partial f}{\partial\boldsymbol{A}}\right)^{\mathrm{T}}\cdot\boldsymbol{A}&\\
+\sum_{i=1}^{3}\boldsymbol{n}_i\boldsymbol{n}_i\cdot\left(\frac{\partial f}{\partial\boldsymbol{n}_i\boldsymbol{n}_i}\right)^{\mathrm{T}}-\sum_{i=1}^{3}\left(\frac{\partial f}{\partial\boldsymbol{n}_i\boldsymbol{n}_i}\right)^{\mathrm{T}}\cdot\boldsymbol{n}_i\boldsymbol{n}_i\Bigg]_\times&=\boldsymbol{0}.
\end{aligned}$$

Because $\boldsymbol{\omega}_0$ is the arbitrary vector we obtain

$$\left[\boldsymbol{A}\cdot\left(\frac{\partial f}{\partial \boldsymbol{A}}\right)^{\mathrm{T}} - \left(\frac{\partial f}{\partial \boldsymbol{A}}\right)^{\mathrm{T}}\cdot\boldsymbol{A} \right. \\ \left. + \sum_{i=1}^{3}\boldsymbol{n}_i\boldsymbol{n}_i\cdot\left(\frac{\partial f}{\partial \boldsymbol{n}_i\boldsymbol{n}_i}\right)^{\mathrm{T}} - \sum_{i=1}^{3}\left(\frac{\partial f}{\partial \boldsymbol{n}_i\boldsymbol{n}_i}\right)^{\mathrm{T}}\cdot\boldsymbol{n}_i\boldsymbol{n}_i\right]_{\times} = \boldsymbol{0}. \tag{56}$$

The partial differential vector equation (56) corresponds to three scalar differential equations. The total number of scalar arguments of the function f is 9 including 6 components of the symmetric tensor $\boldsymbol{A}$ and three parameters (e.g. three Euler angles) characterizing three dyads $\boldsymbol{n}_i\boldsymbol{n}_i$. Each of the scalar partial differential equations in (56) reduces the number of independent arguments by one. Therefore, the total number of independent arguments is 6.

It can be shown that all seven arguments presented by Eqs. (49) or Eqs. (50) satisfies (56). Because only six of them are independent, one functional relation must exist. In the case of the list (50) the functional relation is obvious. Indeed, we can write

$$\tilde{J}_4^2 = \tilde{J}_1\tilde{J}_2\tilde{J}_3. \tag{57}$$

To derive the functional relation for the list (49) one may apply Eqs. (51) to express $\tilde{J}_1, \ldots, \tilde{J}_4$ through $J_1, \ldots, J_7$. The result should be inserted into Eq. (57).

3 Isotropic Failure Criteria

3.1 Equivalent Stress Concept

The strength criteria (based on hypotheses) assume that the mechanical loading states can be characterized, for example, by stresses. There are other suggestions - criteria based on strains or energy and power considerations. Here we will focus our attention mostly on stress formulations.

The dimensioning of structural members is usually carried out from the point of view that materials behave either brittle or ductile. The following hypotheses (sometimes named theories), which correspond the first or the second assumption, are mostly used for strength or yield evaluation (Christensen, 2013; Filonenko-Boroditsch, 1960; Gol'denblat and Kopnov, 1968; Goldenblat and Kopnov, 1971; Pisarenko and Lebedev, 1976; Mālmeisters et al., 1977; Sähn et al., 1993; Yu, 2004)

- the maximum normal stress hypothesis (Rankine hypothesis),
- the maximum shear stress hypothesis (Tresca hypothesis),

- the maximum distorsion energy criterion (von Mises hypothesis), and
- the maximum deviatoric stress hypothesis (Schmidt-Ishlinsky hypothesis).

The first three are as usual implemented in commercial finite element codes as a standard tool. The fourth hypothesis is relatively new. It is sometimes discussed within the theory of plasticity courses.

The stresses in each point of the material or structure are presented by the stress tensor $\boldsymbol{\sigma}$ and, may be, some additional parameters characterizing, for example, the microstructure. For comparison purposes of various stress states the stress tensor cannot be applied since we have to compare a tensor with scalar failure thresholds. That means we should introduce a scalar quantity, for example, the equivalent stress. Let us assume (Altenbach et al., 2014)

$$\sigma_{\mathrm{EQ}} = \sigma_{\mathrm{eq}}(\boldsymbol{\sigma}) + f(\boldsymbol{\nabla}\boldsymbol{\sigma})\, R, \qquad R \geq 0, \tag{58}$$

where σ_{EQ} is some generalization of the classical equivalent stress σ_{eq} (Sect. 1.2) taking into account the microstructure, $\boldsymbol{\nabla}$ is the nabla operator, f denotes an arbitrary scalar-valued function (for example of the quadratic invariant of the stress tensor gradient) and R is a structural parameter, which can be associated with the grain size in gray iron, with the cell size of a hard foam, with the particle size in nanomaterials, etc. This parameter represents the influence of the stress distribution expressed by the stress gradient $\boldsymbol{\nabla}\boldsymbol{\sigma}$. The parameter R is positive-definit and bounded by the minimal dimension of the structural component, e. g. the plate or sheet thickness.

In engineering application it is difficult to identify all parameters in Eq. (58). This is the base for first simplification. Ignoring the microstructure influence in Eq. (58) one get

$$\sigma_{\mathrm{EQ}} = \sigma_{\mathrm{eq}}(\boldsymbol{\sigma}). \tag{59}$$

This implies that the stress state in each point can be described through the stresses at this point only. This formulation has several limitations and must be applied with care if the calculation of stresses is performed for parts with significant stress gradients:

- stress concentration areas,
- load application areas,
- sharp corners, etc.

Nevertheless, the concept of the equivalent stress (59) is widely applicable. This concept allows to compare multi-axial stress states with material parameters estimated in tension tests, e. g. the tensile yield or failure stress σ_+

$$\sigma_{\mathrm{eq}} = \sigma_+. \tag{60}$$

With σ_{eq} strength hypotheses and yield criteria for isotropic materials can be formulated using invariant variables introduced in Sect. 2.1:

- the principal stresses,
- the principal invariants,
- axiatoric-deviatoric invariants,
- cylindrical invariants (Novozhilov's invariants) or
- spherical invariants.

All formulations of invariants are equivalent which means that there is no formal preference of one set of invariants. Note that in the case of incompressible material behavior the first invariant I_1 has no influence on the expression for the strength criteria. In this case set of invariants with I_1 are preferable.

The equivalent stress concept (59) allows to formulate failure criteria, but also general constitutive equations of the material response under multi-axial loading in a compact form using only few parameters. Such formulations are applied in

- elasticity theory (elastic potential) Altenbach et al. (1995a); Ambarcumyan (1982); Lurie (2005); Tsvelodub (2008),
- plasticity theory (plastic potential, yield criterion) Altenbach et al. (1995a); Backhaus (1983); Hill (1948); von Mises (1913, 1928); Prager and Hodge (1954); Skrzypek (1993); Życzkowski (1981),
- creep theory (creep potential) Altenbach et al. (1995a); Betten (2008); Leckie and Hayhurst (1977); Lokoshchenko (2012),
- strength of materials (strength hypothesis or criterion) Altenbach et al. (1995a); Burzyński (1928); Huber (1904); Pisarenko and Lebedev (1976); Yagn (1931); Yu (2004),
- low cyclic fatigue Altenbach et al. (1995a); Lemaitre and Chaboche (1990) and
- phase transformation conditions Levitas and Shvedov (2002); Yao et al. (2005); Pęcherski et al. (2011); Raniecki and Mróz (2008).

Phenomenological yield and failure criteria are widely discussed in the the literature, see Altenbach et al. (1995a, 2014); Chen and Zhang (1991); Christensen (2013); Goldenblat and Kopnov (1971); Mālmeisters et al. (1977); Pisarenko and Lebedev (1976); Sähn et al. (1993); Yu (2004); Życzkowski (1981) among others. They are now a standard tool in the design process.

3.2 Classical Strength Criteria

The three classical models (normal stress hypothesis, Tresca, von Mises) and the model of Schmidt-Ishlinsky represent particular cases of material behavior and are sometimes unable to describe the behavior of materials

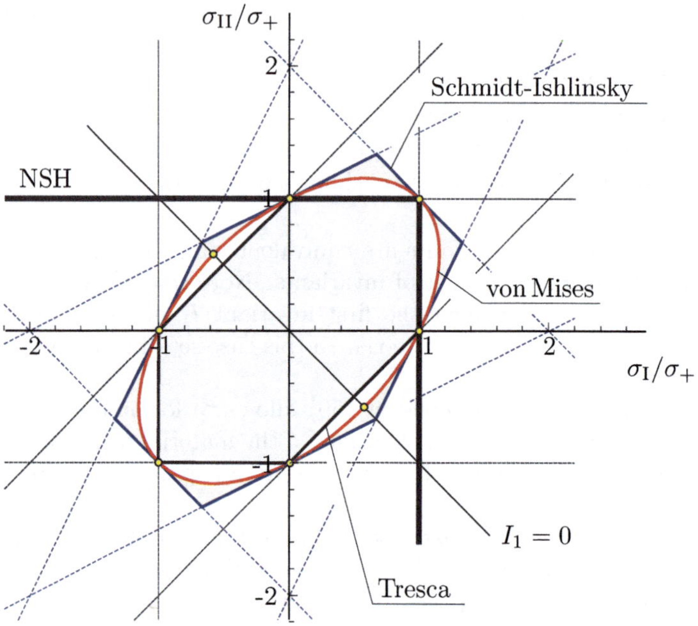

Figure 8. Models for incompressible "ideal ductile" material behavior (von Mises, Tresca and Schmidt-Ishlinsky) and the normal stress hypothesis (NSH) for "absolutely brittle" material behavior in the plane $\sigma_{I} - \sigma_{II}$, $\sigma_{III} = 0$ (after Ishlinsky and Ivlev 2003)

properly. Because of their simplicity they are particulary used in the engineering practice and will be discussed below.

For applied problems the computation can be performed using these models, if no information on the particular material properties is available. The starting point are observations concerning the deformation process. The normal stress hypothesis describes the "absolutely brittle" material behavior, the models of Tresca, von Mises and Schmidt-Ishlinsky - the "ideal ductile" behavior. The four criteria are visualized for the case of plane stress on Fig. 8. Later these criteria will be generalized for intermediate material behaviour. Furthermore these four models are the base for C^0- and C^1-combined models that have been assembled from two or more parts Altenbach et al. (2014).

Normal Stress Hypothesis. The normal stress hypothesis, which was mentioned by Clapeyron, Galilei, Leibniz, Lamé, Maxwell, Navier, Rankine among others, is based on the assumption that the maximum tensile stress is responsible for the failure. In this case the equivalent stress can be expressed as follows

$$\sigma_{\text{eq}} = \max(\sigma_{\text{I}}, \sigma_{\text{II}}, \sigma_{\text{III}}). \tag{61}$$

Equivalent formulation is

$$(\sigma_{\text{I}} - \sigma_{\text{eq}})\,(\sigma_{\text{II}} - \sigma_{\text{eq}})\,(\sigma_{\text{III}} - \sigma_{\text{eq}}) = 0. \tag{62}$$

Equation (62) is a cubic equation with respect to σ_{eq}.

With the help of a parameter identification this equation can be transformed into a third order polynomial of I_1^3, $I_1^2\,\sigma_{\text{eq}}$, $I_1\,\sigma_{\text{eq}}^2$, σ_{eq}^3, $I_2'\,\sigma_{\text{eq}}$ and I_3'. It can be obtained using the model

$$\frac{3\,I_2'\,\sigma_{\text{eq}} + c_3\,I_3'}{1 + 2\,c_3/3^3} = \sigma_{\text{eq}}^3, \tag{63}$$

and the substitution

$$\sigma_{\text{eq}} \to \frac{\sigma_{\text{eq}} - \gamma_1\,I_1}{1 - \gamma_1} \tag{64}$$

with the parameter values

$$c_3 = \frac{3^2}{2}, \qquad \gamma_1 = \frac{1}{3} \tag{65}$$

for the better analysis, unified visualization techniques and systematization. The comparison of the normal stress criterion and the von Mises criterion is given on Fig. 9.

Tresca Hypothesis. The shear stress hypothesis, which was discussed by Coulomb, Guest, Mohr, St. Venant, Tresca among others, supposes that the maximum difference of the principal stresses is relevant for the failure. The hypothesis can be written as follows

$$\tau_{\max} = \frac{1}{2}\max(|\sigma_{\text{I}} - \sigma_{\text{II}}|, |\sigma_{\text{II}} - \sigma_{\text{III}}|, |\sigma_{\text{III}} - \sigma_{\text{I}}|). \tag{66}$$

The equivalent stress can be expressed in this case as

$$\sigma_{\text{eq}} = 2\,\tau_{\max}. \tag{67}$$

In analogy to Eq. (62) an alternative formulation exists

$$(\sigma_{\text{eq}} - |\sigma_{\text{I}} - \sigma_{\text{II}}|)\,(\sigma_{\text{eq}} - |\sigma_{\text{II}} - \sigma_{\text{III}}|)\,(\sigma_{\text{eq}} - |\sigma_{\text{III}} - \sigma_{\text{I}}|) = 0.$$

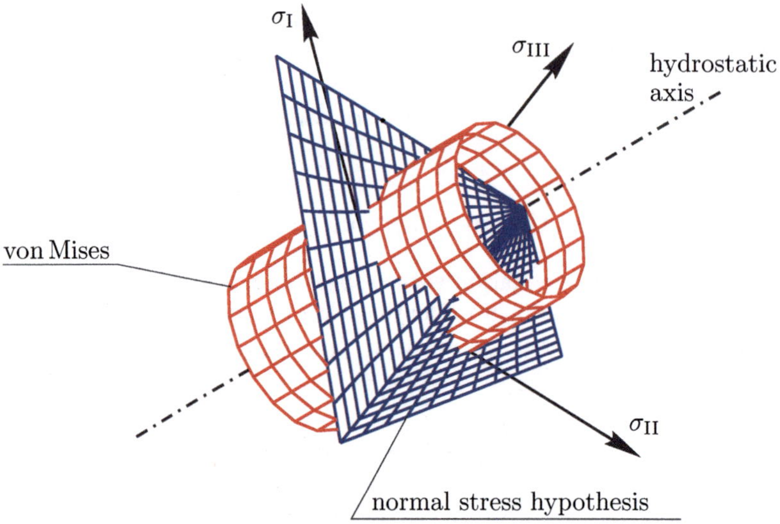

Figure 9. Comparison of the normal stress criterion and the von Mises criterion in the principal stress space.

The Tresca hypothesis can be also expressed by the deviatoric invariants (Prager and Hodge, 1954; Reuss, 1933)

$$\left(I_2' - \sigma_{\mathrm{eq}}^2\right)^2 \left(2^2 I_2' - \sigma_{\mathrm{eq}}^2\right) - 3^3 I_3'^2 = 0, \qquad (68)$$

which is a polynomial of order 6 with respect to σ_{eq}.

Huber-von Mises-Hencky Hypothesis. The distortion energy hypothesis (Huber, 1904; von Mises, 1913; Hencky, 1924)[4], has different interpretations among them that the failure occurs if a critical amount of accumulated distortion energy is achieved

$$\sigma_{\mathrm{eq}}^2 = \frac{3}{2}\boldsymbol{s} \cdot\cdot\, \boldsymbol{s} = \frac{1}{2}\left[(\sigma_{\mathrm{I}} - \sigma_{\mathrm{II}})^2 + (\sigma_{\mathrm{II}} - \sigma_{\mathrm{III}})^2 + (\sigma_{\mathrm{III}} - \sigma_{\mathrm{I}})^2\right] = 3\, I_2'. \qquad (69)$$

This hypothesis is often called von Mises hypothesis but it was independently introduced by Huber (1904), von Mises (1913) and Hencky (1924).

[4]The idea if this criterion was also formulated 1865 in a letter of Maxwell to Lord Kelvin Timoshenko (1953).

Schmidt-Ishlinsky Hypothesis. The criterion of the absolute value of the maximum deviatoric stress, which was introduced by Burzyński (1928); Schmidt (1932); Ishlinsky (1940); Correia de Araújo (1961); Haythornthwaite (1961); Hill (1950) and Yu (1961), is based on the assumption that failure occurs if a critical value of deviatoric components of the stress tensor is achieved

$$\max\left[\left|\sigma_{\mathrm{I}} - \frac{1}{3}I_1\right|, \left|\sigma_{\mathrm{II}} - \frac{1}{3}I_1\right|, \left|\sigma_{\mathrm{III}} - \frac{1}{3}I_1\right|\right] = \frac{2}{3}\sigma_{\mathrm{eq}} \tag{70}$$

or in analogy to Eq. (62)

$$\begin{aligned}&\left[\sigma_{\mathrm{eq}} - \left|\sigma_{\mathrm{I}} - \frac{1}{2}(\sigma_{\mathrm{II}} + \sigma_{\mathrm{III}})\right|\right]\left[\sigma_{\mathrm{eq}} - \left|\sigma_{\mathrm{II}} - \frac{1}{2}(\sigma_{\mathrm{III}} + \sigma_{\mathrm{I}})\right|\right]\\ &\times\left[\sigma_{\mathrm{eq}} - \left|\sigma_{\mathrm{III}} - \frac{1}{2}(\sigma_{\mathrm{I}} + \sigma_{\mathrm{II}})\right|\right] = 0.\end{aligned} \tag{71}$$

This model can be expressed with the deviatoric invariants (Annin, 1999; Yu, 1961) as a polynomial of order 6

$$\left(\frac{3^3}{2^3}I_3' + \frac{3^2}{2^2}I_2'\sigma_{\mathrm{eq}} - \sigma_{\mathrm{eq}}^3\right)\left(\frac{3^3}{2^3}I_3' - \frac{3^2}{2^2}I_2'\sigma_{\mathrm{eq}} + \sigma_{\mathrm{eq}}^3\right) = 0. \tag{72}$$

The models of Tresca, von Mises and Schmidt-Ishlinski are shown for the principal stress space in Fig. 10.

Comparison of the Classical Criteria. Let us perform the basic tests of material testing (Blumenauer, 1996).

- Tension test
  The tension test can be presented mathematically by:

$$\boldsymbol{\sigma} = \sigma\boldsymbol{e}_1\boldsymbol{e}_1.$$

  The deviator in this case is

$$\boldsymbol{s} = \boldsymbol{\sigma} - \frac{1}{3}\boldsymbol{\sigma} \cdot\cdot \boldsymbol{II} = \sigma\left[\frac{2}{3}\boldsymbol{e}_1\boldsymbol{e}_1 - \frac{1}{3}(\boldsymbol{e}_2\boldsymbol{e}_2 + \boldsymbol{e}_3\boldsymbol{e}_3)\right].$$

  Principal stresses can be computed

$$\det(\sigma\boldsymbol{e}_1\boldsymbol{e}_1 - \lambda\boldsymbol{I}) = 0 = (\sigma - \lambda)\lambda^2$$

$$\Rightarrow \sigma_{\mathrm{I}} = \sigma, \quad \sigma_{\mathrm{II}} = \sigma_{\mathrm{III}} = 0.$$

  Finally we get:

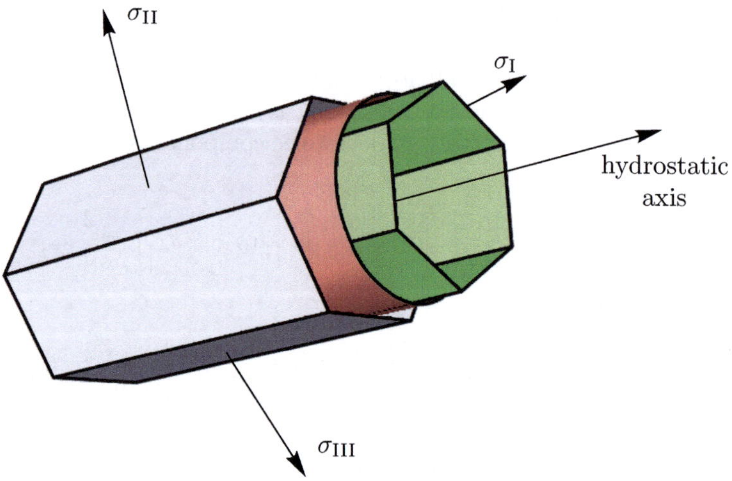

Figure 10. Models of TRESCA, VON MISES, and SCHMIDT-ISHLINSKY in the principal stress space (σ_{I}, σ_{II}, σ_{III}).

– Normal stress criterion

$$\sigma_{\mathrm{eq}} = \sigma,$$

– Tresca criterion

$$2\tau_{\max} = \sigma_{\mathrm{eq}} = |\sigma_{\mathrm{I}} - \sigma_{\mathrm{II}}| = \sigma,$$

– von Mises criterion

$$\sigma_{\mathrm{eq}} = \sigma_{\mathrm{vM}} = \sqrt{\frac{3}{2}\boldsymbol{s} \cdot\cdot\, \boldsymbol{s}} = \sqrt{\frac{3}{2}\sigma^2\left(\frac{4}{9} + \frac{1}{9} + \frac{1}{9}\right)} = \sigma,$$

– Schmidt-Ishlinsky criterion

$$\sigma_{\mathrm{eq}} = \frac{3}{2}\left|\sigma_{\mathrm{I}} - \frac{1}{3}I_1\right| = \frac{3}{2}\left|\sigma - \frac{1}{3}\sigma\right| = \sigma.$$

- Compression test
  The tension test can be presented by:

$$\boldsymbol{\sigma} = -\sigma \boldsymbol{e}_1\boldsymbol{e}_1.$$

The deviator in this case is

$$\boldsymbol{s} = \boldsymbol{\sigma} - \frac{1}{3}\boldsymbol{\sigma} \cdot\cdot\, \boldsymbol{II} = -\sigma\left[\frac{2}{3}\boldsymbol{e}_1\boldsymbol{e}_1 - \frac{1}{3}(\boldsymbol{e}_2\boldsymbol{e}_2 + \boldsymbol{e}_3\boldsymbol{e}_3)\right].$$

Principal stresses can be computed

$$\det(-\sigma \boldsymbol{e}_1\boldsymbol{e}_1 - \lambda \boldsymbol{I}) = 0 = -(\sigma + \lambda)\lambda^2$$

$$\Rightarrow \sigma_{\mathrm{I}} = \sigma_{\mathrm{II}} = 0, \quad \sigma_{\mathrm{III}} = -\sigma.$$

Finally we get:

– Normal stress criterion

$$\sigma_{\mathrm{eq}} = 0,$$

– Tresca criterion

$$2\tau_{\max} = \sigma_{\mathrm{eq}} = |\sigma_{\mathrm{I}} - \sigma_{\mathrm{III}}| = \sigma,$$

– von Mises criterion

$$\sigma_{\mathrm{eq}} = \sigma_{\mathrm{vM}} = \sqrt{\frac{3}{2}\boldsymbol{s} \cdot\cdot \boldsymbol{s}} = \sqrt{\frac{3}{2}\sigma^2\left(\frac{4}{9} + \frac{1}{9} + \frac{1}{9}\right)} = \sigma,$$

– Schmidt-Ishlinsky criterion

$$\sigma_{\mathrm{eq}} = \frac{3}{2}\left|\sigma_{\mathrm{III}} - \frac{1}{3}I_1\right| = \frac{3}{2}\left|-\sigma - \frac{1}{3}\sigma\right| = \sigma$$

- Torsion test
  The torsion test can be presented by

$$\boldsymbol{\sigma} = \tau(\boldsymbol{e}_1\boldsymbol{e}_2 + \boldsymbol{e}_2\boldsymbol{e}_1)$$

The deviator in this case

$$\boldsymbol{s} = \boldsymbol{\sigma} - \frac{1}{3}\boldsymbol{\sigma} \cdot\cdot \boldsymbol{II} = \tau(\boldsymbol{e}_1\boldsymbol{e}_2 + \boldsymbol{e}_2\boldsymbol{e}_1)$$

Principal stresses

$$\det[\tau(\boldsymbol{e}_1\boldsymbol{e}_2 + \boldsymbol{e}_2\boldsymbol{e}_1) - \lambda \boldsymbol{I}] = 0 = -\lambda(\lambda^2 - \tau^2)$$

$$\Rightarrow \sigma_{\mathrm{I}} = \tau, \quad \sigma_{\mathrm{II}} = 0, \quad \sigma_{\mathrm{III}} = -\tau$$

Finally we get:

– Normal stress criterion

$$\sigma_{\mathrm{eq}} = \sigma_{\mathrm{I}} = \tau,$$

– Tresca criterion

$$2\tau_{\max} = \sigma_{\mathrm{eq}} = |\sigma_{\mathrm{I}} - \sigma_{\mathrm{III}}| = |2\tau| = 2\tau,$$

– von Mises criterion

$$\sigma_{\text{eq}} = \sigma_{\text{vM}} = \sqrt{\frac{3}{2}\boldsymbol{s} \cdot\cdot \boldsymbol{s}} = \sqrt{\frac{3}{2}(2\tau^2)} = \sqrt{3}\tau,$$

– Schmidt-Ishlinsky criterion

$$\sigma_{\text{eq}} = \frac{3}{2}|\tau| = \frac{3}{2}\tau.$$

- Hydrostatic pressure test
  The hydrostatic pressure test can be presented by:

$$\boldsymbol{\sigma} = -\sigma \boldsymbol{I}.$$

The deviator in this case is

$$\boldsymbol{s} = -\sigma \boldsymbol{I} + \frac{1}{3}\sigma \boldsymbol{I} \cdot\cdot \boldsymbol{II} = \boldsymbol{0}.$$

Principal stresses can be computed

$$\det(-\sigma \boldsymbol{I} - \lambda \boldsymbol{I}) = 0 = -(\sigma + \lambda)^2$$

$$\Rightarrow \sigma_{\text{I}} = \sigma_{\text{II}} = \sigma_{\text{III}} = -\sigma.$$

Finally we get:
– Normal stress criterion

$$\sigma_{\text{eq}} = 0,$$

– Tresca criterion

$$\tau_{\max} = 0,$$

– von Mises criterion

$$\sigma_{\text{eq}} = 0,$$

– Schmidt-Ishlinsky criterion

$$\sigma_{\text{eq}} = 0.$$

- Conclusion
  The results of the comparison are summarized in Table 1. With respect to the results of analytical solutions presented for the four basic tests the following conclusions can be made:
  – The Tresca criterion, von Mises criterion and Schmidt-Ishlinsky criteria are not sensitive with respect to the loading direction. In this case we assume $\sigma_+ = |\sigma_-|$.

Table 1. Comparison of equivalent stress in the case of the basic tests

Hypothesis	Tension	Compression	Torsion	hydrostatic compression
Tresca	σ	σ	2τ	0
von Mises	σ	σ	$\sqrt{3}\tau$	0
Schmidt-Ishlinsky	σ	σ	$3\tau/2$	0
normal stress	σ	0	τ	0

– If we assume the von Mises criterion the tension test and the torsion test result in the following relation between the critical values for the normal and the shear stress

$$\sigma = \sqrt{3}\tau,$$

which coincides with Eq. (1).

– If we assume the Tresca criterion the tension test and the torsion test result in the following relation between the critical values for the normal and the shear stress

$$\sigma = 2\tau.$$

– If we assume the Schmidt-Ishlinsky criterion the tension test and the torsion test result in the following relation between the critical values for the normal and the shear stress

$$\sigma = \frac{3}{2}\tau.$$

– For hydrostatic pressure the classical criteria give identical results - no failure.

– For compression the normal stress criterion yields no failure.

3.3 Generalization of Classical Criteria

For the generalization of the for classical criteria we need more information on the material behavior. The following nine tests (Fig. 11), which extend the basic tests in Blumenauer (1996), are suggested for the analysis and comparison of the limit surfaces Φ:

- two loadings corresponding to one-dimensional stress states (tension, compression),
- five loadings corresponding to plane stress states (torsion, two balanced plane states, two thin-walled tubular specimens with closed ends under inner and outer pressure) and

- two loadings corresponding to hydrostatic (3D balanced) tension and compression.

All these loading cases have approved verbal formulations and can be considered as basic tests for establishment of generalized criteria predicting the failure of materials or structures. The values relating the respective stresses to σ_+ are introduced in order to obtain

$$k = d = i_{\mathrm{Z}} = u_{\mathrm{D}} = b_{\mathrm{Z}} = b_{\mathrm{D}} = 1 \qquad \text{and} \qquad a_-^{\mathrm{hyd}},\, a_+^{\mathrm{hyd}} \to \infty \tag{73}$$

for the von Mises hypotheses (69). The ratios are presented in Table 2.

The models for incompressible behavior can be compared in the $d-k$-diagram (Altenbach et al., 2014). In this diagram the models of Haythornthwaite (Haythornthwaite, 1960, 1961) and Sayir II (Capurso, 1967; Sayir,

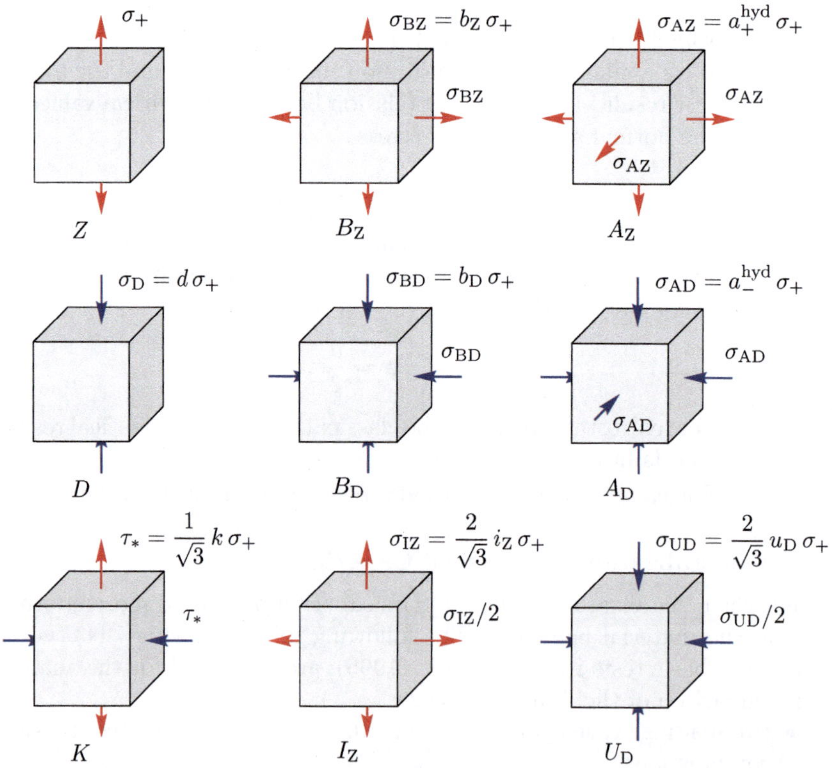

Figure 11. Nine basic tests for material characterization. Some basic features of these tests are given in Table 2.

1970) limit the convex shapes of the surface Φ in the π-plane. For the models of compressible material behavior the $\frac{1}{d} - k-$diagram (Pisarenko and Lebedev, 1976; Altenbach et al., 2014), which allows to represent the properties $d \to \infty$, $k = \sqrt{3}$ of the normal stress hypothesis among others, is recommended. In these diagrams the areas of validity of all criterions and various ideas of generalization can be visualized.

3.4 Standard Criteria

The standard models (the strain hypothesis, the model of Mohr-Coulomb, the model of Pisarenko-Lebedev and the model of Burzyński-Yagn) are frequently applied models for first approximations of measurements: they are easy to handle, can be used to describe different material types (brittle-ductile range) and their parameters can be obtained using simple tests Altenbach et al. (2014). As usual they are two-parameter criteria. In many cases they combine classical criteria and include fitting parameters which express the weight of the different parts of the given standard criterion.

Strain criterion. Strain hypothesis is obtained assuming Hooke's law

$$\epsilon_{\mathrm{eq}} = \frac{\sigma_{\mathrm{eq}}}{E} = \epsilon_{\mathrm{I}} = \frac{1}{E}\left[\sigma_{\mathrm{I}} - \nu(\sigma_{\mathrm{II}} + \sigma_{\mathrm{III}})\right]. \tag{74}$$

The other two equations are obtained by cyclic permutation of indices. The formulation (74) is equivalent to

$$\sigma_{\mathrm{eq}} = \sigma_{\mathrm{I}} - \nu(\sigma_{\mathrm{II}} + \sigma_{\mathrm{III}}).$$

The model contains the following special cases
- the normal stress hypothesis (Rankine, 1876) with $\nu = 0$,
- the Mariotte hypothesis (Mariotte, 1700).

The limit cases are the following surfaces Φ
- triangular prism in the principal stress space with $\nu = 1/2$ and
- plane through point Z orthogonal to hydrostatic axis with $\nu = -1$.

Mohr-Coulomb criterion. The model is introduced on the phenomenological basis with the help of the analysis of the slip of geological and granular materials (Coulomb, 1776; Mohr, 1900a,b). It arises as equations, which are obtained by permutation of indices in

$$\left[\sigma_{\mathrm{I}} - \frac{1}{d}\,\sigma_{\mathrm{II}} - \sigma_{\mathrm{eq}}\right]\left[\sigma_{\mathrm{I}} - \frac{1}{d}\,\sigma_{\mathrm{III}} - \sigma_{\mathrm{eq}}\right]\left[\sigma_{\mathrm{II}} - \frac{1}{d}\,\sigma_{\mathrm{III}} - \sigma_{\mathrm{eq}}\right] = 0. \tag{75}$$

This model yields

Table 2. Basic stress states, relations, normalized coordinates in the principal stress space, normalized axiatoric-deviatoric invariants, the stress angle θ and the angle ψ.

Loading	Stress	Label	Relation	$\left(\frac{\sigma_{\mathrm{I}}}{\sigma_+}, \frac{\sigma_{\mathrm{II}}}{\sigma_+}, \frac{\sigma_{\mathrm{III}}}{\sigma_+}\right)$	$\frac{I_1}{\sigma_+}$	$\frac{\sqrt{3I_2'}}{\sigma_+}$	$\frac{3\sqrt[3]{I_3'/2}}{\sigma_+}$	θ	ψ
				Basic stress states					
Tension	σ_+	Z	1	$(1, 0, 0)$	1	1	1	0	$\frac{\pi}{4}$
Compression	σ_-	D	d	$(-d, 0, 0)$	$-d$	d	d	$\frac{\pi}{3}$	$-\frac{\pi}{4}$
Torsion	τ_*	K	k	$\left(\frac{k}{\sqrt{3}}, -\frac{k}{\sqrt{3}}, 0\right)$	0	k	0	$\frac{\pi}{6}$	$\frac{\pi}{2}$
Biaxial tension	σ_{BZ}	B_{Z}	b_{Z}	$(b_{\mathrm{Z}}, b_{\mathrm{Z}}, 0)$	$2\,b_{\mathrm{Z}}$	b_{Z}	$-b_{\mathrm{Z}}$	$\frac{\pi}{3}$	$\arctan\frac{1}{2}$
Biaxial compression	σ_{BD}	B_{D}	b_{D}	$(-b_{\mathrm{D}}, -b_{\mathrm{D}}, 0)$	$-2\,b_{\mathrm{D}}$	b_{D}	b_{D}	0	$-\arctan\frac{1}{2}$
Inner pressure	σ_{IZ}	I_{Z}	i_{Z}	$\left(\frac{2}{\sqrt{3}}\,i_{\mathrm{Z}}, \frac{1}{\sqrt{3}}\,i_{\mathrm{Z}}, 0\right)$	$\sqrt{3}\,i_{\mathrm{Z}}$	i_{Z}	0	$\frac{\pi}{6}$	$\frac{\pi}{6}$
Outer pressure	σ_{UD}	U_{D}	u_{D}	$\left(-\frac{2}{\sqrt{3}}\,u_{\mathrm{D}}, -\frac{1}{\sqrt{3}}\,u_{\mathrm{D}}, 0\right)$	$-\sqrt{3}\,u_{\mathrm{D}}$	u_{D}	0	$\frac{\pi}{6}$	$-\frac{\pi}{6}$
Hydrostatic tension	σ_{AZ}	A_{Z}	a_+^{hyd}	$\left(a_+^{\mathrm{hyd}}, a_+^{\mathrm{hyd}}, a_+^{\mathrm{hyd}}\right)$	$3\,a_+^{\mathrm{hyd}}$	0	0	-	0
Hydrostatic compression	σ_{AD}	A_{D}	a_-^{hyd}	$\left(-a_-^{\mathrm{hyd}}, -a_-^{\mathrm{hyd}}, -a_-^{\mathrm{hyd}}\right)$	$-3\,a_-^{\mathrm{hyd}}$	0	0	-	π

- with $d \to \infty$ to the normal stress hypothesis (Rankine, 1876) and
- with $d = 1$ to the model of Tresca (Tresca, 1868).

The relation $d \geq 2$ is recommended if computations involving so called fatigue limits should be performed.

Pisarenko-Lebedev criterion. The model is presented by (Pisarenko and Lebedev, 1968)

$$\frac{1}{d}\sqrt{\frac{1}{2}[(\sigma_{\rm I}-\sigma_{\rm II})^2+(\sigma_{\rm II}-\sigma_{\rm III})^2+(\sigma_{\rm III}-\sigma_{\rm I})^2]}+\left(1-\frac{1}{d}\right)\max[\sigma_{\rm I},\,\sigma_{\rm II},\,\sigma_{\rm III}]=\sigma_{\rm eq} \tag{76}$$

with $d \geq 1$. This is a linear combination of the equivalent stresses after the normal stress hypothesis ($d \to \infty$) and the VON MISES hypothesis ($d = 1$). The model of Sdobyrev (1959) follows with $d = 2$.

Burzyński-Yagn criterion This model (Burzyński, 1928, 1929a,b; Yagn, 1931) belongs to the rotationally symmetric model and is a function of two parameters

$$3\,I_2' = \frac{\sigma_{\rm eq} - \gamma_1\,I_1}{1-\gamma_1}\,\frac{\sigma_{\rm eq} - \gamma_2\,I_1}{1-\gamma_2}. \tag{77}$$

The values k and d compute to

$$d = \frac{1}{1-\gamma_1-\gamma_2}, \qquad k^2 = \frac{1}{(1-\gamma_1)(1-\gamma_2)}. \tag{78}$$

The position of the hydrostatic node at tension one gets from

$$a_+^{\rm hyd} = \frac{1}{3\gamma_1}. \tag{79}$$

The model (77) represents the general equation of a second order surface of revolution about the hydrostatic axis. Some important special case are presented in Table 3.

Let us make some conclusions for the standard criteria:

- The models differ by the symmetry type in the π-plane and by the power of stresses n. The rotationally symmetric model (77) has the stress power $n = 2$. The strain hypothesis and the model of Pisarenko-Lebedev has the stress power $n = 3$ and the model of Mohr-Coulomb - $n = 6$. The last three models have a trigonal symmetry in the π-plane.
- The models can be characterized by the shape of the meridian line. The strain hypothesis, the hypothesis of Mohr-Coulomb, of Pisarenko-Lebedev, of von Mises and of Drucker-Prager have a straight line as

Table 3. Settings for the rotationally symmetric model of BURZYŃSKI-YAGN.

Three-dimensional Representation	Parameter	Quelle
cylinder	$\gamma_1 = \gamma_2 = 0$	von Mises (1913)
cone	$\gamma_1 = \gamma_2 \in]0, 1[$	Drucker and Prager (1952); Mirolyubov (1953)
paraboloid	$\gamma_1 \in]0, 1[, \gamma_2 = 0$	Burzyński (1928, 1929a,b); Torre (1947); Balandin (1937)
ellipsoid centered of symmetry plane $I_1 = 0$	$\gamma_1 = -\gamma_2 \in]0, 1[$	Beltrami (1885)
ellipsoid	$\gamma_1 \in]0, 1[, \gamma_2 < 0$	Schleicher (1926, 1928)
hyperboloid of two sheets	$\gamma_1 \in]0, 1[, \gamma_2 \in]0, \gamma_1[$	Burzyński (1928, 2009); Yagn (1931, 1933)
hyperboloid of one sheet centered of symmetry plane $I_1 = 0$	$\gamma_1 = -\gamma_2 = a\imath \quad (\imath = \sqrt{-1}, a \in \mathbb{R})$	Kuhn (1980)
hyperboloid of one sheet	$\gamma_{1,2} = b \pm a\imath \quad (\imath = \sqrt{-1}, a, b \in \mathbb{R})$	Gol'denblat and Kopnov (1968); Filonenko-Boroditsch (1960); Filin (1975)

the meridian that means a linear dependency of I_1. The model (77) has additional to the straight line curvilinear meridians: parabola, hyperbola and ellipse.

- The form in the π-plane and the inclination of the meridian of the models of Mohr-Coulomb and Pisarenko-Lebedev depends on a single parameter. This is a restriction for the fitting of experimental data.
- There is a need of additional criteria since the classical and the presented here standard criteria does not allow to characterize all advanced materials.

4 Mathematical Formulations of Criteria

In the case of phenomenological models some mathematical framework based on the combination of invariants, etc. can be applied for the formulation. The aim is to establish a general expression for the equivalent stress which includes classical and other models as special cases. The following formulation ideas are suggested in the literature and will be discussed below:

- Criteria of Altenbach-Zolochevsky I and II (Altenbach et al., 1993; Altenbach and Zolochevsky, 1994; Altenbach et al., 1995a,b),
- Model in terms of the integrity basis, and
- Models based on the stress deviator.

4.1 Criterion of Altenbach-Zolochevsky I

The formulation of the criterion of Altenbach-Zolochevsky I is given in Altenbach et al. (1993, 1994)

$$\sigma_{\mathrm{eq}} = \sigma_{\mathrm{vM}} \left(\lambda_1 \sin\varphi + \lambda_2 \cos\varphi + \lambda_3\right) + I_1 \left(\lambda_4 + \lambda_5 \sin\varphi + \lambda_6 \cos\varphi\right). \quad (80)$$

This is a combination of the first invariant of the stress tensor I_1, the function of the second invariant of the stress deviator $\sigma_{\mathrm{vM}} = \sqrt{3I_2'}$ and the stress angle φ. In the original paper the following definition of the stress angle is used, cf. (2) and Novozhilov (1951a)

$$\sin 3\varphi = -\frac{3\sqrt{3}}{2}\frac{I_3'(\boldsymbol{s})}{I_2'(\boldsymbol{s})^{3/2}}, \qquad |\varphi| \leq \frac{\pi}{6}.$$

The identification of the λ_i parameters can be performed with the help of tests. It is necessary to compute I_1, σ_{vM} and φ for each loading and to insert into Eq. (80).

1. Experiments (the tests 1a - 1c belongs to the basic tests in material testing Blumenauer 1996, the tests 1d - 1f are introduced to include the influence of the multiaxiality of the stress state)

a) uniaxial tension

$$\sigma_{11} = \sigma_+, \tag{81}$$

b) uniaxial compression

$$\sigma_{11} = -\sigma_-, \tag{82}$$

c) torsion

$$\sigma_{12} = \tau_*, \tag{83}$$

d) thin tubular specimen with closed ends under inner pressure p

$$\sigma_{11} = \frac{\sigma_+}{2}, \qquad \sigma_{22} = \sigma_+ \tag{84}$$

with

$$\sigma_+ = \frac{pR}{t},$$

where R is the radius and t is the thickness of the specimen,

e) balanced two-dimensional tension in a tubular specimen under inner pressure p^* and tension force F^*

$$\sigma_{11} = \frac{\sigma_{\mathrm{t}}^*}{2} + \frac{F^*}{A}, \qquad \sigma_{22} = \sigma_{\mathrm{t}}^* \tag{85}$$

with the constraint

$$\sigma_{11} = \sigma_{22} = \sigma^*$$

and

$$\sigma_{\mathrm{t}}^* = \frac{p^* R^*}{t^*},$$

where R^* is the radius, t^* is the thickness and A the cross section area of the specimen,

f) tension test in a hydrostatic pressure chamber

$$\sigma_{11} = \frac{F}{A} - q, \qquad \sigma_{22} = \sigma_{22} = -q \tag{86}$$

with the constraint that the first invariant $I_1 = 0$ that means

$$\sigma_{11} = \frac{2}{3}\sigma^{**}, \qquad \sigma_{22} = \sigma_{33} = -\frac{1}{3}\sigma^{**},$$

F is the tension force, q is the hydrostatic pressure and A the cross section area of the specimen.

2. Analytical solutions

a) uniaxial tension

$$I_1 = \sigma_+, \qquad \sigma_{\mathrm{vM}} = \sigma_+, \qquad \varphi = -\frac{\pi}{6}, \tag{87}$$

$$-\lambda_1 + \sqrt{3}\lambda_2 + 2\lambda_3 + 2\lambda_4 - \lambda_5 + \sqrt{3}\lambda_6 = 2, \tag{88}$$

b) uniaxial compression ($d = |\sigma_-|/\sigma_+$)

$$I_1 = -\sigma_-, \qquad \sigma_{\mathrm{vM}} = \sigma_-, \qquad \varphi = \frac{\pi}{6}, \tag{89}$$

$$\lambda_1 + \sqrt{3}\lambda_2 + 2\lambda_3 - 2\lambda_4 - \lambda_5 - \sqrt{3}\lambda_6 = \frac{2}{d}, \tag{90}$$

c) torsion ($k = \sqrt{3}\tau^*/\sigma_+$)

$$I_1 = 0, \qquad \sigma_{\mathrm{vM}} = \sqrt{3}\tau^* = k\sigma_+, \qquad \varphi = 0, \tag{91}$$

$$\lambda_2 + \lambda_3 = \frac{1}{k}, \tag{92}$$

d) tubular specimen under inner pressure p

$$I_1 = \frac{3}{2}\sigma_{\mathrm{t}}, \qquad \sigma_{\mathrm{vM}} = \frac{\sqrt{3}}{2}\sigma_{\mathrm{t}}, \qquad \varphi = 0, \tag{93}$$

$$\sqrt{3}\lambda_2 + \sqrt{3}\lambda_3 + 3\lambda_4 + 3\lambda_6 = 2\frac{\sigma_+}{\sigma_{\mathrm{t}}}, \tag{94}$$

e) balanced two-dimensional tension in a tubular specimen under inner pressure p^* and tension force F^*

$$I_1 = 2\sigma^*, \qquad \sigma_{\mathrm{vM}} = \sigma^*, \qquad \varphi = \frac{\pi}{6}, \tag{95}$$

$$\lambda_1 + \sqrt{3}\lambda_2 + 2\lambda_3 + 4\lambda_4 + 2\lambda_5 + 2\sqrt{3}\lambda_6 = 2\frac{\sigma_+}{\sigma^*}, \tag{96}$$

f) tension test in a hydrostatic pressure chamber

$$I_1 = 0, \qquad \sigma_{\mathrm{vM}} = \sigma^{**}, \qquad \varphi = -\frac{\pi}{6}, \tag{97}$$

$$-\lambda_1 + \sqrt{3}\lambda_2 + 2\lambda_3 = 2\frac{\sigma_+}{\sigma^{**}}. \tag{98}$$

From the Eqs. (88), (90), (92), (94), (96) and (98) now the values of the λ_i are following:

$$
\begin{aligned}
\lambda_1 &= \frac{1}{3}\left(\frac{2}{d} - 3\frac{\sigma_+}{\sigma^{**}} + \frac{\sigma_+}{\sigma^*}\right), \\
\lambda_2 &= -\frac{1}{3(2-\sqrt{3})}\left(\frac{2}{d} - \frac{2}{k} + 3\frac{\sigma_+}{\sigma^{**}} + \frac{\sigma_+}{\sigma^*}\right), \\
\lambda_3 &= \frac{1}{3(2-\sqrt{3})}\left(\frac{2}{d} - \frac{\sqrt{3}}{k} + 3\frac{\sigma_+}{\sigma^{**}} + \frac{\sigma_+}{\sigma^*}\right), \\
\lambda_4 &= \frac{1}{3(2-\sqrt{3})}\left(3 - \frac{1}{d} + \frac{1}{k} - 2\sqrt{3}\frac{\sigma_+}{\sigma_\mathrm{t}} - 3\frac{\sigma_+}{\sigma^{**}} + \frac{\sigma_+}{\sigma^*}\right), \\
\lambda_5 &= -\frac{1}{3}\left(3 + \frac{1}{d} - 3\frac{\sigma_+}{\sigma^{**}} - \frac{\sigma_+}{\sigma^*}\right), \\
\lambda_6 &= -\frac{1}{3(2-\sqrt{3})}\left(3 - \frac{1}{d} + \frac{2}{\sqrt{3}k} - 4\frac{\sigma_+}{\sigma_\mathrm{t}} - 3\frac{\sigma_+}{\sigma^{**}} + \frac{\sigma_+}{\sigma^*}\right).
\end{aligned}
\tag{99}
$$

It is obvious that the criterion (80) has limited possibilities to fit experimental results since there are only the material parameters $\lambda_i (i = 1, \ldots, 6)$. Let us present the main special cases of this 6-parameter criterion:

- 1-parameter criteria
  - von Mises criterion (von Mises, 1913)

$$\lambda_3 = 1, \qquad \lambda_1 = \lambda_2 = \lambda_4 = \lambda_5 = \lambda_6 = 0$$

  - Tresca criterion (Tresca, 1868)

$$\lambda_2 = \frac{2\sqrt{3}}{3}, \qquad \lambda_1 = \lambda_3 = \lambda_4 = \lambda_5 = \lambda_6 = 0$$

  - Mariotte criterion (normal strain criterion with $\nu = 1/2$) (Mariotte, 1700)

$$\lambda_1 = -\frac{1}{2}, \qquad \lambda_2 = \frac{\sqrt{3}}{2}, \qquad \lambda_3 = \lambda_4 = \lambda_5 = \lambda_6 = 0$$

  - Johnson or normal stress criterion (Johnson, 1960)

$$\lambda_1 = -\lambda_4 = -\frac{1}{3}, \qquad \lambda_2 = \frac{\sqrt{3}}{3}, \qquad \lambda_3 = \lambda_5 = \lambda_6 = 0$$

  - Sdobyrev criterion (Sdobyrev, 1959)

$$\lambda_1 = -\lambda_4 = -\frac{1}{6}, \qquad \lambda_2 = \frac{\sqrt{3}}{6}, \qquad \lambda_3 = \frac{1}{2}, \qquad \lambda_5 = \lambda_6 = 0$$

- 2-parameter criteria ($d = \sigma_+/|\sigma_-|$ or $\chi \in [0,1]$)
  - Mohr-Coulomb criterion (Coulomb, 1776; Mohr, 1900a,b)

$$\lambda_1 = -\lambda_4 = \frac{1}{3}\left(\frac{1}{d} - 1\right), \quad \lambda_2 = \frac{\sqrt{3}}{3}\left(\frac{1}{d} + 1\right),$$
$$\lambda_3 = \lambda_5 = \lambda_6 = 0$$

  - Drucker-Prager criterion (Drucker and Prager, 1952)

$$\lambda_3 = \frac{1}{2}\left(1 + \frac{1}{d}\right), \qquad \lambda_4 = \frac{1}{2}\left(1 - \frac{1}{d}\right),$$
$$\lambda_1 = \lambda_2 = \lambda_5 = \lambda_6 = 0$$

  A similar criterion was introduced by Mirolyubov (1953) (see also (Pisarenko and Lebedev, 1976; Tsybulko and Romanenko, 2008)
  - Pisarenko-Lebedev criterion (Pisarenko and Lebedev, 1968)

$$\lambda_1 = -\lambda_4 = \frac{1}{3}\left(\frac{1}{d} - 1\right), \qquad \lambda_2 = \frac{\sqrt{3}}{3}\left(1 - \frac{1}{d}\right),$$
$$\lambda_3 = \frac{1}{d}, \qquad \lambda_5 = \lambda_6 = 0$$

  - Sandel criterion (Sandel, 1919)

$$\lambda_2 = \frac{\sqrt{3}}{3}\left(1 + \frac{1}{d}\right), \qquad \lambda_4 = \frac{1}{2}\left(1 - \frac{1}{d}\right),$$
$$\lambda_1 = \lambda_3 = \lambda_5 = \lambda_6 = 0$$

  - incompressible Edelman-Drucker criterion (Edelman and Drucker, 1951; Koval'chuk, 1981)

$$\lambda_2 = \frac{2\sqrt{3}}{3}(1 - \chi), \qquad \lambda_3 = \chi, \qquad \lambda_1 = \lambda_3 = \lambda_5 = \lambda_6 = 0$$

- 3-parameter criteria
  - Paul criterion (Paul, 1961)

$$\lambda_1 = \frac{1}{3}(2a_2 - a_1 - a_3), \qquad \lambda_2 = \frac{\sqrt{3}}{3}(a_1 - a_3),$$
$$\lambda_4 = \frac{1}{3}(a_1 + a_2 + a_3), \qquad \lambda_3 = \lambda_5 = \lambda_6 = 0$$

- incompressible Tsvelodub criterion (Tsvelodub, 1991)

$$\lambda_1 \neq 0, \qquad \lambda_2 \neq 0, \qquad \lambda_3 \neq 0, \qquad \lambda_4 = \lambda_5 = \lambda_6 = 0$$

- 4-parameter criteria
  - Birger criterion (Birger, 1977)

$$\lambda_1 = \frac{1}{3}(2a_2 - a_1 - a_3), \qquad \lambda_2 = \frac{\sqrt{3}}{3}(a_1 - a_3), \qquad \lambda_3 = a_4,$$
$$\lambda_4 = \frac{1}{3}(a_1 + a_2 + a_3), \qquad \lambda_5 = \lambda_6 = 0$$

  - Tarasenko criterion (Tarasenko, 1957)

$$\lambda_2 \neq 0, \qquad \lambda_4 \neq 0, \qquad \lambda_5 \neq 0, \qquad \lambda_6 \neq 0, \qquad \lambda_1 = \lambda_3 = 0$$

4.2 Criterion of Altenbach-Zolochevsky II

The starting point of the formulation of this isotropic criterion are the following invariants (Sect. 2.1)

$$J_1 = \boldsymbol{\sigma} \cdot\cdot \boldsymbol{I}, \qquad J_2 = \boldsymbol{\sigma} \cdot\cdot \boldsymbol{\sigma}, \qquad J_3 = \boldsymbol{\sigma} \cdot (\boldsymbol{\sigma} \cdot\cdot \boldsymbol{I}). \tag{100}$$

These invariants can be combined and one gets a new linear (Σ_1), quadratic (Σ_2) and cubic (Σ_3) invariant

$$\Sigma_1 = \mu_1 J_1, \qquad \Sigma_2^2 = \mu_2 J_1^2 + \mu_3 J_2, \qquad \Sigma_3^3 = \mu_4 J_1^3 + \mu_5 J_1 J_2 + \mu_6 J_3. \tag{101}$$

Finally, the equivalent stress can be introduced as follows

$$\sigma_{\mathrm{eq}} = \alpha \Sigma_1 + \beta \Sigma_2 + \gamma \Sigma_3. \tag{102}$$

Here $\mu_i (i = 1, \ldots, 6)$ are material parameters, which should be estimated from some basic tests. In Altenbach and Zolochevsky (1994) the following tests are suggested:

- uniaxial tension with measurements in the
  - longitudinal direction,
  - transverse direction,
- uniaxial compression,
- torsion with measurements of
  - shear,
  - longitudinal direction,
- hydrostatic pressure

α, β, γ characterize the weight of each invariant expression (101). Note that the criterion (102) has more possibilities to approximate real tests in comparison with the criterion (80).

It is easy to show that the von Mises criterion can be obtained from Eqs. (100) - (102). At first one has to assume the following values

$$\alpha = \gamma = 0, \qquad \beta = 1$$

The values μ_2 and μ_3 can be estimated as follows. The von Mises criterion (69) is given by

$$\sigma_{\rm vM} = \sqrt{\frac{3}{2}\boldsymbol{s} \cdot\cdot \boldsymbol{s}}, \qquad \boldsymbol{s} = \boldsymbol{\sigma} - \frac{1}{3}\boldsymbol{\sigma} \cdot\cdot \boldsymbol{I}.$$

The quadratic invariant Σ_2 leads to

$$\Sigma_2^2 = \mu_2 I_1^2 + \mu_3 I_2 = \mu_2 (\boldsymbol{\sigma} \cdot\cdot \boldsymbol{I})^2 + \mu_3 \boldsymbol{\sigma} \cdot\cdot \boldsymbol{\sigma} = \left(\mu_2 + \frac{1}{3}\mu_3\right) (\boldsymbol{\sigma} \cdot\cdot \boldsymbol{I})^2 + \mu_3 \boldsymbol{s} \cdot\cdot \boldsymbol{s}$$

The comparison with the von Mises criterion results in

$$\mu_3 = \frac{1}{3}, \qquad \mu_2 = -\frac{1}{2}$$

In the same way other special cases can be deduced. For example, the criterion of Burzyński (1928) and Yagn (1931) follows with $\gamma = 0$ and the criterion of Spitzig et al. (1975) with $\alpha \neq 0$, $\beta \neq 0$, $\gamma \neq 0$ and $\mu_2 = \mu_4 = \mu_5 = 0$.

4.3 Model in Terms of the Integrity Basis

This model results from the invariants I_1, $(I_2')^{1/2}$ and I_3', cf. Betten and Borrmann (1988); Betten (2001); Desai (1980); Schur and Grunsky (1968) that means at first scalar valued functions of a given order are formulated:

$$\begin{aligned} S_1 &= a_1\, I_1 + b_1\, (I_2')^{1/2}, \\ S_2 &= a_2\, I_1^2 + b_2\, I_2', \\ S_3 &= a_3\, I_1^3 + b_3\, (I_2')^{3/2} + c_3\, I_3 + d_3\, I_1\, I_2 + e_3\, I_1^2\, (I_2')^{1/2}, \\ &\ldots \end{aligned} \tag{103}$$

The sum of the S_i with the same power n yields

$$(S_1)^n + (S_2)^{n/2} + (S_3)^{n/3} + \ldots = \sigma_{\rm eq}^n. \tag{104}$$

The choice of integer exponents n/i is recommended for the terms S_i: $n = 1, 2, 3, 6, 9$ and 12 are preferred values to obtain known criteria. With

$n = 1$ and the first three functions S_i one gets the criterion of Altenbach-Zolochevsky II (102), which was shown in Kolupaev (2006). The advantage of this model is that the equivalent stress $\sigma_{\rm eq}$ can be expressed explicitly.

Another modification one gets if the weight $\sigma_{\rm eq}^{n-i}$ for S_i is introduced

$$\sigma_{\rm eq}^{n-1} S_1 + \sigma_{\rm eq}^{n-2} S_2 + \sigma_{\rm eq}^{n-3} S_3 + \ldots + \sigma_{\rm eq} S_{n-1} + S_n = \sigma_{\rm eq}^n. \tag{105}$$

By this way we get the same power of the stresses in each term (Kolupaev, 2006). The exponent $n > 1$ and the terms in (105) can be selected in such a manner that an analytical solution is possible with respect to $\sigma_{\rm eq}$. Equation (77) is an example of a quadratic equation. Models which are given cubic, bi-cubic and tri-quadratic equations are

$$S_1 \sigma_{\rm eq}^2 + S_2 \sigma_{\rm eq} + S_3 = \sigma_{\rm eq}^3, \tag{106}$$

$$S_2 \sigma_{\rm eq}^4 + S_4 \sigma_{\rm eq}^2 + S_6 = \sigma_{\rm eq}^6, \tag{107}$$

$$S_3 \sigma_{\rm eq}^3 + S_6 = \sigma_{\rm eq}^6. \tag{108}$$

More examples are discussed in Kolupaev (2006). Disadvantages of this approach can be summarized as follows:

- increasing number of parameters,
- difficult convexity limits for the parameters,
- the influence of I_1 can be not separated from the influence of I'_2 and I'_3 and
- missing geometrical interpretation of the parameters.

4.4 Models based on the Invariants of the Stress Deviator

The functions of the invariants of the stress deviator can be defined as follows

$$\begin{aligned} S'_2 &= b_2 I'_2, \\ S'_3 &= b_3 (I'_2)^{3/2} + c_3 I_3, \\ S'_4 &= b_4 (I'_2)^2 + f_4 (I'_2)^{1/2} I_3, \\ &\ldots \end{aligned} \tag{109}$$

The sum of S'_i with the same power results in

$$(S'_2)^{n/2} + (S'_3)^{n/3} + (S'_4)^{n/4} + \ldots = \sigma_{\rm eq}^n \tag{110}$$

and

$$\sigma_{\rm eq}^{n-2} S'_2 + \sigma_{\rm eq}^{n-3} S'_3 + \ldots + \sigma_{\rm eq} S'_{n-1} + S'_n = \sigma_{\rm eq}^n, \tag{111}$$

cf. Eqs. (104) and (105).

Another possibility is

$$\sigma_{\text{eq}}^{n-2\,m_2}\,(S_2')^{m_2}+\sigma_{\text{eq}}^{n-3\,m_3}\,(S_3')^{m_3}+\ldots+\sigma_{\text{eq}}\,(S_{n-1}')+(S_n')=\sigma_{\text{eq}}^n. \quad (112)$$

The formulations (110), (111) and (112) yield in various models of incompressible material behavior, for example Dodd and Naruse (1989); Freudenthal and Gou (1969); Iyer and Lissenden (2003); Maitra et al. (1973); Sayir (1970); Spitzig et al. (1975).

Multiplicative combinations of various S_i' are possible, for example,

$$(S_2')^{(n-j)/2}\,S_j'=\sigma_{\text{eq}}^n. \quad (113)$$

This equation results in the cosine ansatz models on the base of the stress angle (18):

$$(3\,I_2')^{\frac{n}{2}}\,\frac{1+c_3\cos 3\theta+c_6\cos^2 3\theta}{1+c_3+c_6}=\sigma_{\text{eq}}^n \quad (114)$$

with

$$d^n=\frac{1+c_3+c_6}{1-c_3+c_6},\qquad k^n=1+c_3+c_6. \quad (115)$$

The following summary can be given:

- The formulation of the models with the deviatoric basis (109) should be preferred since they are simpler in comparison with models on the basis of Eq. (103).
- In the case of rational functions of I_3' (functions of I_3' with integer power) one gets convex shapes in the π-plane.
- The compressible generalization can be performed using the substitution on the base of I_1.

5 Compressibility and Incompressibility

The modeling of incompressible material behavior is based on neglecting of the first invariant

$$\Phi(I_2',I_3',\sigma_{\text{eq}})=0$$

or

$$\Phi(I_2',\theta,\sigma_{\text{eq}})=0.$$

Such an idealization cannot be accepted in many practical cases. It is an acceptable concept, especially if $I_1\leq 0$.

The incompressible models can be extended to compressible ones:

- by a linear substitution (straight meridian) (Sayir, 1970)

$$\sigma_{\mathrm{eq}} \to \frac{\sigma_{\mathrm{eq}} - \gamma_1 I_1}{1 - \gamma_1}, \quad \gamma_1 \in [0, 1[, \tag{116}$$

- by a quadratic substitution (77)

$$\sigma_{\mathrm{eq}}^2 \to \frac{\sigma_{\mathrm{eq}} - \gamma_1 I_1}{1 - \gamma_1} \frac{\sigma_{\mathrm{eq}} - \gamma_2 I_1}{1 - \gamma_2}, \quad \gamma_1 \in [0, 1[, \tag{117}$$

- and a generalized substitution

$$\sigma_{\mathrm{eq}} \to \sqrt[j+l+m]{\left(\frac{\sigma_{\mathrm{eq}} - \gamma_1 I_1}{1 - \gamma_1}\right)^j \left(\frac{\sigma_{\mathrm{eq}} - \gamma_2 I_1}{1 - \gamma_2}\right)^l \sigma_{\mathrm{eq}}^m}, \quad \gamma_1 \in [0, 1[. \tag{118}$$

The parameters $\gamma_1 \in [0, 1[$ and γ_2 describe the position of the hydrostatic nodes. In the case of materials which do not fail at hydrostatic pressure (steel, brass, etc.) one gets $\gamma_2 \in [0, \gamma_1[$. Otherwise for materials which fail at hydrostatic pressure (hard foams, ceramics, sintered materials, etc.) $\gamma_2 < 0$. The integer powers $j \geq 1$, $l \geq 0$ and $m \geq 0$ wich describe the curvature of the meridian. The meridian with $l = m = 0$ is a straight line and with $l = 0$ - a parabola.

6 Anisotropic Failure Criteria

With the development of some technologies, for example metal sheet rolling, and advanced materials like glass-reinforced plastics, the classical or improved isotropic criteria should be modified. The aim of the modification is to take into account the anisotropy of the material behavior. There are several suggestions in the literature (Ashkenazi, 1965; Goldenblat and Kopnov, 1965; Hashin, 1980; Mālmeisters et al., 1977; Schürmann, 2007; Christensen, 2013; Ganczarski and Skrzypek, 2013). In many papers it is mentioned that the first anisotropic criterion was presented by Hill (1948), but it is easy to proof that von Mises (1928) has made the first and more generalized suggestion for an anisotropic failure criterion.

The formulation of anisotropic criteria can be performed similar to the isotropic criteria:

- intuitive extension of classical or improved isotropic criteria or
- mathematical formulations.

In contrast to the isotropic criteria here we start with the mathematical formulation which is known as the tensor polynomial failure criterion. It seems that this criterion at first was presented in Goldenblat and Kopnov (1965).

6.1 Tensor Polynomial Failure Criterion

Goldenblat and Kopnov (1965) published the following criterion[5]

$$(\boldsymbol{A}\cdot\cdot\,\boldsymbol{\sigma})^{\alpha} + (\boldsymbol{\sigma}\cdot\cdot\,{}^{(4)}\boldsymbol{B}\cdot\cdot\,\boldsymbol{\sigma})^{\beta} + \left[(\boldsymbol{\sigma}\cdot\cdot\,{}^{(6)}\boldsymbol{C}\cdot\cdot\,\boldsymbol{\sigma})\cdot\cdot\,\boldsymbol{\sigma}\right]^{\gamma} + \ldots \leq 1. \qquad (119)$$

Here $\boldsymbol{A}, {}^{(4)}\boldsymbol{B}, {}^{(6)}\boldsymbol{C}, \ldots$ are strength tensors of different orders, α, β, γ are fitting parameters. The Goldenblat-Kopnov criterion is a mathematical formulation, which is an extension of the suggestion of von Mises (1928). It is not possible to apply this criterion in practice since the number of parameters in Eq. (119) is increasing dramatically with the increase of the number of terms in this criterion. It is easy to show that $\boldsymbol{A}$ contains 6 parameters in the general case of anisotropy (assuming only that the stress tensor $\boldsymbol{\sigma}$ is symmetrically). ${}^{(4)}\boldsymbol{B}$ contains 21 parameters similar to the Hookean tensor in the anisotropic elasticity, taking all symmetries into account ${}^{(6)}\boldsymbol{C}$ has 56 linear independent parameters.

It is obvious that nobody can establish enough tests for the estimation of these parameters. The number of parameters slightly decreases assuming incompressibility or plan stress state. But even in these cases there are no examples of parameter identification presented in the literature.

With respect of these difficulties in Goldenblat and Kopnov (1965) the general criterion was limited to linear and quadratic terms

$$\boldsymbol{A}\cdot\cdot\,\boldsymbol{\sigma} + \sqrt{\boldsymbol{\sigma}\cdot\cdot\,{}^{(4)}\boldsymbol{B}\cdot\cdot\,\boldsymbol{\sigma}} \leq 1. \qquad (120)$$

and the plane stress state was assumed. This results in a significant decrease of the number of strength parameters in $\boldsymbol{A}$ and ${}^{(4)}\boldsymbol{B}$. For materials with identical properties at tension and compression $\boldsymbol{A} = \boldsymbol{0}$ and we get the von Mises theory (von Mises, 1928). Examples of the estimation of the coordinates of the tensors are presented in Goldenblat and Kopnov (1965) and Mālmeisters et al. (1977).

6.2 Modified Altenbach-Zolochevsky Criterion

In Altenbach et al. (1995a) the anisotropic extension of the isotropic model Altenbach-Zolochevsky II (Sect. 4.2) is discussed. Instead of the general form of the criterion of Goldenblat and Kopnov (1965) the following equivalent stress expression is suggested

$$\sigma_{\text{eq}} = \alpha\Sigma_1 + \beta\Sigma_2 + \gamma\Sigma_3. \qquad (121)$$

[5]The basics of tensor calculus are given, for example, in Altenbach (2012); Lebedev et al. (2010); Lurie (2005).

Here α, β, γ are again weight coefficients and the $\Sigma_1, \Sigma_2, \Sigma_3$ now denotes

$$\Sigma_1 = \boldsymbol{a} \cdot\cdot \boldsymbol{\sigma}, \qquad \Sigma_2^2 = \boldsymbol{\sigma} \cdot\cdot {}^{(4)}\boldsymbol{b} \cdot\cdot \boldsymbol{\sigma}, \qquad \Sigma_3^3 = \boldsymbol{\sigma} \cdot\cdot \left(\boldsymbol{\sigma} \cdot\cdot {}^{(6)}\boldsymbol{c} \cdot\cdot \boldsymbol{\sigma}\right).$$

It is obvious that the three strength tensors $\boldsymbol{a}, {}^{(4)}\boldsymbol{b}, {}^{(6)}\boldsymbol{c}$ contain 819 coordinates ($\boldsymbol{a}$ - 9, ${}^{(4)}\boldsymbol{b}$ - 81, ${}^{(6)}\boldsymbol{c}$ - 729). From the assumption that the stress tensor is symmetrically we get a reduction up to 83 coordinates ($\boldsymbol{a}$ - 6, ${}^{(4)}\boldsymbol{b}$ - 21, ${}^{(6)}\boldsymbol{c}$ - 56). But even for this case hardly can be introduced enough independent tests for the estimation of the stiffness parameters.

Further reduction we get for special cases of the anisotropy:

- Orthotropy
  Now we have 32 coordinates ($\boldsymbol{a}$ - 3, ${}^{(4)}\boldsymbol{b}$ - 9, ${}^{(6)}\boldsymbol{c}$ - 20).
- Transversal-Isotropy
  In this case we have 16 coordinates ($\boldsymbol{a}$ - 2, ${}^{(4)}\boldsymbol{b}$ - 5, ${}^{(6)}\boldsymbol{c}$ - 9).
- Isotropy
  In this case we have 6 coordinates ($\boldsymbol{a}$ - 1, ${}^{(4)}\boldsymbol{b}$ - 2, ${}^{(6)}\boldsymbol{c}$ - 3).

Remark 6.1. From the last item we can make a helpful conclusion: if we have a classical isotropic material only 2-parameter criteria can give an adequate and full description of the failure behavior. Introducing additional restrictions like the incompressibility assumption a one-parameter criterion can give acceptable results. If we have more effects (for example the strength differential effect), then we need as a minimum a 2-parameter criterion for incompressible materials, otherwise a 3-parameter criterion.

Remark 6.2. The Goldenblat-Kopnov criterion (120) and the modified Altenbach-Zolochevsky (121) result in identical formulation if

- (120) is restricted by three terms,
- in (120) are taken $\alpha = 1, \beta = \frac{1}{2}, \gamma = \frac{1}{3}$, and
- in (121) are taken $\alpha = \beta = \gamma = 1$.

Remark 6.3. It should be mentioned that the applicability range of anisotropic criteria is not well investigated. In this sense a recently published paper Ganczarski and Skrzypek (2014) is highly recommended for further reeding.

6.3 Other Approaches

Considering anisotropic material behavior other proposals for failure criteria are suggested in the literature. One approach is based on the mapping of the anisotropic continuum to the isotropic model continuum. The approach is discussed with respect to anisotropic material behavior which can be described with the help of a potential in Betten (1976, 1981, 1982). This approach is effective - but the number of material parameters is very huge.

Special cases with respect to composites and metallic materials are widely discussed in the literature. Here want to give only references to Christensen (2013) and the chapters 2 and 5 of F. Barlat and R. Talreja in this lecture notes.

7 Conclusion

The correctness of any hypothesis can be verified by experimental data. On the other hand, there are not enough accurate data at combined stress states. For example, the scattering of the measurements allows the fitment of different models by the same experimental sets. From this it follows that the uniqueness of the choice of a criterion cannot be established - there are no sufficient conditions for the choice.

Let us introduce some points for the surface Φ in the principal space:

- From the isotropy assumption it follows the trigonal or hexagonal symmetry of the surface in the π-plane. The rotational symmetry can be obtained for models with smooth surfaces as an interim solution.
- Assumption of convexity in the π-plane and meridian plane is not necessary for all failure criteria.

Considering the big number of models suggested, up to now there are no physical statements for the shape of the surface Φ. The models can be only formulated empirically. Various models were recently proposed, but they are valid only in particular cases. The generalized models like

- Altenbach-Zolochevsky I (Sect. 4.1),
- Altenbach-Zolochevsky II (Sect. 4.2),
- models in terms of the integrity basis (Sects 4.3 and 4.4),
- tensorial polynomial criterion (Sect. 6.1),
- modified Altenbach-Zolochevsky criterion (Sect. 6.2)

allow a simple classification, but they are developed within different mathematical frameworks.

One of the most efficient methods is the restriction of the permissible forms of the surface in the stress space. There are some plausibility conditions:

- reliability and trustworthiness of the predictions,
- simple and confident application,
- clear geometric background,
- physical background,
- minimal number of parameters,
- dimensionless parameters,

- continuous differentiability even for limit surfaces,
- continuous differentiability in the hydrostatic nodes,
- containing well-known hypotheses to $\sigma_{\rm eq}$,
- wide range of convex shapes in the π-plane,
- continuous-differentiable change of the shape of the surface in the π-plane for intersections $I_1 = $ const.,
- account of the medial stress,
- dependence of the models for Φ of all three invariants,
- only rational functions of the invariants I_1 and I'_3, and
- maximum of the stress power not higher than 12.

From these items we conclude that the following models are very efficient:

- the Unified Strength Theory (UST) of Yu (2004); Kolupaev et al. (2013b):
  The model is piecewise linear with straight meridians and can be fitted in a satisfying manner for materials, if the hydrostatic pressure is not restricted. It can be recommended for plane stress states. The generalization for hydrostatic tension can be performed with the cut-off of the normal stress hypothesis.
- the cosine ansatz of (114) with the substitution of (118) if $n = 2, 3, 6$: The curvature of the meridian is independent from the shape of the π-plane. During the fitting procedure the convexity of the π-plane should be controlled. With $l = m$=0 some solutions of the UST can be approximated.
- the linear combination of the models of Sayir and Haythornthwaite, Sect. 3.2, (Altenbach et al., 2014) with the substitution of (118) if $n = 2, 3, 6$:
  This model can present the majority of shapes in the π-plane. With the substitution of (118) various types of meridians can be realized. With $l = m = 0$ various solutions of the UST can be approximated.

The topic (development of limit criteria) is still in the focus of research and there are up to now many publications in this field, like Brencich and Gambarotta (2001); Cuntze (2013); Donato and Bianchi (2012); Ehlers and Avci (2013); Ehlers and Scholz (2007); Ghorbel (2008); Heyder et al. (2002); Karaoulanis (2013); Lian et al. (2013); Liu et al. (2012); Mortara (2009); Penasa et al. (2014); Yoon et al. (2014); Zapara et al. (2012); Zhang et al. (2014) etc..

Bibliography

H. Altenbach. *Kontinuumsmechanik - Eine Einführung in die materialunabhängigen und materialabhängigen Gleichungen.* Springer, 2012.

H. Altenbach and A.A. Zolochevsky. Eine energetische Variante der Theorie des Kriechens und der Langzeitfestigkeit für isotrope Werkstoffe mit komplizierten Eigenschaften. *ZAMM*, 74(3):189–199, 1994.

H. Altenbach, U. Lauschke, and A. Zolochevsky. Ein verallgemeinertes Versagenskriterium und seine Gegenüberstellung mit Versuchsergebnissen. *ZAMM*, 73(4-5):T 372–T 375, 1993.

H. Altenbach, J. Murín, M. Dutko, and A. Zoločevsky. O možnosti zovšeobecnenia klasickych pevnostných hypotéz (On the possibilities of generalization of classical srength criteria, in Sloviakian). *Strojnícky Časopis*, 45(5):418–427, 1994.

H. Altenbach, J. Altenbach, and A. Zolochevsky. *Erweiterte Deformationsmodelle und Versagenskriterien der Werkstoffmechanik.* Deutscher Verlag für Grundstoffindustrie, 1995a.

H. Altenbach, J. Altenbach, and A. Zolochevsky. A generalized constitutive equation for creep of polymers at multi-axial loading. *Mechanics of Composite Materials*, 31(6):511–518, 1995b.

H. Altenbach, A. Bolchoun, and V.A. Kolupaev. Phenomenological yield and failure criteria. In H. Altenbach and A. Öchsner, editors, *Plasticity of Pressure-Sensitive Materials*, pages 49–152. Springer, Heidelberg, 2014.

S.A. Ambarcumyan. *Multimodulus Elasticity Theory (Razno-modul'naja teorija uprugosti, in Russ.).* Nauka, Moscow, 1982.

B.D. Annin. Theory of ideal plasticity with a singular yield surface. *J. of Applied Mechanics and Technical Physics*, 40(2):347–353, 1999.

E. K. Ashkenazi. Problems of the anisotropy of strength. *Mechanics of Polymers*, 1(2):60–70, 1965.

C. Bach. *Elastizität und Festigkeit.* Springer, 1902.

G. Backhaus. *Deformationsgesetze.* Akademie-Verlag, Berlin, 1983.

P.P. Balandin. On the strength hypotheses (K voprosu o gipotezakh prochnosti, in Russ.). *Vestnik inzhenerov i tekhnikov*, 1:19–24, 1937.

A.J.C. Barré de Saint-Venant. Théorie et equations générales du mouvement non permanent des eaux courantes. *Comptes Rendus des Séances de l'Académie des Sciences*, 73:147–154, 1871.

E. Beltrami. Sulle condizioni di resistenza dei corpi elastici. *Il Nuovo Cimento*, 18(1):145–155, 1885.

J. Besson, D. Steglich, and W. Brocks. Modelling of plain strain ductile rupture. *Intern. J. Plasticity*, 19:47–67, 2003.

J. Betten. Ein Beitrag zur Invariantentheorie in der Plastomechanik anisotroper Werkstoffe. *ZAMM*, 56:557–559, 1976.

J. Betten. Zur Aufstellung der Stoffgleichungen in der Kriechmechanik anisotroper körper. *Rhelogica Acta*, 20(6):527–535, 1981.

J. Betten. Zur Aufstellung einer Integritätsbasis für Tensoren zweiter und vierter Stufe. *ZAMM*, 62:T274–275, 1982.

J. Betten. *Tensorrechnung für Ingenieure.* B.G. Teubner, Stuttgart, 1987.

J. Betten. *Kontinuumsmechanik.* Springer, Berlin, 2nd edition, 2001.

J. Betten. *Creep Mechanics.* Springer, Berlin, 2008.

J. Betten and M. Borrmann. Der Poynting-Effekt als Ursache einer werkstoffbedingten Anisotropie. *Forschung im Ingenieurwesen*, 54(1):16–18, 1988.

I.A. Birger. On a criterion for fracture and plasticity (Ob odnom kriterii razrushenija i plastichnosti, in Russ.). *Mechanika tverdogo tela, Izvestija Akademii Nauk SSSR*, (4):143–150, 1977.

B. Bischoff-Beiermann and O. Bruhns. A physically motivated set of invariants and tensor generators in the case of transverse isotropy. *Int. J. Eng. Sci.*, 32:1531–1552, 1994.

H. Blumenauer. *Werkstoffprüfung.* Dt. Verl. für Grundstoffindustrie, Leipzig, 1996.

J.P. Boehler, editor. *Application of Tensor Functions in Solid Mechanics*, CISM Courses and Lectures no. 292, Wien-New York, 1987. Springer.

A. Bower. *Applied Mechanics of Solids.* CRC Press, Boca Raton, 2010.

A. Brencich and L. Gambarotta. Isotropic damage model with different tensile-compressive response for brittle materials. *Int. J. Solids Struct.*, 38:5865–5892, 2001.

O. Bruhns, H. Xiao, and A. Meyers. On representation of yield functions for crystals, quasicrystals and transversely isotropic solids. *Eur. J. Mech. A/Solids*, 18:47–67, 1999.

W. Burzyński. *Studjum nad hipotezami wytężenia.* Akademia Nauk Technicznych, Lwów, 1928.

W. Burzyński. Über die Anstrengungshypothesen. *Schweizerische Bauzeitung*, 94(21):259–262, 1929a.

W. Burzyński. Über die Anstrengungshypothesen. *Schweizerische Bauzeitung*, 95(7):87–88, 1929b.

W. Burzyński. Selected passages from Włodzimierz Burzyński's doctoral dissertation "Study on material effort hypotheses" printed in Polish by the Academy of Technical Sciences, Lwów, 1928, 1-192. *Engng Trans. Polish Academy of Sciences*, 57:3–4, 127–157, 2009.

M. Capurso. Yield conditions for incompressible isotropic and orthotropic materials with different yield stress in tension and compression. *Meccanica*, 2(2):118–125, 1967.

W.F. Chen and H. Zhang. *Structural Plasticity - Theory, Problems, and CAE Software.* Springer, New York, 1991.

R.M. Christensen. *The Theory of Materials Failure.* University Press, Oxford, 2013.

C.A. Coulomb. Essai sur une application des regles des maximis et minimis a quelquels problemes de statique relatifs, a la architecture. *Mem. Acad. Roy. Div. Sav.*, 7:343–387, 1776.

R. Courant. *Methods of Mathematical Physics*, volume 2. Partial Differential Equations. Wiley Interscience Publication, New York, 1989.

F. Correia de Araújo. *Elasticidade e plasticidade.* Imprensa Portuguesa, Porto, 1961.

R.G. Cuntze. Comparison between experimental and theoretical results using Cuntze's "failure mode concept" model for composites under triaxial loadings - Part B of the second world-wide failure exercise. *Journal of Composite Materials*, 47(6-7):893–924, 2013.

C.S. Desai. A general basis for yield, failure and potential functions in plasticity. *Int. J. for Numerical and Analytical Methods in Geomechanics*, 4 (4):361–375, 1980.

B. Dodd and K. Naruse. Limitation on isotropic yield criteria. *Int. J. of Mechanical Sciences*, 31(7):511–519, 1989.

G.H.B. Donato and M. Bianchi. Pressure dependent yield criteria applied for improving design practices and integrity assessments against yielding of engineering polymers. *J. Materials Research and Technology*, 1(1): 2–7, 2012.

D.C. Drucker and W. Prager. Soil mechanics and plastic analysis or limit design. *Quarterly of Appl. Mathematics*, 10:157–165, 1952.

F. Edelman and D.C. Drucker. Some extensions of elementary plasticity theory. *J. of the Franklin Institute*, 251(6):581–605, 1951.

W. Ehlers and O. Avci. Stress-dependent hardening and failure surfaces of dry sand. *International Journal for Numerical and Analytical Methods in Geomechanics*, 37(8):787–809, 2013.

W. Ehlers and B. Scholz. An inverse algorithm for the identification and the sensitivity analysis of the parameters governing micropolar elasto-plastic granular material. *Archive of Applied Mechanics*, 77(12):911–931, 2007.

V.A. Eremeyev and W. Pietraszkiewicz. Local symmetry group in the general theory of elastic shells. *J. Elasticity*, 85(2):125–152, 2006.

V.A. Eremeyev, L.P. Lebedev, and H. Altenbach. *Foundations of Micropolar Mechanics.* SpringerBriefs in Applied Sciences and Technology. Springer, Heidelberg, 2013.

A.P. Filin. *Applied Mechanics of Solid Deformable Bodies (in Russ.: Prikladnaja mechanika tverdogo deformiruemogo tela)*, volume 1. Nauka, Moscow, 1975.

M.M. Filonenko-Borodich. *Theory of Elasticity.* P. Noordhoff W. N., Groningen, 1960.

M.M. Filonenko-Boroditsch. *Festigkeitslehre*, volume 1. Technik, Berlin, 1960.

I. Finnie and W.R. Heller. *Creep of Engineering Materials.* McGraw-Hill, New York, 1959.

A.M. Freudenthal and R.F. Gou. Second order effects in the theory of plasticity. *Acta Mech.*, 8(1):34–52, 1969.

A. Ganczarski and J. Skrzypek. *Mechanics of advanced materials (Mechanika nowoczesnykh materiałów, in Polish).* Politechnika Krakowska, Kraków, 2013.

A.W. Ganczarski and J.J. Skrzypek. Constraints on the applicability range of Hill's criterion: strong orthotropy or transverse isotropy. *Acta Mech.*, 225(9):2563–2582, 2014.

E. Ghorbel. A viscoplastic constitutive model for polymeric materials. *International Journal of Plasticity*, 24(11):2032 – 2058, 2008.

I.I. Goldenblat and V.A. Kopnov. Strengt of glass-reinforced plastics in the complex stress state. *Mechanics of Polymers*, 2:54–59, 1965.

I.I. Gol'denblat and V.A. Kopnov. *Yield and Strength Criteria for Structural Materials (Kriterii prochnosti i plastichnosti konstrukzionnych materialov, in Russ.).* Mashinostroenie, Moscow, 1968.

I.I. Goldenblat and V.A. Kopnov. General theory of strengt criteria of isotropic and anisotropic materials (Obshaja teorija kriteriev prochosti isotropnykh i anizotropnykh materialov, in Russ.). *Problemy Prochnosti*, 5:65–69, 1971.

F. Grashof. *Theorie der Elasticität und Festigkeit.* Gaertner, Berlin, 1878.

A.L. Gurson. Continuum theory of ductile rupture by void nucleation and growth: Part I. Yield criteria and flow rules for porous ductile media. *Trans. ASME. J. of Engineering Materials and Technology*, 99:215, 1977.

Z. Hashin. Failure criteria for unidirectional fiber composites. *Trans ASME. J. Appl. Mech.*, 47(2):329–334, 1980.

R.M. Haythornthwaite. Mechanics of triaxial tests for soil. *Proc. ASCE. J. Soil Mech. and Foundations Div. SM5*, 86:35–62, 1960.

R.M. Haythornthwaite. Range of yield condition in ideal plasticity. *Proc. ASCE. J. Eng. Mech. Div. EM6*, 87:117–133, 1961.

H. Hencky. Zur Theorie plastischer Deformationen und der hierdurch im Material hervorgerufenen Nachspannungen. *ZAMM*, 4(4):323–334, 1924.

M. Heyder, A. Öchsner, and S. Ströhla. Bewertung von Spannungszuständen bei der experimentellen Bestimmung des Schubmoduls an zellularen Materialien. *Materialwissenschaft und Werkstofftechnik*, 33 (2):85–89, 2002.

R. Hill. A theory of the yielding and plastic flow of anisotropic metals. *Proc. Roy. Soc. London*, 193:281–297, 1948.

R. Hill. On the inhomogeneous deformation of a plastic lamina in a compression test. *Philosophical Magazine, Series 7*, 41(319):733–744, 1950.

M.T. Huber. Specific strain work as a measurment of material effort (właściwa praca odkształcenia jako miara wytężenia materyału, in Polish). *Czasopismo Techniczne*, 22:34–40, 49–50, 61–62, 80–81, 1904.

A. Yu. Ishlinsky. Hypothesis of strength of shape change (Gipoteza prochnosti formoizmenenija, in Russ.). *Uchebnye Zapiski Moskovskogo Universiteta, Mekhanika*, 46:104–114, 1940.

A.Yu. Ishlinsky and D.D. Ivlev. *Mathematical Theory of Plasticity (Matematicheskaja teorija plastichnosti, in Russ.)*. Fizmatlit, Moscow, 2003.

M. Itskov. *Tensor Algebra and Tensor Analysis for Engineers*. Springer, Berlin, 2007.

S.K. Iyer and C.J. Lissenden. Multiaxial constitutive model accounting for the strength-differential in Inconel 718. *Int. J. of Plasticity*, 19:2055–2081, 2003.

A.E. Johnson. Complex stress creep of metals. *J. Met. Reviews*, 20:477–506, 1960.

F.E. Karaoulanis. Implicit numerical integration of nonsmooth multisurface yield criteria in the principal stress space. *Archives of Computational Methods in Engineering*, 20(3):263–308, 2013.

V.A. Kolupaev. *Dreidimensionales Kriechverhalten von Bauteilen aus unverstärkten Thermoplasten*. PhD thesis, Martin-Luther-Universität Halle-Wittenberg, Halle, 2006.

V.A. Kolupaev, W. Becker, H. Massow, and D. Dierkes. Auslegung von Probekörpern aus Hartschaum zur Ermittlung der biaxialen Zugfestigkeit. *Forschung im Ingenieurwesen*, 2013a. doi: DOI 10.1007/s10010-014-0175-9.

V.A. Kolupaev, M.H. Yu, and H. Altenbach. Visualisation of the Unified Strength Theory. *Archive of Appl. Mechanics*, 83:1061–1085, 2013b.

B.I. Koval'chuk. On the criterion for limit state of some hull steels under complex loading states at normal and above-normal temperatures (O kriterii predel'nogo sostojanija nekotorich korpusnich stalej v uslovijach slozhnogo naprjazhennogo sostojanija pri kompatnoj i povishennich temperaturach, in Russ.). *Problemi Prochnosti*, (5):10–15, 1981.

P. Kuhn. *Grundzüge einer allgemeinen Festigkeitshypothese, Auszug aus Antrittsvorlesung des Verfassers vom 11. Juli 1980 "Vom Konstrukteur und den Festigkeitshypothesen"*. Inst. für Maschinenkonstruktionslehre, Karlsruhe, 1980.

A. Lagzdiņš and V. Tamužs. Tensorial representation of the orientation distribution function of internal structure elements for heterogeneous solids. *Mathematics and Mechanics of Solids*, 1:193–205, 1996.

A. Lagzdiņš, V. Tamužs, G. Teters, and A. Krēgers. *Orientational Averaging in Mechanics of Solids*. Longman, London, 1992.

A.Zh. Lagzdin and V.P. Tamuzh. Construction of a phenomenological theory of fracture of anisotropic media. *Mechanics of Polymers*, 7(4):563–571, 1971.

L.P. Lebedev, M.J. Cloud, and V.A. Eremeyev. *Tensor Analysis with Applications in Mechanics*. World Scientific, Singapore, 2010.

F.A. Leckie and D.R. Hayhurst. Constitutive equations for creep rupture. *Acta Metall.*, 25:1059–1070, 1977.

J. Lemaitre and J.L. Chaboche. *Mechanics of Solid Materials*. Cambridge University Press, Cambridge, 1990.

V.I. Levitas and L.K. Shvedov. Low-pressure phase transformation from rhombohedral to cubic bn: Experiment and theory. *Phys. Rev. B*, 65: 104109, Feb 2002.

Q.M. Li. Strain energy density failure criterion. *International J. of Solids and Structures*, 38:69977013, 2001.

J. Lian, M. Sharaf, F. Archie, and S. Münstermann. A hybrid approach for modelling of plasticity and failure behaviour of advanced high-strength steel sheets. *International Journal of Damage Mechanics*, 22(2):188–218, 2013.

M. Liu, Y. Gao, and H. Liu. A nonlinear drucker-prager and matsuoka-nakai unified failure criterion for geomaterials with separated stress invariants. *International Journal of Rock Mechanics and Mining Sciences*, 50:1 – 10, 2012.

A.M. Lokoshchenko. Long-term strength of metals in complex stress state (a survey). *Mechanics of Solids*, 47(3):357–372, 2012.

A.I. Lurie. *Theory of Elasticity*. Springer, Berlin, 2005.

M. Maitra, K. Majumdar, and A. Das. Unified plastic yield criterion for ductile solids. *AIAA J.*, 11(10):1428–1429, 1973.

A. Mālmeisters, V. Tamužs, and G. Teters. *Mechanik der Polymerwerkstoffe*. Akademie-Verlag, Berlin, 1977.

E. Mariotte. *Traité du mouvement des eaux et des autres corps fluides*. J. Jambert, Paris, 1700.

I.N. Mirolyubov. On the generalization of the strengt theory based on the octaedral stresses in the case of brittle materials (K voprosu ob obobshchenii teorii prochnosti oktaedricheskikh kasatel'nyh naprjazhenij na khrupkie materialy, in Russ.). *Trudy Leningradskogo Technologicheskogo Instituta*, pages 42–52, 1953.

O. Mohr. Welche Umstände bedingen die Elastizitätsgrenze und den Bruch eines Materials. *Z. VDI*, 45:1524–1530, 1900a.

O. Mohr. Welche Umstände bedingen die Elastizitätsgrenze und den Bruch eines Materials. *Z. VDI*, 46:1572–1577, 1900b.

G. Mortara. A hierarchical single yield surface for frictional materials. *Computers and Geotechnics*, 36(6):960 – 967, 2009.

K. Naumenko and H. Altenbach. *Modeling of Creep for Structural Analysis.* Springer, Heidelberg, 2007.

A. Needleman and V. Tvergaard. An analysis of ductile rupture in notched bars. *J. Mech. Phys. Solids*, 32:461–490, 1984.

A. Needleman and V. Tvergaard. An analysis of ductile rupture at a crack tip. *J. Mech. Phys. Solids*, 35:151–183, 1987.

F.E. Neumann. *Vorlesungen über die Theorie der Elastizität der festen Körper und des Lichtäthers.* B. G. Teubner, Leipzig, 1885.

V.V. Novozhilov. On the connection between stresses and strains in a nonlinear-elastic continuum (o svjazi mezhdu naprjazhenijami i deformazijami v nelinejno-uprugoj srede, in Russ.). *Prikladnaja Matematika i Mekhanika*, XV(2):183–194, 1951a.

V.V. Novozhilov. On the principles of the statical analysis of the experimental results for isotropic materials (O prinzipakh obrabotki rezultatov staticheskikh ispytanij izotropnykh materialov, in Russ.). *Prikladnaja Matematika i Mekhanika*, XV(6):709–722, 1951b.

N.S. Ottosen and M. Ristinmaa. *The Mechanics of Constitutive Modeling.* Elsevier Science, London, 2005.

P. Paufler. *Physikalische Kristallographie.* Akademie Verlag, Berlin, 1986.

B. Paul. A modification of the Coulomb-Mohr theory of fracture. *J. of Materials Science, Ser. E*, 28(2):259–268, 1961.

R. B. Pęcherski, P. Szeptyński, and M. Nowak. An extension of Burzyński hypothesis of material effort accounting for the third invariant of stress tensor. *Arch. Metallurgy and Materials*, 56(2):503–508, 2011.

M. Penasa, A. Piccolroaz, L. Argani, and D. Bigoni. Integration algorithms of elastoplasticity for ceramic powder compaction. *Journal of the European Ceramic Society*, 34(11):2775 – 2788, 2014.

G.S. Pisarenko and A.A. Lebedev. On the shape of a limit surface for a mechanical strength criterion (O forme predel'noj poverchnosti mechanicheskogo kriterija prochnosti, in Russ.). *Prikladnaja Mechanika*, 4 (3):45–50, 1968.

G.S. Pisarenko and A.A. Lebedev. *Deformation and Strength of Materials under Complex Stress State (Deformirovanie i prochnost' materialov pri slozhnom naprjazhennom sostojanii, in Russ.).* Naukowa Dumka, Kiev, 1976.

W. Prager and P. Hodge. *Theorie ideal plastischer Körper.* Springer, Wien, 1954.

D. Radaj. *Festigkeitsnachweise 1 (Grundverfahren) und 2 (Sonderverfahren).* Dt. Verl. für Schweißtechnik, Düsseldorf, 1974.

B. Raniecki and Z. Mróz. Yield or martensitic phase transformation conditions and dissipation functions for isotropic, pressure-insensitive alloys exhibiting SD effect. *Acta Mech.*, 195:81–102, 2008.

W.J.M. Rankine. *Manual of Applied Mechanics.* Griffin, London, 1876.

J. Résal. *Résistance des matériaux.* Librairie polytechnique, Baudry & cie., Paris, 1898.

A. Reuss. Vereinfachte Beschreibung der plastischen Formänderungsgeschwindigkeiten bei Voraussetzung der Schubspannungsfließbedingung. *ZAMM*, 13(5):356–360, 1933.

G. Rousselier. Finite constitutive strain relations including ductile fracture ramming. In S. Nemat-Nasser, editor, *Three-Dimensional Constitutive Relations and Ductile Fracture*, pages 331–355. North Holland, Amsterdam, 1981.

G. Rousselier. The Rousselier model for porous metal plasticity and ductile fracture. In J. Lemaitre, editor, *Handbook of Materials Behavior Models*, pages 436–445. Academic Press, San Diego, 2001a.

G. Rousselier. Dissipation in porous metal plasticity and ductile fracture. *J. Mech. Phys. Solids*, 49:1727–1746, 2001b.

J. Rychlewski and J. Zhang. On representation of tensor functions: a review. *Advances in Mechanics*, 14:75–94, 1991.

S. Sähn, H. Göldner, K.-F. Fischer, and J. Nickel. *Bruch- und Beurteilungskriterien in der Festigkeitslehre.* Fachbuchverbuchverlag, Leipzig-Kln, 1993.

G.D. Sandel. *Über die Festigkeitsbedingungen: ein Beitrag zur Lösung der Frage der zulässigen Anstrengung der Konstruktionsmaterialen.* PhD thesis, TeH, Stuttgart, 1919.

J. Sauter and N. Wingerter. *Neue und alte statische Festigkeitshypothesen*, volume 191 of *Fortschritts-Berichte VDI Reihe 1: Konstruktionstechnik/Maschinenelemente.* VDI, Düsseldorf, 1990.

M. Sayir. Zur Fließbedingung der Plastizitätstheorie. *Ing. Arch*, 39:414–432, 1970.

H. Schade and K. Neemann. *Tensoranalysis.* de Gruyter, Berlin, 2. edition, 2006.

F. Schleicher. Der Spannungszustand an der Fließgrenze (Plastizitätsbedingung). *Ztschr. f. Math. und Mech.*, 6(3):199–216, 1926.

F. Schleicher. Über die Sicherheit gegen Überschreiten der Fliessgrenze bei statischer Beanspruchung. *Der Bauingenieur*, 9(15):253–261, 1928.

R. Schmidt. Über den Zusammenhang von Spannungen und Formänderungen im Verfestigungsgebiet. *Ing. Arch*, 3(3):215–235, 1932.

I. Schur and H. Grunsky. *Vorlesungen über Invariantentheorie: Die Grundlehren der mathematischen Wissenschaften in Einzelndarstellungen.* Springer, Berlin, 1968.

H. Schürmann. *Konstruieren mit Faser-Kunststoff-Verbunden.* Springer, Berlin, 2007.

V.P. Sdobyrev. Criterion for the long term strength of some heat-resistant alloys at a multiaxial loading (Kriterij dlitelnoj prochnosti dlja nekotorykh zharoprochnykh splavov pri slozhnom naprjazhennom sostojanii, in Russ.). *Izvestija Akademii Nauk SSSR, Otdelenie tekhnicheskikh Nauk, Mechanika i Mashinostroenie*, (6):93–99, 1959.

J. J. Skrzypek. *Plasticity and Creep: Theory, Examples and Problems*. CRC Press, Boca Raton, 1993.

A.J.M. Spencer. Isotropic polynomial invariants and tensor functions. In J. Boehler, editor, *Applications of Tensor Functions in Solid Mechanics*, CISM Lecture Notes no. 292, pages 141–169. Springer, Wien-New York, 1987.

W. Spitzig, R. Sober, and O. Richmond. Pressure dependence of yielding and associated volume expansion in tempered martensite. *Acta Metall.*, 23(7):885–893, 1975.

V.P. Tamuzh and A.Zh. Lagzdyn'sh. A variant of the phenomenological theory of fracture. *Mechanics of Polymers*, 4(4-6):493–500, 1968.

I.I. Tarasenko. On the criteria of brittle strength of materials (O kriterijach khrupkoj prochnosti materialov, in Russ.). *Sbornik nauchnykh trudov, Leningradskij inzhenerno-stroitelnij institut*, 26:161–168, 1957.

S. P Timoshenko. *History of Strength of Materials: With a Brief Account of the History of Theory of Elasticity and Theory of Structure*. McGraw-Hill, New York, 1953.

S.P. Timoshenko and J.N. Goodier. *Theory of Elasticity*. McGraw-Hill, New York, 1987.

C. Torre. Einfluss der mittleren Hauptnormalspannung auf die Fließ- und Bruchgrenze. *Österreichisches Ingenieur-Archiv*, I(4/5):316–342, 1947.

H. Tresca. Mémoire sur l'ecoulement des corps solides. *Mémoires Pres. par Div. Savants*, 18:733–799, 1868.

I.Yu. Tsvelodub. *Stability postulate and its applications in the creep theory for metallic materials (Postulat ustojchivosti i ego prilozhenija v teorii polzuchesti metallicheskich materialov, in Russ.)*. Institut Gidromechaniki, Novosibirsk, 1991.

I.Yu. Tsvelodub. Multimodulus elasticity theory. *J. of Applied Mechanics and Technical Physics*, 49(1):129–135, 2008.

A. E. Tsybulko and E. A. Romanenko. Natural criterion of the limit tolerance state of materials differently resisting stretching and compression. *J. of Machinery Manufacture and Reliability*, 37(1):34–77, 2008.

V. Tvergaard. Influence of voids on shear band instabilities under plane strain conditions. *Int. J. Fracture*, 17:389–407, 1981.

V. Tvergaard. On localization in ductile materials containing spherical voids. *Int. J. Fracture*, 18:237–252, 1982.

V. Tvergaard and A. Needleman. Analysis of the cup-cone fracture in a round tensile bar. *Acta Metall.*, 32:157–169, 1984.

W. Voigt. *Lehrbuch der Krystallphysik.* B.G. Teubner, Leipzig, 1910.

R. von Mises. Mechanik des festen Körpers im plastischen deformablen Zustand. *Nachrichten der Königlichen Gesellschaft der Wissenschaften Göttingen, Mathematisch-physikalische Klasse*, pages 589–592, 1913.

R. von Mises. Mechanik der plastischen Formänderung von Kristallen. *ZAMM*, 8:161–185, 1928.

Yu. I. Yagn. *Strength of Materials: Theory and Problems (Soprotivlenie materialov: teorja i zadachnik, in Russ.).* Kubuch, Leningrad, 1933.

Yu.I. Yagn. New methods of strength prediction (Novye metody rascheta na prochnost', in Russ.). *Vestnik inzhenerov i tekhnikov*, 6:237–244, 1931.

Y.-P. Yao, D. Sun, and H. Matsuoka. *SMP Criterion-Based Unified Constitutive Model for Geomaterials*, chapter 13, pages 333–357. ASCE, 2005.

J.W. Yoon, Y. Lou, J. Yoon, and M.V. Glazoff. Asymmetric yield function based on the stress invariants for pressure sensitive metals. *International Journal of Plasticity*, 56:184 – 202, 2014.

M.-H. Yu. General behaviour of isotropic yield function (in Chinese). *Scientific and Technological Research Paper of Xi'an Jiaotong University*, pages 1–11, 1961.

M.-H. Yu. *Unified Strength Theory and its Applications.* Springer, Berlin, 2004.

M. Zapara, N. Tutyshkin, W.H. Müller, and R. Wille. Constitutive equations of a tensorial model for ductile damage of metals. *Continuum Mechanics and Thermodynamics*, 24(4-6):697–717, 2012.

S. Zhang, B. Song, X. Wang, D. Zhao, and X. Chen. Deduction of geometrical approximation yield criterion and its application. *Journal of Mechanical Science and Technology*, 28(6):2263–2271, 2014.

Q.S. Zheng. Theory of representations of tensor functions. *Appl. Mech. Rev.*, 47(11):545–587, 1994.

P.A. Zhilin. The modified theory of the tensor symmetry and tensor invariants (Modifizirovannaya teoriya simmetrii tenzorov i tenzornykh invariantov, in Russ.). *Izvestiya Vuzov. Severo-Kavkazkii Region. Estestvennye Nauki*, Special Issue:176–195, 2003.

P.A. Zhilin. *Rational Continuum Mechanics (Racional'naya mekhanika sploshnykh sred, in Russ.).* Polytechnic University Publisher, St. Petersburg, 2012.

M. Życzkowski. *Combined Loadings in the Theory of Plasticity.* PWN-Polish Scientific Publ., Warszawa, 1981.

Constitutive Description of Isotropic and Anisotropic Plasticity for Metals

Frédéric Barlat [*†] and Myoung-Gyu Lee [*]
[*] Graduate Institute of Ferrous Technology,
Pohang University of Science and Technology, Republic of Korea
[†] Currently at Institute for Virtual Manufacturing (IVP),
ETH Zürich, Switzerland

Abstract Modeling of the plastic behavior for isotropic and anisotropic metals is the topic of this article. The motivation for such work is briefly introduced. Then, a description of the main features of plastic deformation in metals at different scales is summarized to prepare the subsequent choices and assumptions made in the next sections. The main properties of the Cauchy stress tensor are reviewed because it is the main variable for plastic yielding. The description of plasticity for isotropic metals is discussed, which includes the yield condition, the flow rule and strain hardening. Then, the generalization of the concepts to plastic anisotropy, which is particularly important to the case of metal sheets and plates, is outlined. Finally, the influence of the constitutive description for plasticity on failure is briefly discussed using two examples.

1 Motivation

Numerical simulations based on the finite element (FE) approach are very useful to optimize manufacturing processes and predict product performances. Accurate results are achievable if sufficient consideration is given to the choice of key features, including type of mesh, boundary conditions and material constitutive behavior. The latter, in particular the plastic behavior of metals, is the topic of this paper. In plasticity, multi-scale modeling has been instrumental for understanding the relationship between macroscopic properties and microstructural features at different scales and has been successfully applied for material design (McDowell, 2010). Philosophically, multi-scale is a very comprehensive and interpretive approach to constitutive modeling. However, in many instances, it does not address very well practical needs when simple, yet accurate, material models with time-efficient implementations in commercial finite element codes are required. This is a domain where continuum descriptions are still very powerful.

H. Altenbach, T. Sadowski (Eds.), *Failure and Damage Analysis of Advanced Materials*,
CISM International Centre for Mechanical Sciences
DOI 10.1007/978-3-7091-1835-1_2

The goal of this article is to describe how to model plasticity for metallic materials at the continuum scale while keeping, in an approximate way, some aspects of the microstructure in the formulation. In order to develop macroscopic constitutive models that are relevant to metals and capture in a simplified way their structure, it is important to understand the relationship between macroscopic plastic properties and microscopic deformation mechanisms. These two aspects will be briefly discussed in Sect. 2. Section 3 will review the main properties of the stress tensor. The constitutive modeling of plasticity for isotropic and anisotropic metals will be introduced in Sects 4 and 5, respectively. Finally, applications of constitutive models to the prediction of failure will be illustrated by two examples in Sect. 5. The first deals with plastic flow localization during sheet metal forming while the second is about fracture toughness in thick plates.

2 Description and Modeling of Plasticity

2.1 Plasticity at Macro-Scale

Aspects of the plastic deformation of metals and alloys at low and moderate strain rates and temperatures, subjected to monotonic loading or to a few load cycles, are briefly discussed here. The reader is referred to textbooks (Dieter, 1988; Meyers and Chawla, 2009) or more advanced publications on the topic (Krausz and Krausz, 1996; Kocks et al., 1998; Raabe et al., 2004) for more details. The stress-strain behavior of metals and alloys at low strain is almost always reversible and linear. The elastic range however, is bounded by the yield limit, the stress above which permanent or inelastic deformations occur. In the plastic range, the flow stress usually increases with the total amount of plastic dissipation or a corresponding measure of accumulated plastic strain. This flow stress becomes the new yield stress if the material is unloaded and reloaded.

For fully dense metals, it is considered that plastic deformation occurs without any volume change and that hydrostatic pressure has virtually no influence on yielding. For material containing some porosity, the pressure independence assumption is no longer valid. Experiments conducted at high confinement pressure showed that, though very small, a pressure effect is quantifiable (Spitzig and Richmond, 1984) and can explain the strength differential (SD) effect for high strength steels. The SD effect corresponds to the difference between tension and compression yield stresses when both tests are conducted independently from an annealed state.

The Bauschinger effect is a common feature in metals and alloys that occurs when a material is deformed up to a given strain, unloaded and reloaded in the reverse direction, typically, tension followed by compression. The new yield stress after strain reversal is lower than the flow stress before unloading from the first deformation step. This phenomenon is a consequence of the building up of a back-stress field, i.e., a self-equilibrated stress in the matrix that remains when the material is freed from external loads. Bauschinger and SD effects are two different phenomena.

The flow stress of a metal usually decreases when the testing temperature increases. At low homologous temperatures, time has usually a very small influence on the flow stress and plasticity in general. However, at higher temperatures, strain rates effects are important. In fact, it has been observed that strain rate and temperature have similar effects on plasticity. Raising the temperature under which an experiment is carried out is equivalent to decreasing the strain rate. Temperature has another influence on plasticity. When subjected to a constant stress smaller than the yield limit, a material can deform by creep. A similar phenomenon, called relaxation, corresponds to a decrease in the applied stress when the strain is held constant.

Finally, solid state transformations can occur in materials due to an applied stress. These transformations lead to phase changes under stresses that are lower than the yield stress of either phase and can induce plasticity.

2.2 Plasticity at Micro-Scale

Commercial metals and alloys used in manufacturing are usually polycrystalline. They are composed of numerous grains, each with a given lattice orientation with respect to macroscopic axes. At low homologous temperature, metals and alloys deform by dislocation glide and by twinning on given crystallographic planes and directions, which produce microscopic shear deformations. Therefore, the distribution of grain orientations, the crystallographic texture, plays an important role in plasticity. Because of the geometrical nature of slip and twinning deformations, strain incompatibilities arise between grains and produce short-range residual stresses, which are partly responsible for the Bauschinger effect. Slip results in a gradual lattice rotation as deformation proceeds while twinning leads to abrupt changes in lattice orientation. The number of available slip systems determines the nature of the deformation mechanisms. Body-centered cubic (BCC) and face-centered cubic (FCC) materials tend to deform by slip because of the large number of available slip systems. However, hexagonal close-packed (HCP) materials, in which the number of potential slip systems is limited, generally tend to twin as an alternate mechanism to accommodate an imposed deformation. After slip, dislocations accumulate at microstructural obstacles, including twinned regions, and increase the slip resistance for further deformation, leading to strain hardening with its characteristic stress-strain curve.

At higher temperature, more slip systems may be available to accommodate the deformation. Atomic diffusion is also another mechanism that affects plastic deformation at high temperature and contributes to creep. In addition, grain boundary sliding becomes more significant. For instance, superplastic forming occurs mainly by grain boundary sliding. In this case, the grain size and shape are important parameters.

Commercial materials contain second-phases or intermetallic particles. These phases are present in materials by design in order to control either the microstructure such as the grain size or mechanical properties such as strength. However, some amounts of second-phases are undesired. In any case, the presence of these non-homogeneities alters the material behavior because of their differences in elastic properties with the matrix as it happens in composite materials, or because of their strong interactions with dislocations. In both cases, these effects produce incompatibility stresses that contribute to the Bauschinger effect.

The mechanisms of failure intrinsic to metals and alloys are plastic flow localization and fracture. Localization tends to occur in the form of shear bands, either micro-bands, which tend to be crystallographic, or macro-bands which are not. Ductile fracture is generally the result of mechanisms of void nucleation, growth and coalescence. The associated micro-porosity leads to volume changes although the matrix is plastically incompressible, and hydrostatic pressure affects the material behavior. At low homologous temperature, second-phases are principally the sites of damage. The stress concentration around these phases lead to void nucleation, and growth occurs by plasticity. Coalescence is the result of plastic flow micro-localization of the ligaments between voids. At higher temperature, where creep becomes dominant, cavities nucleate at grain boundaries by various mechanisms including grain boundary sliding and vacancy concentration.

2.3 Constitutive Modeling

As briefly reviewed above, plasticity of metals involves many aspects at scales spanning from atomic to macroscopic. Thus, it is necessary to limit the discussion to very focused areas only, mostly rate-independent anisotropic plasticity. In addition, the Bauschinger effect will be ignored.

As a general framework, the classical elements of the theory of plasticity are considered. For an isotropic material subjected to uniaxial tension at given temperature and strain rate, plastic yielding occurs when the stress σ reaches a critical value, the yield stress σ_{y}. This can be represented mathematically by the yield condition

$$\Phi = \sigma - \sigma_{\mathrm{y}} = 0 \tag{1}$$

Then, due to the accumulation of dislocations during deformation, this critical value increases, i.e., the material work-hardens. For an incompressible plastic material, the strain in any direction orthogonal to the tensile axis is $-1/2$ times that in the longitudinal direction, which defines the flow behavior in uniaxial tension. More generally, the yield condition is assumed to depend on the stress, temperature Θ and a number of state variables X_i, which describe the microstructure explicitly or implicitly

$$\Phi\left(\sigma, \Theta, X_i\right) = 0 \tag{2}$$

When the yield condition is fulfilled, the strain rate (or strain increment) is also a function of the stress, temperature and state variables. It defines the so-called state equation

$$\dot{\varepsilon} = \dot{\varepsilon}\left(\sigma, \Theta, X_i\right) \tag{3}$$

Since the material microstructure changes during plastic deformation, the state variables obey the following evolution equations

$$\dot{X}_i = \dot{X}_i\left(\sigma, \Theta, X_i\right) \tag{4}$$

For multiaxial loading, the stress is no longer defined by a scalar but by a tensor $\boldsymbol{\sigma}$. The state variables may be as numerous as necessary for a given problem and, in addition, defined by tensors. However, only one or two scalars will be considered in this article. The yield condition defines a surface in a six-dimensional space, which evolves as plastic deformation proceeds. This generalizes the notion of strain hardening in uniaxial tension. Finally, Eq. (3) is, in general, a tensorial relationship, which defines the strain rate tensor and extends the flow behavior explained above for uniaxial tension.

3 Stress Tensor

3.1 Representation

The Cauchy stress tensor is expressed in a set of mutually orthogonal unit base vectors $(\boldsymbol{e}_1, \boldsymbol{e}_2, \boldsymbol{e}_3)$ as

$$\boldsymbol{\sigma} = \sigma_{ij}\boldsymbol{e}_i \otimes \boldsymbol{e}_j, \tag{5}$$

where Einstein summation convention on repeated indices applies. The operator $\otimes$ represents the tensor (or open) product. This operation is defined with the scalar (or dot) product by operating on a third vector, namely,

$$\left(\boldsymbol{e}_i \otimes \boldsymbol{e}_j\right) \cdot \boldsymbol{e}_k = \boldsymbol{e}_i \otimes \boldsymbol{e}_j \cdot \boldsymbol{e}_k = \left(\boldsymbol{e}_j \cdot \boldsymbol{e}_k\right) \boldsymbol{e}_i = \delta_{jk}\boldsymbol{e}_i \tag{6}$$

where δ_{jk} denotes the Kronecker's symbol. Alternatively,

$$\boldsymbol{e}_k \cdot (\boldsymbol{e}_i \otimes \boldsymbol{e}_j) = \delta_{ki}\boldsymbol{e}_j \tag{7}$$

The traction vector, i.e., the stress vector $\boldsymbol{t}_i$ applied at a point of a solid on an infinitesimal surface of normal $\boldsymbol{e}_i$ is

$$\boldsymbol{t}_i = \boldsymbol{e}_i \cdot \boldsymbol{\sigma} = \boldsymbol{e}_i \cdot (\sigma_{pq}\boldsymbol{e}_p \otimes \boldsymbol{e}_q) = \sigma_{pq}\delta_{ip}\boldsymbol{e}_q = \sigma_{iq}\boldsymbol{e}_q \tag{8}$$

The stress tensor may also be represented by its generic component, σ_{ij}, or by the following matrix form

$$[\boldsymbol{\sigma}] = \begin{bmatrix} \sigma_{11} & \sigma_{12} & \sigma_{31} \\ \sigma_{12} & \sigma_{22} & \sigma_{23} \\ \sigma_{31} & \sigma_{23} & \sigma_{33} \end{bmatrix} \tag{9}$$

Note that, because moment equilibrium requirements (not discussed here), the stress tensor and the corresponding matrix are symmetric.

3.2 Transformations

If the tensor components in a certain set of base vectors $(\boldsymbol{e}_1, \boldsymbol{e}_2, \boldsymbol{e}_3)$ are known, the components in a different reference frame, say $(\hat{\boldsymbol{e}}_1, \hat{\boldsymbol{e}}_2, \hat{\boldsymbol{e}}_3)$, as illustrated in Fig. 1a, may be calculated. For this, let the matrix $\boldsymbol{R}$ rotate the set of base vectors $(\boldsymbol{e}_1, \boldsymbol{e}_2, \boldsymbol{e}_3)$ into $(\hat{\boldsymbol{e}}_1, \hat{\boldsymbol{e}}_2, \hat{\boldsymbol{e}}_3)$, i.e.,

$$\hat{\boldsymbol{e}}_i = R_{ij}\boldsymbol{e}_j \tag{10}$$

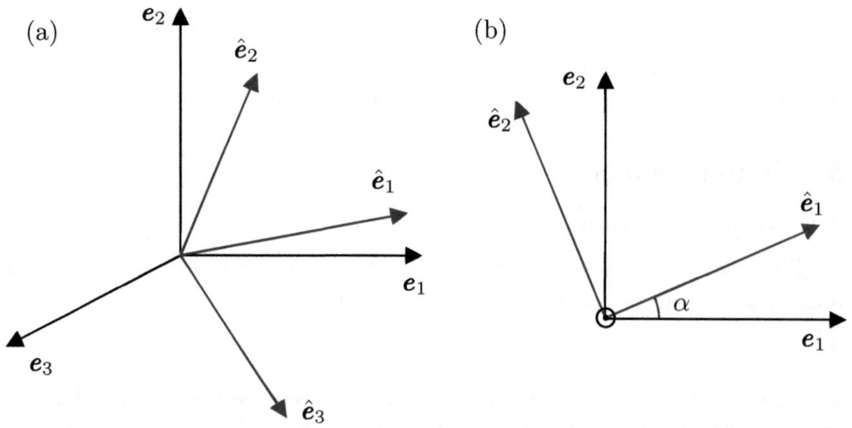

Figure 1. Reference bases made of mutually orthogonal unit vectors in 3D (a) and 2D (b) spaces.

Alternatively, this rotation may be expressed in the following matrix form

$$\begin{bmatrix} \hat{\boldsymbol{e}}_1 \\ \hat{\boldsymbol{e}}_2 \\ \hat{\boldsymbol{e}}_3 \end{bmatrix} = [\boldsymbol{R}] \begin{bmatrix} \boldsymbol{e}_1 \\ \boldsymbol{e}_2 \\ \boldsymbol{e}_3 \end{bmatrix} \tag{11}$$

Then, the stress tensor $\boldsymbol{\sigma}$ expressed in the two reference frames is

$$\boldsymbol{\sigma} = \hat{\sigma}_{ij}\hat{\boldsymbol{e}}_i \otimes \hat{\boldsymbol{e}}_j = \hat{\sigma}_{ij} R_{im}\boldsymbol{e}_m \otimes R_{jn}\boldsymbol{e}_n = \hat{\sigma}_{ij} R_{im} R_{jn}\boldsymbol{e}_m \otimes \boldsymbol{e}_n = \sigma_{mn}\boldsymbol{e}_m \otimes \boldsymbol{e}_n \tag{12}$$

Thus, the relationships between the components of a tensor expressed in two base vector sets are

$$\sigma_{mn} = R_{im} R_{jn} \hat{\sigma}_{ij} \quad \text{or, in matrix notation,} \quad [\sigma] = [\boldsymbol{R}]^{\mathrm{T}} [\hat{\sigma}] [\boldsymbol{R}] \tag{13}$$

Alternatively,

$$\hat{\sigma}_{mn} = R_{mi} R_{nj} \sigma_{ij} \quad \text{or, in matrix notation,} \quad [\hat{\sigma}] = [\boldsymbol{R}] [\sigma] [\boldsymbol{R}]^{\mathrm{T}} \tag{14}$$

The relationships in Eqs. (13) and (14) are very useful, particularly when studying the behavior of anisotropic materials.

Exercise 1

Plane stress is defined by the following matrix representation of the stress tensor,

$$[\boldsymbol{\sigma}] = \begin{bmatrix} \sigma_{11} & \sigma_{12} & 0 \\ \sigma_{12} & \sigma_{22} & 0 \\ 0 & 0 & 0 \end{bmatrix}$$

Calculate the components of $[\hat{\boldsymbol{\sigma}}]$ when $[\boldsymbol{R}]$ is defined by a rotation about the axis $\boldsymbol{e}_3$ and the angle α (Fig. 1b).

Hint: Use Eq. (11) to define $[\boldsymbol{R}]$ and Eq. (14).

Answer of exercise 1

$$\begin{aligned} \hat{\sigma}_{11} &= \sigma_{11}\cos^2\alpha + \sigma_{22}\sin^2\alpha + 2\sigma_{12}\cos\alpha\sin\alpha, \\ \hat{\sigma}_{22} &= \sigma_{11}\sin^2\alpha + \sigma_{22}\cos^2\alpha - 2\sigma_{12}\cos\alpha\sin\alpha, \\ \hat{\sigma}_{12} &= (\sigma_{22} - \sigma_{11})\sin\alpha\cos\alpha + \sigma_{12}(\cos^2\alpha - \sin^2\alpha) \end{aligned}$$

Exercise 2

Assume that $[\hat{\boldsymbol{\sigma}}]$ is represented in matrix form by

$$[\hat{\boldsymbol{\sigma}}] = \begin{bmatrix} \sigma_1 & 0 & 0 \\ 0 & \sigma_2 & 0 \\ 0 & 0 & 0 \end{bmatrix}$$

Using the results in the exercise above, calculate the maximum shear stress among all possible angles α.

Hint: Application of exercise 1 leads to $\hat{\sigma}_{12} = (\sigma_2 - \sigma_1)\sin\alpha\cos\alpha$.

Answer of exercise 2

The shear stress is maximum when $\alpha = \pi/4$.

3.3 Invariants

In order to introduce invariants of a tensor, it is useful to make an analogy with the case of a vector, for instance a force. A force is physically well defined by its intensity and direction although its components depend on the base vectors in which it is expressed. However, the force intensity (length of the vector) and direction are invariant because they are independent of the reference base vectors. Similarly, a tensor is a physical quantity for which invariants can be defined. The Cauchy stress tensor has three independent invariants but various combinations of such invariants are possible. A number of sets of stress tensor invariants are compared in Życzkowski (1981). The so-called principal invariants σ_k may be determined by solving the characteristic equation $\det(\boldsymbol{\sigma} - \sigma_k \boldsymbol{I}) = 0$, where $\boldsymbol{I}$ is the second order unit tensor ($I_{ij} = \delta_{ij}$ the Kronecker symbol). The expressions of these invariants I_1, I_2 and I_3 are

$$\begin{aligned} I_1 &= \sigma_{11} + \sigma_{22} + \sigma_{33}, \\ I_2 &= \sigma_{23}^2 + \sigma_{31}^2 + \sigma_{12}^2 - \sigma_{22}\sigma_{33} - \sigma_{33}\sigma_{11} - \sigma_{11}\sigma_{22}, \\ I_3 &= 2\sigma_{23}\sigma_{31}\sigma_{12} + \sigma_{11}\sigma_{22}\sigma_{33} - \sigma_{11}\sigma_{23}^2 - \sigma_{22}\sigma_{31}^2 - \sigma_{33}\sigma_{12}^2 \end{aligned} \tag{15}$$

Exercise 3

Develop $\det(\boldsymbol{\sigma} - \sigma_k \boldsymbol{I}) = 0$ to determine the characteristic polynomial equation and find the values of the principal invariants.

Hint: The determinant of $[\boldsymbol{\sigma} - \sigma_k \boldsymbol{I}]$ should be of the form

$$-\sigma_k^3 + I_1\sigma_k^2 + I_2\sigma_k + I_3$$

Answer of exercise 3

$$\begin{aligned} |\boldsymbol{\sigma} - \sigma_k \boldsymbol{I}| = & - \sigma_k^3 \\ & + \sigma_k^2(\sigma_{11} + \sigma_{22} + \sigma_{33}) \\ & - \sigma_k(\sigma_{11}\sigma_{22} + \sigma_{22}\sigma_{33} + \sigma_{33}\sigma_{11} - \sigma_{12}^2 - \sigma_{23}^2 - \sigma_{31}^2) \\ & + \sigma_{11}\sigma_{22}\sigma_{33} + 2\sigma_{12}\sigma_{23}\sigma_{31} - \sigma_{11}\sigma_{23}^2 - \sigma_{22}\sigma_{31}^2 - \sigma_{33}\sigma_{12}^2 = 0 \end{aligned}$$

In addition to the three principal invariants, the "angular invariant" may be introduced by the formula

$$\theta = \arccos\left[\frac{1}{2}\left(2I_1^3 + 3I_1I_2 + 2I_3\right)\left(I_1^2 + I_2\right)^{-3/2}\right] \tag{16}$$

from which, the ordered principal stresses ($\sigma_1 \geq \sigma_2 \geq \sigma_3$) are given by

$$\begin{aligned} \sigma_1 &= 2\sqrt{I_1^2 + I_2}\cos\left(\frac{\theta}{3}\right) + I_1, \\ \sigma_2 &= 2\sqrt{I_1^2 + I_2}\cos\left(\frac{\theta + 4\pi}{3}\right) + I_1, \\ \sigma_3 &= 2\sqrt{I_1^2 + I_2}\cos\left(\frac{\theta + 2\pi}{3}\right) + I_1 \end{aligned} \tag{17}$$

by solving for the roots of a cubic equation. Note that the principal stresses are invariant as well. Finally, three principal directions may generally be associated with the three principal values.

For a plane stress state, where three components are equal to zero, the principal values of the stress tensor are, more simply

$$\begin{aligned} \sigma_1 &= \frac{1}{2}\left[\sigma_{11} + \sigma_{22} + \sqrt{(\sigma_{11} - \sigma_{22})^2 + 4\sigma_{12}^2}\right], \\ \sigma_2 &= \frac{1}{2}\left[\sigma_{11} + \sigma_{22} - \sqrt{(\sigma_{11} - \sigma_{22})^2 + 4\sigma_{12}^2}\right] \end{aligned} \tag{18}$$

3.4 Deviator

The first invariant, I_1, is the trace of the Cauchy stress tensor and is very often replaced by the mean stress $\sigma_\mathrm{m} = I_1/3$. Then, the stress deviator s_{ij} can be defined from the stress tensor as

$$s_{ij} = \sigma_{ij} - \sigma_\mathrm{m}\delta_{ij} \tag{19}$$

The principal invariants of the stress deviator J_1, J_2 and J_3 can be calculated using Eq. (15) but applied to the components of the stress deviator instead of the stress tensor

$$\begin{aligned} J_1 &= s_{11} + s_{22} + s_{33} = 0, \\ J_2 &= s_{23}^2 + s_{31}^2 + s_{12}^2 - s_{22}s_{33} - s_{33}s_{11} - s_{11}s_{22}, \\ J_3 &= 2s_{23}s_{31}s_{12} + s_{11}s_{22}s_{33} - s_{11}s_{23}^2 - s_{22}s_{31}^2 - s_{33}s_{12}^2 \end{aligned} \tag{20}$$

Exercise 4

Show that

a. $J_2 = \frac{1}{2}s_{ij}s_{ij}$,

b. $J_2 = s_{23}^2 + s_{31}^2 + s_{12}^2 + \frac{(s_{22}-s_{33})^2 + (s_{33}-s_{11})^2 + (s_{11}-s_{22})^2}{6}$,

c. $J_2 = \sigma_{23}^2 + \sigma_{31}^2 + \sigma_{12}^2 + \frac{(\sigma_{22}-\sigma_{33})^2 + (\sigma_{33}-\sigma_{11})^2 + (\sigma_{11}-\sigma_{22})^2}{6}$

Hint: From Eq. (20), $J_2 = J_2' + J_2''$ with $J_2' = -s_{11}s_{22} - s_{22}s_{33} - s_{33}s_{11}$ and $J_2'' = s_{23}^2 + s_{31}^2 + s_{12}^2$. First, using $s_{22} = -s_{33} - s_{11}$, etc., show that $2J_2' = s_{11}^2 + s_{22}^2 + s_{33}^2$, which leads to the solution *a*. Second, adding J_2' on both side of the previous relationship, i.e. $3J_2' = s_{11}^2 + s_{22}^2 + s_{33}^2 + J_2'$, $3J_2'$ can be developed and regrouped appropriately, leading to *b.* and, using Eq. (19), to *c.*

Answer of exercise 4

$$J_2' = s_{11}(s_{33} + s_{11}) + s_{22}(s_{11} + s_{22}) + s_{33}(s_{22} + s_{33})$$

and, after developing

$$J_2' = s_{11}^2 + s_{22}^2 + s_{33}^2 - J_2'$$

or

$$2J_2' = s_{11}^2 + s_{22}^2 + s_{33}^2$$

Therefore,

$$2J_2 = 2J_2' + 2J_2'' = s_{11}^2 + s_{22}^2 + s_{33}^2 + 2s_{12}^2 + 2s_{23}^2 + 2s_{31}^2 = s_{ij}s_{ij}$$

which answers a.

$$3J_2' = s_{11}^2 + s_{22}^2 + s_{33}^2 + J_2' = s_{11}^2 + s_{22}^2 + s_{33}^2 - s_{11}s_{22} - s_{22}s_{33} - s_{33}s_{11}$$

which can be rearranged as

$$3J_1' = \frac{(s_{11} - s_{22})^2 + (s_{22} - s_{33})^2 + (s_{33} - s_{11})^2}{2}$$

This expression is used in

$$3J_2 = 3J_2' + 3J_2''$$

to answer *b.* and, with Eq. (19), *c.*

Exercise 5

Obtain the expressions of the principal invariants of the stress deviator as a function of the invariants of the stress tensor.
Hint: From Eq. (15), $I_2 = I_2' + I_2''$ with $I_2' = -\sigma_{11}\sigma_{22} - \sigma_{22}\sigma_{33} - \sigma_{33}\sigma_{11}$ and $I_2'' = \sigma_{23}^2 + \sigma_{31}^2 + \sigma_{12}^2$. First, using Eq. (19), show that $I_2' = J_2' - I_1^3/3$. This leads to J_2. Similarly, using Eqs. (15) and (19), find J_3.

Answer of exercise 5

$$\begin{aligned} J_1 &= 0, \\ J_2 &= I_2 + \frac{1}{3} I_1^2 \\ J_3 &= I_3 + \frac{1}{3} I_1 I_2 - \frac{2}{27} I_1^3 \end{aligned}$$

The above equations indicate that $J_1 = 0$ while J_2 is never negative. The principal values of the stress deviator, s_1, s_2 and s_3 are also invariants.

The behavior of an isotropic material should not depend on the reference base vectors in which the stress tensor is expressed. Therefore, the constitutive behavior of an isotropic material must be expressed by invariants. For instance, the yield condition can be written as a function of the three independent principal invariants of the stress tensor I_1, I_2 and I_3 or any other invariant or combinations of invariants, in particular, the three principal stresses. For isotropic materials that are independent of the mean stress σ_m, yielding may be written as a function of the two independent principal invariants J_2 and J_3 of the stress deviator or by the three principal deviatoric stresses among which, two only are independent. The use of principal values to define the plastic behavior of isotropic materials is preferred in the present context for reasons that will be given later.

The plastic behavior of metallic materials is highly non-linear. As a result, strains are calculated incrementally as described by Eq. (3). These plastic strain increments form a tensor with properties that are similar to those of the stress tensor. Therefore, these properties are not discussed any longer. For simple deformation modes, these increments may add to calculate the plastic strain. For complex stress states though, this requires a more advanced formalism, i.e., finite strains. However, this formalism is not needed in this article.

4 Isotropic Plasticity

A large number of yield conditions for many types of materials have been proposed over the last century and extensive reviews can be found, for example, in Życzkowski (1981); Yu (2002); Altenbach et al. (2014). In this chapter, only the most relevant yield conditions for isotropic metals are discussed.

4.1 Isosensitive Materials

The yield conditions presented in this section are not affected by a simultaneous change of the signs of all the stress components. In other words, these criteria predict equal absolute values of the yield stresses in tension and compression. Such materials, for which the corresponding yield surface exhibits symmetry about the origin, were called "isosensitive" by Życzkowski (1981).

Fully Dense Materials. In uniaxial tension, plastic deformation initiates when the stress reaches a critical value. For multiaxial loading, the yield condition is defined as

$$\Phi = \phi\left(\sigma_{ij}\right) - \phi_0 = 0 \tag{21}$$

where ϕ_0 is a certain constant. ϕ itself, which operates on the Cauchy stress tensor σ_{ij}, is called the yield function. The oldest condition for plastic yielding, proposed by Tresca (1864), expresses that plastic flow initiates when the maximum shear stress (see exercise 2 as an example) reaches a critical value, σ_s, the simple shear yield stress

$$\frac{\sigma_1 - \sigma_3}{2} = \sigma_s \tag{22}$$

This defines the yield surface as a hexagon in the principal stress space or in the π-plane. This yield function does not depend on the mean stress and can be written simply using the principal deviatoric stresses as well. The representation of the Tresca criterion as a function of principal stress invariants is more complex. The most common yield condition used for general purposes may be expressed with the second deviatoric stress invariant

$$3J_2 = \sigma_y^2 \tag{23}$$

or, in expanded form (with principal stresses only)

$$\frac{(\sigma_2 - \sigma_3)^2 + (\sigma_3 - \sigma_1)^2 + (\sigma_1 - \sigma_2)^2}{2} = \sigma_y^2 \tag{24}$$

Here, σ_y denotes the uniaxial tension yield stress. This yield condition is usually attributed to von Mises (1913) who proposed it as a convenient approximation of Eq. (22). However, Huber proposed it as early as 1904 as a condition of constant distortion energy, see Zyczkowski (1981); Engel (1994). The von Mises yield surface is a cylinder in the principal stress space and does not depend on the mean stress. In a similar way as Tresca, it can be written as a function of the principal deviatoric stresses as well.

Exercise 6

Express the Tresca and von Mises yield criteria using the principal deviatoric stresses.
Hint: Use Eq. (19).

Answer of exercise 6

Tresca criterion:

$$\frac{s_1 - s_3}{2} = \sigma_s$$

von Mises yield criterion:

$$\frac{(s_2 - s_3)^2 + (s_3 - s_1)^2 + (s_1 - s_2)^2}{2} = \sigma_y^2$$

Many experimental data showed that the measured yield surface of metals is located between the Tresca and Huber-Mises predictions. Taylor and Quinney (1931) reported extensive test results in which isotropic copper and steel tubes were loaded in combined tension and torsion. These authors found that the data were located between the two criteria. These findings suggested that the third invariant J_3 should be included in the expression of the yield criterion. Drucker (1949) proposed the following form

$$J_2^3 - cJ_3^2 = \sigma_s^6 \tag{25}$$

Assuming the material coefficient $c = 0$, the von Mises yield condition is recovered (with $3\sigma_s^2 = \sigma_y^2$), whereas direct transition to Tresca is not possible. Hershey (1954) proposed another description for the yield surface, which includes Tresca or von Mises as a particular case,

$$\phi = |\sigma_1 - \sigma_2|^a + |\sigma_2 - \sigma_3|^a + |\sigma_3 - \sigma_1|^a = 2\sigma_y^a \quad (26)$$

For the exponent $a = 2$ or $a = 4$, Equation (26) reduces to von Mises, whereas for $a = 1$ and in the limiting case $a \to \infty$, it leads to the Tresca yield condition. The yield function in Eq. (26) was developed as a good approximation of self-consistent crystal plasticity calculations. It was suggested by Hosford (1972) again for a similar reason. This author showed that, with exponents of 6 and 8, this equation leads to an almost perfect representation of the isotropic BCC and FCC yield surfaces, respectively, calculated with the Taylor (1938) full-constraint crystal plasticity model (Fig. 2). Karafillis and Boyce (1993) generalized Eq. (26) with the following form (denoted KB)

$$\begin{aligned} \phi &= (1-c)\left\{|s_1 - s_2|^a + |s_2 - s_3|^a + |s_3 - s_1|^a\right\} \\ &+ c\frac{3^a}{2^{a-1}+1}\left\{|s_1|^a + |s_2|^a + |s_3|^a\right\} = 2\sigma_y^a \end{aligned} \quad (27)$$

Exercise 7

Express the Hershey yield criterion with the principal deviatoric stresses and the KB criterion using the principal Cauchy stresses.
Hint: Use Eq. (19).

Answer of exercise 7

Hershey yield criteria:

$$|s_1 - s_2|^a + |s_2 - s_3|^a + |s_3 - s_1|^a = 2\sigma_y^a$$

KB yield criteria:

$$\begin{aligned} &(1-c)(|\sigma_1 - \sigma_2|^a + |\sigma_2 - \sigma_3|^a + |\sigma_3 - \sigma_1|^a) \\ &+c\frac{3^a}{2^{a-1}+1}\left(\frac{|2\sigma_1 - \sigma_2 - \sigma_3|^a}{3} + \frac{|2\sigma_2 - \sigma_3 - \sigma_1|^a}{3} + \frac{|2\sigma_3 - \sigma_1 - \sigma_2|^a}{3}\right) = 2\sigma_y^a \end{aligned}$$

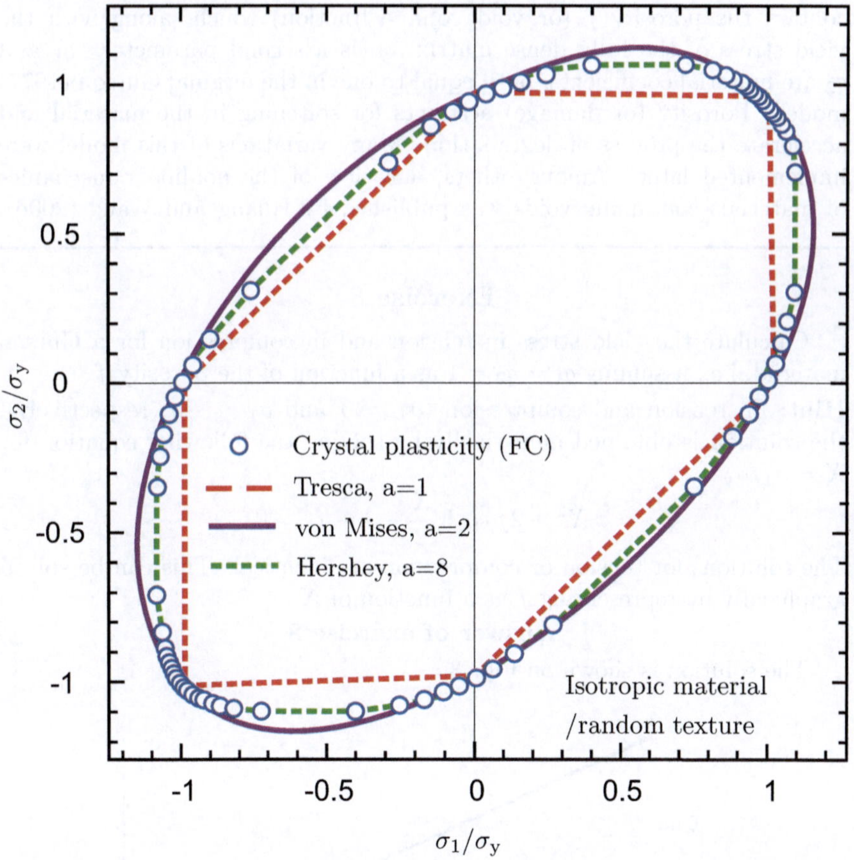

Figure 2. Isotropic yield surfaces obtained by Tresca, von Mises and Hershey, and compared to that determined by crystal plastic calculations using the full constraint (FC) model for an isotropic FCC material (random distribution of grain orientations).

Material with Porosity. To specifically account for porosity as the main mechanism of degradation, constitutive models, such as the model proposed by Gurson (1977) and later extended by Tvergaard (1982)

$$\Phi = \frac{3J_2}{\sigma_y^2} + 2q_1 f \cosh\left(\frac{q_2 I_1}{2\sigma_y}\right) - 1 + (q_1 f)^2 = 0 \tag{28}$$

contain the porosity f (or void volume fraction) which, along with the yield stress of the fully dense matrix σ_y, is a second parameter. q_1 and q_2 are material coefficients, both equal to one in the original Gurson (1977) model. Porosity (or damage) accounts for softening in the material and accelerate the process of degradation. Many variations of this model were implemented later. Among others, a review of the nonlinear mechanics of materials containing voids was published by Huang and Wang (2006).

Exercise 8

Calculate the yield stress in tension and in compression for a Gurson material, i.e., assuming $q_1 = q_2 = 1$ as a function of the porosity f.

Hint: In tension and compression ($\sigma_1 = \sigma$ and $\sigma_1 = -\sigma$, respectively), the solution is obtained numerically by solving the following equation for $X = \sigma_1/\sigma_y$

$$X^2 + 2f \cosh\frac{X}{2} - 1 + f^2 = 0$$

The solutions for tension or compression are identical. This can be solved graphically by representing f as a function of X.

Answer of exercise 8

The solution is shown on Fig. 3

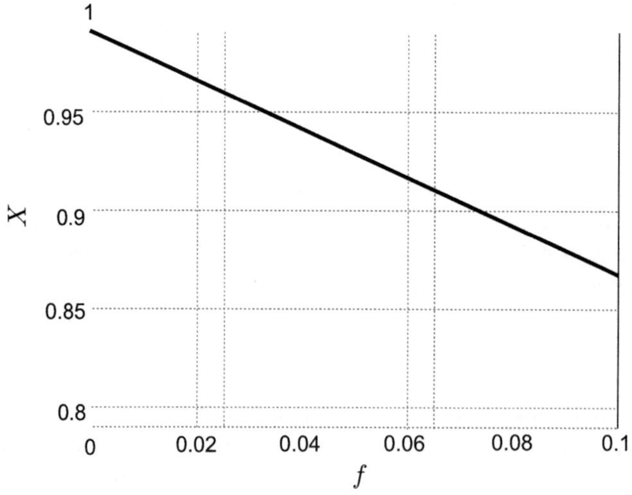

Figure 3. Solution of exercise 8.

4.2 Anisosensitive Yield Conditions

If yielding is affected by a simultaneous change of the signs of all the stress components, in other words, if the yield stresses in tension and compression are different, the material is called "anisosensitive" by Życzkowski (1981). This may be due to different deformation mechanisms in tension and compression or to the effect of the mean stress.

Pressure-Independent Materials. Although their plastic behaviour is independent of the mean stress, HCP materials are anisosensitive because their deformation mechanisms, slip and twinning, are different in tension and compression. Cazacu and Barlat (2004) and Cazacu et al. (2006) proposed two yield criteria for these materials. One is expressed with the principal invariants of the stress deviator

$$J_2^{3/2} - cJ_3 = \sigma_{\mathrm{s}}^3 \tag{29}$$

which corresponds to a small variation of the Drucker yield criterion in Eq. (25). The other is expressed as a function of the principal deviatoric stresses Cazacu et al. (2006)

$$\phi = ||s_1| - ks_1|^a + ||s_2| - ks_2|^a + ||s_3| - ks_3|^a = 2\sigma_{\mathrm{y}}^a \tag{30}$$

in which k controls the tension-compression asymmetry. For both material descriptions, the strength-differential effect (SD) can be characterized by the uniaxial compression-to-tension yield stress ratio

$$\frac{\sigma_{\mathrm{c}}}{\sigma_{\mathrm{t}}} = \left(\frac{3\sqrt{3} - 2c}{3\sqrt{3} + 2c}\right)^{1/3} = \left(\frac{2^{a-1}\,|1-k|^a + |1+k|^a}{2^{a-1}\,|1+k|^a + |1-k|^a}\right)^{1/a}, \tag{31}$$

which is equal to one when c or k are equal to zero.

Exercise 9

Use Eqs (29) and (30) to obtain Eq. (31)

Hint: First, find the deviatoric stresses and invariants (exercises 4 and 5) for tension ($\sigma_1 = \sigma_{\mathrm{t}}$) and compression ($\sigma_1 = \sigma_{\mathrm{c}}$). Then, apply the two criteria.

Answer of exercise 9

Eq. (31)

As indicated by Fig. 4, the yield function in Eq. (30) can reproduce

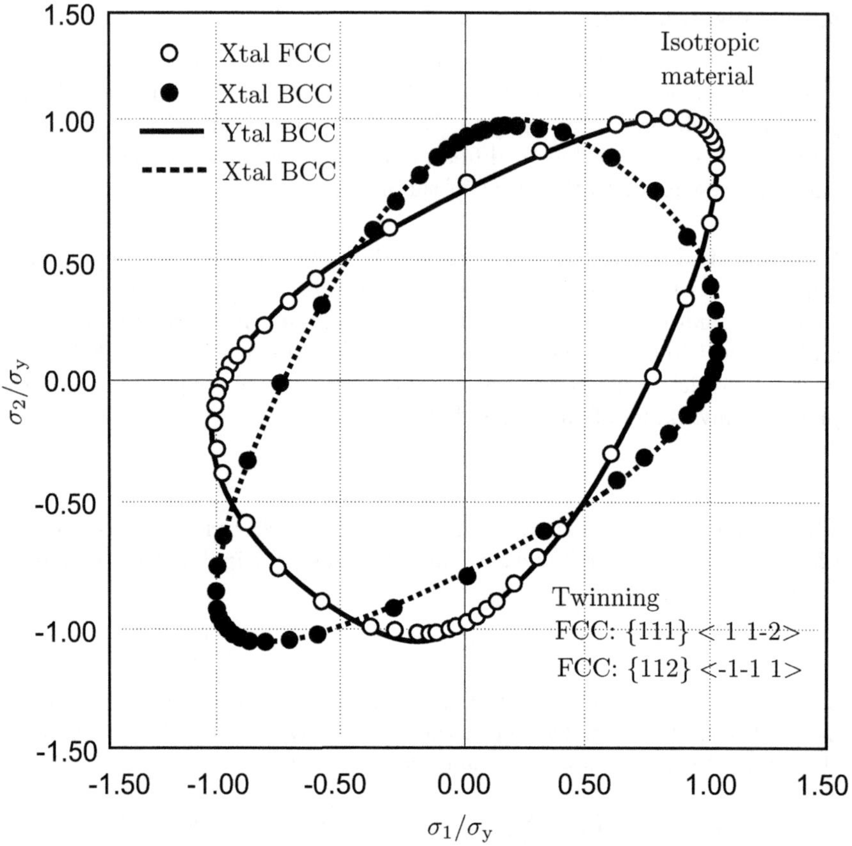

Figure 4. Isotropic yield surfaces for FCC and BCC isotropic polycrystals deformed by twinning and calculated with the full constraint (FC) crystal plasticity model or by the yield condition of Eq. (30) (reprinted from Cazacu and Barlat 2004, with permission from Elsevier).

almost perfectly the yield surfaces of randomly oriented FCC and BCC polycrystals deforming solely by twinning and computed either with full constraint (Hosford and Allen, 1973) or visco-plastic self-consistent (Lebensohn and Tomé, 1993) crystal plasticity models. Both approaches capture the strength-differential (SD) effect with a ratio of compressive to tensile yield stresses larger than one for FCC materials and smaller than one for BCC materials.

Pressure-Dependent Materials. Many metals are almost isosensitive but, for instance, cast iron and geomaterials, which are sensitive to the mean stress, do not evidently belong to this group. A relatively simple but sufficiently general yield condition for mean stress-dependent materials was proposed by Burzyński (1929)

$$\phi = bJ_2 + cI_1^2 + dI_1 = \phi_0 \tag{32}$$

where b, c and d are material coefficients. A particular case of this formulation was employed in Spitzig and Richmond (1984) who measured a small influence of the means stress on the plastic flow behaviour of fully dense high strength steels, thus, explaining the S-D effect in these materials.

4.3 Flow Rule

In the classical theory of plasticity (Hill, 1950), the strain increment derives from a potential. Traditionally, this potential is identified with the yield function and the corresponding relationship between stress and strain increment is called the associated flow rule (AFR). Non-associated flow rules (NAFR), in which the plastic potential and the yield surface are different, have also been used in earlier (Mroz, 1963) and more recent works (Stoughton, 2002; Stoughton and Yoon, 2009) but, for metals, there are arguments in favour of the AFR.

Bishop and Hill (1951a,b) showed that, for a polycrystal crystal obeying the Schmid law (Schmid and Boas, 1935), i.e., dislocation glide occurs when the resolved shear stress on a slip system reaches a critical value, the resulting yield surface was convex and its normal was collinear to the strain increment. This was demonstrated without any assumption about the interaction mode between grains or the uniformity of the deformation gradient. In addition, Hecker (1976) reviewed a number of multiaxial experiments, which verified to a fair degree of approximation, these assumptions about normality and convexity. Therefore, the assumption of normality between the yield surface and the strain increment is believed to be a good approximation for metals although it is, of course, possible to find deviations. Mathematically, the associated flow rule means that the yield function becomes a potential for the plastic strain increments ($\mathrm{d}\varepsilon_{ij} = \dot{\varepsilon}_{ij}\mathrm{d}t$), which, mathematically, translates as follows

$$\mathrm{d}\varepsilon_{ij} = \mathrm{d}\lambda \frac{\partial \phi}{\partial \sigma_{ij}} \tag{33}$$

where $\mathrm{d}\lambda$ is the plastic multiplier, a scalar. Note that the above relationship is consistent with the general formulation in Eq. (3).

Exercise 10

Express the associated flow rule, Eq. (33), as a function of the deviatoric stress components for a fully dense metal.

Answer of exercise 10

Use the composition of a derivation to find

$$\mathrm{d}\varepsilon_{ij} = \mathrm{d}\lambda \frac{\partial \phi}{\partial s_{ij}}$$

The results of Bishop and Hill (1951a,b) have other consequences, in particular for the yield function, which must be convex with respect to the stress components. The convexity is verified if the Hessian matrix associated with the yield function

$$H_{ij} = \frac{\partial^2 \phi}{\partial \sigma_i \partial \sigma_j} \tag{34}$$

is positive semi-definite, i.e., if the principal values are non-negative (Rockafellar, 1972). This condition is much easier to check when the yield function is expressed with the principal stresses. This is the reason why this type of formulations has been given a preference in the present article. All of the isotropic yield functions defined in the above section are convex in a certain parameter range that can be established using Eq. (34).

Exercise 11

Show that Hershey's yield function in plane stress (only σ_1 and σ_2 are non-zero) is convex for $a = 4$.

Answer of exercise 11

The Hessian matrix is only 2×2. The signs of the corresponding principal values can be obtained from the sum and the product of these values.

4.4 Strain Hardening

One State Variable. As plastic deformation proceeds, the material hardens, which is captured by an evolution of the yield surface. The basic information concerning strain hardening is given by the tension test. Therefore, for multiaxial stress states, most theories of plastic hardening attempt to employ that information and "transfer" this reference stress-strain curve, $\sigma_r = h(\varepsilon_r)$, to the general three-dimensional case. In general, the reference curve is that measured in uniaxial tension. However, other choices are possible.

The simplest approach of this type is to formulate an equivalent or effective stress and make it equal to σ_r. A widely used method is to start with the von Mises yield condition and substitute the reference stress σ_r for the yield stress σ_y. Therefore, it is possible to define the von Mises equivalent stress as follows

$$\sigma_e = (3J_2)^{1/2} = \sigma_r \tag{35}$$

Exercise 12

Give different expressions of the von Mises effective stress σ_e.

Answer of exercise 12

Use the results of exercise 4:

$$\sigma_e = \sqrt{\frac{3}{2} s_{ij} s_{ij}}$$

$$\sigma_e = \sqrt{3(s_{23}^2 + s_{31}^2 + s_{12}^2) + \frac{(s_{22} - s_{33})^2 + (s_{33} - s_{11})^2 + (s_{11} - s_{22})^2}{2}}$$

$$\sigma_e = \sqrt{3(\sigma_{23}^2 + \sigma_{31}^2 + \sigma_{12}^2) + \frac{(\sigma_{22} - \sigma_{33})^2 + (\sigma_{33} - \sigma_{11})^2 + (\sigma_{11} - \sigma_{22})^2}{2}}$$

More generally, many yield functions are homogeneous of a certain degree "a" with respect to the Cauchy or deviatoric stress components, i.e., $\phi(\alpha\sigma_{ij}) = \alpha^a \phi(\sigma_{ij})$. For a number of applications, it is convenient to represent these functions as homogeneous functions of the first degree, which can be done without any difficulty. For example, in the case of Tresca, it is readily satisfied. For von Mises, it is obtained by replacing the second invariant by its square root as in Eq. (35) and, for the Hershey yield function, by replacing ϕ by $\phi^{1/a}$. It is therefore always possible to rewrite the yield condition with an effective stress $\bar{\sigma}$ (homogeneous function of first degree) equal to the reference flow stress σ_r. The definition of the associated effective strain may result from the plastic work equivalence

$$\mathrm{d}\bar{\varepsilon} = \frac{\sigma_{ij}\mathrm{d}\varepsilon_{ij}}{\bar{\sigma}} \tag{36}$$

which, of course, leads to $\bar{\varepsilon} = \varepsilon$, i.e., the longitudinal strain, for uniaxial tension if this is the reference loading curve.

Exercise 13

The von Mises effective strain increment $d\varepsilon_{\mathrm{e}}$ can be calculated from Eq. (36), i.e., $\mathrm{d}\varepsilon_{\mathrm{e}} = \mathrm{d}\bar{\varepsilon}$ with the corresponding associated flow rule. Express the effective strain increment $\mathrm{d}\varepsilon_{\mathrm{e}}$ as a function of the incremental strain components.

Answer of exercise 13

From Eq. (36) and the associated flow rule, express $\mathrm{d}\varepsilon_{\mathrm{e}}$ only with deviatoric plastic strain increments (note that the trace of the plastic strain increment tensor is zero). After simplification, the result should be

$$\mathrm{d}\varepsilon_{\mathrm{e}} = \sqrt{\frac{2}{3}\mathrm{d}\varepsilon_{ij}\mathrm{d}\varepsilon_{ij}}$$

Therefore, the yield condition can be finally rewritten as

$$\Phi = \bar{\sigma}\left(\sigma_{ij}\right) - \sigma_{\mathrm{r}}\left(\bar{\varepsilon}\right) = 0 \tag{37}$$

This type of approach is called isotropic hardening (or isotropic work-hardening). The associated yield surface expands in stress space as plastic deformation accumulates. Note that the reference stress needs not be uniaxial but if another state is chosen, Eq. (36) provides a rational way to define the effective strain. The advantage of a convex function of first degree is that, combined with the yield condition (37), it results in $\mathrm{d}\lambda = \mathrm{d}\bar{\varepsilon}$ in the associated flow rule of Eq. (33). In the context of the general formulation introduced by Eqs (2) to (4), isotropic work-hardening is controlled by one state variable only, which can be the specific plastic work, the reference stress σ_{r} or, equivalently, $\bar{\varepsilon}$. The evolution equation for $\bar{\varepsilon}$ is given in Eq. (36). Alternatively, the evolution equation for σ_{r} is

$$\mathrm{d}\sigma_{\mathrm{r}} = \theta\left(\bar{\varepsilon}\right)\mathrm{d}\bar{\varepsilon} \tag{38}$$

where $\theta\left(\bar{\varepsilon}\right) = \mathbf{d}\sigma_{\mathrm{r}}/\mathrm{d}\bar{\varepsilon}$ is the rate of strain hardening.

The reference curve can be defined by $\theta\left(\bar{\varepsilon}\right)$ or directly by σ_{r} under different mathematical forms, among which, the Swift power law (Swift, 1952)

$$\sigma_{\mathrm{r}}\left(\bar{\varepsilon}\right) = K\left(\varepsilon_0 + \bar{\varepsilon}\right)^n \tag{39}$$

and the Voce saturation law (Voce, 1955)

$$\sigma_{\rm r}\left(\bar{\varepsilon}\right) = \sigma_s - \left(\sigma_s - \sigma_{\rm y}\right) \exp\left(-\bar{\varepsilon}/\bar{\varepsilon}_0\right) \tag{40}$$

The materials coefficients $K, \varepsilon_0, n, \sigma_{\rm s}, \sigma_{\rm y}$ and $\bar{\varepsilon}_0$ can be determined by curve fitting. The Swift law is called Hollomon law (Hollomon, 1945) when $\varepsilon_0 = 0$.

Two State Variables. In the Gurson model, Eq. (28), the state variable $\sigma_{\rm r}$, characterizing the strength of the fully dense matrix, may be substituted for $\sigma_{\rm y}$. In addition, the porosity f is also a state variable with its specific evolution, namely

$$\dot{f} = (1 - f)\,\dot{\varepsilon}_{kk} \tag{41}$$

which expresses the conservation of matter. Thus, this model contains two state variables. It was shown by Leblond et al. (1995), though, that there is no interaction between the work-hardening of the fully dense matrix and the porosity growth. In other words, the evolution of f is totally independent of the reference hardening curve $\sigma_{\rm r}$. However, the study of a hollow sphere under deviatoric and hydrostatic loads indicates that the porosity growth should depend on strain hardening. Thus, Leblond et al. (1995) modified the Gurson model to capture this effect, i.e.,

$$\Phi\left(\sigma_{\rm e}, \sigma_{\rm m}, f, \sigma_{\rm r}', \sigma_{\rm r}''\right) = \left(\frac{\sigma_{\rm e}}{\sigma_{\rm r}'}\right)^2 + 2q_1 f \cosh\left(\frac{3q_2\sigma_{\rm m}}{2\sigma_{\rm r}''}\right) - 1 - q_1^2 f^2 = 0 \tag{42}$$

where $\sigma_{\rm r}'$ and $\sigma_{\rm r}''$ are expressed by integrals suggested by the study of the hollow sphere under hydrostatic and deviatoric loading. Since the integrals are not simple to evaluate, an approximation was proposed in Karabin et al. (2009), to capture this effect in a more phenomenological manner by simply assuming

$$\begin{aligned} \sigma_{\rm r}' &= \sigma_{\rm r}\left(\bar{\varepsilon}'\right), \\ \sigma_{\rm r}'' &= \sigma_{\rm r}\left(\bar{\varepsilon}''\right) \end{aligned} \tag{43}$$

and

$$\begin{aligned} \dot{\bar{\varepsilon}}' &= \dot{\bar{\varepsilon}}\left(1 - 1/\xi'\right), \\ \dot{\bar{\varepsilon}}'' &= \dot{\bar{\varepsilon}}\left(1 - 1/\xi''\right) \end{aligned} \tag{44}$$

In the above relationships, $\dot{\bar{\varepsilon}}$ is the effective strain rate of the fully dense matrix, and ξ' and ξ'' are two constant coefficients. Thus, according to the relationships (43) and (44), this model still contains two independent state variables $\bar{\varepsilon}$ and f. This simple approach was able to capture the strain hardening effect on porosity growth as shown by Fig. 5. The porosity evolution for the original Gurson model is described by the solid lines, irrespective of the strain hardening index, n. However, the model proposed by Leblond

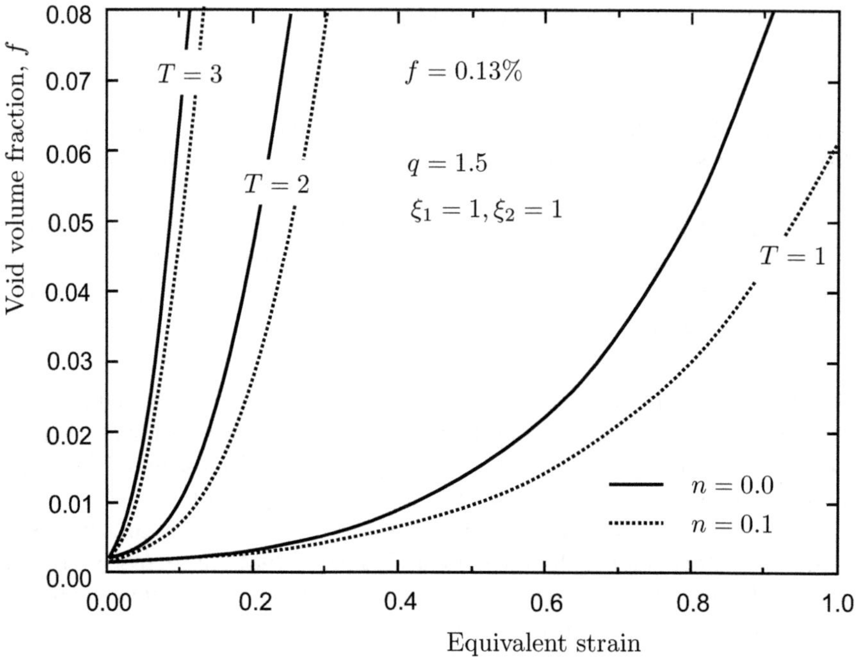

Figure 5. Porosity evolution for two values of strain hardening exponent n in the Swift law. Relevant coefficients are given in the figure. The stress triaxiality parameter is defined as $T = \sigma_m/\sigma_e$ (reprinted from Karabin et al. 2009, with permission from Springer).

et al. (1995), and simplified as explained above, indicates that a higher strain hardening exponent decreases the porosity growth rate significantly. This effect will be used in Sect. 6.2 on fracture toughness for thick plates.

4.5 Temperature and Strain Rate Effects

As mentioned before, temperature (Θ) and strain rate ($\dot{\bar{\varepsilon}}$) are also affecting the plastic behavior significantly. This is not discussed in detail in this article but a simple way to include strain rate effects is to take the flow stress as

$$\sigma_r = h(\bar{\varepsilon}) \left(\dot{\bar{\varepsilon}}/\dot{\bar{\varepsilon}}_0\right)^m \tag{45}$$

The hardening $h(\bar{\varepsilon})$ curve (for instance Voce or Swift) is determined at the reference strain rate $\dot{\bar{\varepsilon}}_0$. m, the strain rate sensitivity parameter, is obtained by conducting tests at different rates or by performing rate jumps.

The effects of both strain rate and temperature may be taken into account, for instance, by the Johnson and Cook (1983) description of the reference flow stress

$$\sigma_{\rm r} = \left(A + B\bar{\varepsilon}^N\right)\left[1 + C\ln\left(\frac{\dot{\bar{\varepsilon}}}{\dot{\bar{\varepsilon}}_0}\right)\right]\left[1 - \left(\frac{\Theta - 298}{\Theta_{\rm melt} - 298}\right)^M\right] \tag{46}$$

where A, B, C, N and M are materials coefficients, $\dot{\bar{\varepsilon}}_0$ the reference strain rate and $\Theta_{\rm melt}$ the melting temperature.

Many experimental studies showed that the yield surface evolution is more complex than the expansion resulting from the simple isotropic work-hardening assumption, even when temperature and strain rate effects are not considered. However, it works reasonably well at first approximation to solve plasticity problems, in particular, when the material deforms under proportional loading. Other forms of hardening, including yield surface translation, distortion and rotation are available in the literature (Życzkowski, 1981; Chaboche, 2008) but these more advanced methods are out of the scope of the present article.

5 Anisotropic Yield Functions

5.1 Classical Approach

For anisotropic materials, the classical theory of plasticity is very similar to that for isotropic materials described above. The yield condition is described by Eq. (37) and the flow rule by Eq. (33). In the latter, $\mathrm{d}\lambda = \mathrm{d}\bar{\varepsilon}$ as defined by Eq. (36) for a first degree homogeneous yield function. In addition, it is assumed that the anisotropic yield surface expands isotropically by work-hardening. The main difference is that, due to the dependence of plastic properties on the loading direction, the yield function does not depend on stress tensor invariants and has to be expressed in a specific reference frame attached to the material $\boldsymbol{e}_x, \boldsymbol{e}_y$ and $\boldsymbol{e}_z$. For instance, these directions are the rolling (RD), transverse (TD) and normal (ND) directions for a rolled sheet or plate.

In an anisotropic yield condition, the stress invariants should be replaced by simultaneous (common) invariants of the stress and material (or structural) tensors (see Życzkowski, 1981). The substitution of the stresses by their transformed values into well-known isotropic yield conditions allows for a direct generalization of these conditions for anisotropic materials. The first anisotropic yield condition was proposed by von Mises (1928)

$$A_{ijkl}\sigma_{ij}\sigma_{kl} = 1 \quad \text{with} \quad A_{ijkl} = A_{jikl} = A_{ijlk} = A_{klij} \tag{47}$$

The associated symmetry conditions reduce the number of anisotropy components to 21. Further, the additional requirement of independence of Eq. (47) on the mean stress σ_m reduces the number of plastic moduli to 15. But in the most simplified form, this yield condition provides a generalization of the von Mises isotropic criterion. For an orthotropic material, Hill (1948) proposed a specific case of the above relationship in the following form

$$\begin{aligned} & F\left(\sigma_{yy}-\sigma_{zz}\right)^2+G\left(\sigma_{zz}-\sigma_{xx}\right)^2+H\left(\sigma_{xx}-\sigma_{yy}\right)^2 \\ + \; & 2L\sigma_{yz}^2+2M\sigma_{zx}^2+2N\sigma_{xy}^2=\bar{\sigma}^2 \end{aligned} \tag{48}$$

where F, G, H, L, M and N are anisotropy coefficients. Note that the subscripts x, y and z for the stress components indicate that the yield function is expressed in the material symmetry axes $\boldsymbol{e}_x, \boldsymbol{e}_y$ and $\boldsymbol{e}_z$, e.g., the RD, TD and ND for a sheet. Other yield functions were proposed later by Hill (1979, 1990a,b) but, by far, have never been used as much as that represented by Eq. (48).

Exercise 14

For plane stress, i.e.,

$$[\boldsymbol{\sigma}]=\begin{bmatrix} \sigma_{xx} & \sigma_{xy} & 0 \\ \sigma_{xy} & \sigma_{yy} & 0 \\ 0 & 0 & 0 \end{bmatrix},$$

express Hill's yield criterion (Hill, 1948) and write all the components of the strain increment assuming the associated flow rule.

Hint: Apply associated flow rule, Eq. (33), to Hill's yield function, Eq. (48). Note that for shear, the operation leads to twice the value of the shear increment because the term σ_{xy} in the yield function is in fact

$$\frac{\sigma_{xy}+\sigma_{yx}}{2}$$

Answer of exercise 14

$$\phi=F\sigma_{yy}^2+G\sigma_{xx}^2+H(\sigma_{xx}-\sigma_{yy})^2+2N\sigma_{xy}^2=\bar{\sigma}^2$$

$$\begin{aligned} \mathrm{d}\varepsilon_{xx} &= 2\mathrm{d}\lambda[(G+H)\sigma_{xx}-H\sigma_{yy}], \\ \mathrm{d}\varepsilon_{yy} &= 2\mathrm{d}\lambda[(H+F)\sigma_{yy}-H\sigma_{xx}], \\ \mathrm{d}\varepsilon_{zz} &= -(\mathrm{d}\varepsilon_{xx}+\mathrm{d}\varepsilon_{yy}), \\ \mathrm{d}\varepsilon_{xy} &= 2N\mathrm{d}\lambda\sigma_{xy} \end{aligned}$$

Exercise 15

For plane stress, typically the stress state in sheet metal during a forming process, the uniaxial yield stresses in the rolling direction (RD), transverse direction (TD) and at 45° from the RD are σ_0, σ_{45}, and σ_{90}. Moreover, the yield stress in balanced biaxial tension is such that $\sigma_{xx} = \sigma_{yy} = \sigma_b$. Express the coefficients of Hill's yield criterion (Hill, 1948) as a function of σ_0, σ_{45}, and σ_{90} and σ_b assuming that the reference stress is the uniaxial yield stress in the rolling direction (the formulae of Sect. 3.2 may be needed).

Hint: Write four relationships from the plane stress version of Eq. (48) for each of the four loading conditions. Then, solve a system of four equations with four unknowns.

Answer of exercise 15

$$
\begin{aligned}
2F &= \left(\frac{\sigma_0}{\sigma_{90}}\right)^2 + \left(\frac{\sigma_0}{\sigma_b}\right)^2 - 1, \\
2G &= \left(\frac{\sigma_0}{\sigma_b}\right)^2 + 1 - \left(\frac{\sigma_0}{\sigma_{90}}\right)^2, \\
2H &= 1 + \left(\frac{\sigma_0}{\sigma_{90}}\right)^2 - \left(\frac{\sigma_0}{\sigma_b}\right)^2, \\
2N &= 4\left(\frac{\sigma_0}{\sigma_{45}}\right)^2 - \left(\frac{\sigma_0}{\sigma_b}\right)^2
\end{aligned}
$$

5.2 Tensor Representation

A general way of extending isotropic yield functions to the case of anisotropic materials is to modify the invariants in a way that is compatible with the symmetry of the material. Then, these modified invariants can be used in the isotropic yield function. For instance, assuming that plastic flow is insensitive to hydrostatic pressure, the yield condition for an isotropic material can be expressed with a function of the second and third invariants of the stress deviator J_2 and J_3, respectively. The theory of representation of tensor functions (e.g., Boehler (1978); Liu (1982); Wang (1970); Betten (1988) Rogers 1990) provides a framework in which Cazacu and Barlat (2001, 2003) proposed generalizations J_2^o and J_3^o of the stress deviator invariants. For orthotropic symmetry, i.e., for rolled plates or sheets, J_2^o and J_3^o are required to be homogeneous functions of degree two and three in stresses, respectively, that reduce to J_2 and J_3 for isotropic conditions, are insensitive to pressure, and are invariant to any transformation belonging to the symmetry group of the material. These conditions are met only for specific forms of J_2^o and J_3^o. It was shown that any yield function expressed with these modified invariants contain 17 independent anisotropy coefficients. In Cazacu and Barlat (2001), this approach was used to extend the isotropic yield criterion (Drucker, 1949), Eq. (25), by replacing the two invariants J_2 and J_3 by their modified forms J_2^o and J_3^o.

5.3 Linear Transformation Approach

The advantage of this approach is that it reduces simply to a yield function of the type given in Sect. 4. Moreover, the convexity is relatively easy to verify.

One Transformation. Based on linear transformations of the stress tensor, a method suitable for convex formulations was developed (Barlat et al., 1991, 2003, 2005; Karafillis and Boyce, 1993). This approach is detailed below for incompressible materials for which a linear transformation is performed on the stress deviator, s, leading to the transformed tensor $\tilde{s}$

$$\tilde{s}_{ij} = C_{ijkl} s_{kl} \tag{49}$$

C, a fourth order tensor, contains the anisotropy coefficients, accounts for the macroscopic symmetries of the material, and reduces to the identity tensor for isotropic materials. In contrast with Sobotka (1969) and Boehler and Sawczuk (1970) who were using the transformed components directly in the formulation, in the present theory, an isotropic yield function ϕ can be generalized to anisotropy by replacing the principal deviatoric stress components by the principal values $\tilde{s}_1, \tilde{s}_2$ and $\tilde{s}_3$ of the transformed stress deviator $\tilde{s}$. For a general tensor, the components $\tilde{s}_{ij}$ are first determined using Eq. (49). Then, the transformed invariants $\tilde{I}_1, \tilde{I}_2$ and $\tilde{I}_3$ are calculated as in Eq. (15) for these transformed tensors. Finally, the principal values of $\tilde{s}$ are obtained using the general equations provided in Eq. (17) by substituting the invariants $\tilde{I}_k$ for I_k.

Logan and Hosford (1980) showed that a restricted extension of Eq. (26) to anisotropic materials leads to results consistent with crystal plasticity calculations for textured polycrystals with $a = 6$ and $a = 8$ for BCC and FCC structures, respectively. Barlat et al. (1991) proposed a complete extension of Eq. (26) using one linear transformation. Later, Karafillis and Boyce (1993) started with a more general isotropic form to further extend

Eq. (26)

$$\begin{aligned} \phi &= (1-c)\left\{|\tilde{s}_1-\tilde{s}_2|^a+|\tilde{s}_2-\tilde{s}_3|^a+|\tilde{s}_3-\tilde{s}_1|^a\right\} \\ &+ c\frac{3^a}{2^{a-1}+1}\left\{|\tilde{s}_1|^a+|\tilde{s}_2|^a+|\tilde{s}_3|^a\right\}=2\bar{\sigma}^a \end{aligned} \tag{50}$$

where c is a coefficient.

Two Transformations. It is also possible to use two or more linear transformations, as long as the yield function is isotropic with respect to the $3N$ variables $\tilde{s}_k^{(1)},\ldots,\tilde{s}_k^{(N)}$ (for N linear transformations). For instance, two linear transformations on the stress deviator lead to the tensors $\tilde{s}'$ and $\tilde{s}''$ defined as

$$\begin{aligned} \tilde{s}'_{ij} &= C'_{ijkl}s_{kl} \\ \tilde{s}''_{ij} &= C''_{ijkl}s_{kl} \end{aligned} \tag{51}$$

The principal value of each transformed tensor can be calculated in the same way as with one transformation. Applied to plane stress, the two transformations, in matrix representation, reduce to

$$\begin{aligned} \begin{bmatrix} \tilde{s}'_{xx} \\ \tilde{s}'_{yy} \\ \tilde{s}'_{xy} \end{bmatrix} &= \begin{bmatrix} C'_{11} & C'_{12} & 0 \\ C'_{21} & C'_{22} & 0 \\ 0 & 0 & C'_{66} \end{bmatrix} \begin{bmatrix} s_{xx} \\ s_{yy} \\ s_{xy} \end{bmatrix}, \\ \begin{bmatrix} \tilde{s}''_{xx} \\ \tilde{s}''_{yy} \\ \tilde{s}''_{xy} \end{bmatrix} &= \begin{bmatrix} C''_{11} & C''_{12} & 0 \\ C''_{21} & C''_{22} & 0 \\ 0 & 0 & C''_{66} \end{bmatrix} \begin{bmatrix} s_{xx} \\ s_{yy} \\ s_{xy} \end{bmatrix} \end{aligned} \tag{52}$$

leading to the anisotropic yield function called Yld2000-2d in Barlat et al. (2003)

$$\phi = |\tilde{s}'_1-\tilde{s}'_2|^a + |2\tilde{s}''_2+\tilde{s}''_1|^a + |2\tilde{s}''_1+\tilde{s}''_2|^a = 2\bar{\sigma}^a \tag{53}$$

This expression reduces to Hershey's in Eq. (26) when the tensor $\boldsymbol{C}'$ and $\boldsymbol{C}''$ are both set to the identity tensor, i.e., $\tilde{\boldsymbol{s}}' = \tilde{\boldsymbol{s}}'' = \boldsymbol{s}$. Out of the ten dependent anisotropy coefficients in the formulation, two can be set to zero ($C'_{12} = C'_{21} = 0$). Other plane stress yield functions (Aretz, 2004, 2005; Banabic et al., 2005), which were proposed later, were shown to reduce to Yld2000-2d when the coefficients are properly set (Barlat et al., 2007). Other yield functions based on linear transformations were suggested more recently (e.g., Comsa and Banabic, 2008; Plunkett et al., 2008).

Exercise 16

Show that Eq. (53) reduces to Hershey's yield condition in plane stress for an isotropic material.

Answer of exercise 16

For an isotropic material, $\tilde{\boldsymbol{s}}' = \tilde{\boldsymbol{s}}'' = \tilde{\boldsymbol{s}}$ and $\mathbf{C}' = \mathbf{C}'' = \mathbf{C}$.

For a general stress case, Bron and Besson (2004) extended the Karafillis and Boyce (1993) yield function with two linear transformations by choosing the isotropic form as follows

$$\begin{aligned} \phi &= \frac{1}{2}\left\{|\tilde{s}_1' - \tilde{s}_2'|^{a_1} + |\tilde{s}_2' - \tilde{s}_3'|^{a_1} + |\tilde{s}_3' - \tilde{s}_1'|^{a_1}\right\}^{1/a_1} \\ &+ \frac{3^{a_2}}{2^{a_2}+2}\left\{|\tilde{s}_1''|^{a_2} + |\tilde{s}_2''|^{a_2} + |\tilde{s}_3''|^{a_2}\right\}^{\frac{1}{a_2}} = \bar{\sigma} \end{aligned} \tag{54}$$

where a_1 and a_2 are two constant exponents. Barlat et al. (2005) proposed a generalization of Hershey's yield function, denoted Yld2004-18p,

$$\begin{aligned} \phi &= |\tilde{s}_1' - \tilde{s}_1''|^a + |\tilde{s}_1' - \tilde{s}_2''|^a + |\tilde{s}_1' - \tilde{s}_3''|^a + |\tilde{s}_2' - \tilde{s}_1''|^a + |\tilde{s}_2' - \tilde{s}_2''|^a \\ &+ |\tilde{s}_2' - \tilde{s}_3''|^a + |\tilde{s}_3' - \tilde{s}_1''|^a + |\tilde{s}_3' - \tilde{s}_2''|^a + |\tilde{s}_3' - \tilde{s}_3''|^a = 4\bar{\sigma}^a \end{aligned} \tag{55}$$

which is an isotropic and convex function with respect to its arguments. Beside their associated isotropic generator, the most significant difference between the models associated with Eqs. (54) and (55) is the form of the linear transformation. Each of them contains six independent coefficients in Bron and Besson (2004) but nine in Barlat et al. (2005). In the latter, the matrix representations of the anisotropy tensors C' and C'' are

$$\begin{aligned} C' &= \begin{bmatrix} 0 & -c_{12}' & -c_{13}' & 0 & 0 & 0 \\ -c_{21}' & 0 & -c_{23}' & 0 & 0 & 0 \\ -c_{31}' & -c_{32}' & 0 & 0 & 0 & 0 \\ 0 & 0 & 0 & c_{44}' & 0 & 0 \\ 0 & 0 & 0 & 0 & c_{55}' & 0 \\ 0 & 0 & 0 & 0 & 0 & c_{66}' \end{bmatrix}, \\ C'' &= \begin{bmatrix} 0 & -c_{12}'' & -c_{13}'' & 0 & 0 & 0 \\ -c_{21}'' & 0 & -c_{23}'' & 0 & 0 & 0 \\ -c_{31}'' & -c_{32}'' & 0 & 0 & 0 & 0 \\ 0 & 0 & 0 & c_{44}'' & 0 & 0 \\ 0 & 0 & 0 & 0 & c_{55}'' & 0 \\ 0 & 0 & 0 & 0 & 0 & c_{66}'' \end{bmatrix} \end{aligned} \tag{56}$$

In Eq. (56), the relationship $\tilde{s}_1 + \tilde{s}_2 + \tilde{s}_3 = 0$ does not hold for either transformation, but since the yield function is still expressed as a function of s_1, s_2 and s_3 through Eq. (51), it is independent of the mean stress. These models described above are appropriate for materials that do not exhibit the SD effect, i.e., when tension and compression yield stresses are identical.

Exercise 17

Show that Eq. (55) reduces to Hershey's yield condition for an isotropic material.

Answer of exercise 17

Similar as exercise 16.

The Hershey type of model, which leads to yield surfaces with rounded corner was validated experimentally by Tozawa and Nakamura (1967, 1972) and Tozawa (1978) for many cubic materials. The Yld2004-18p model was validated on a binary Al-Mg alloy as shown in Fig. 6 (Barlat et al., 1997). This figure represents the yield locus of this alloy as measured experimentally, calculated with the Taylor-Bishop-Hill crystal plasticity model (TBH model, see Taylor, 1938; Bishop and Hill, 1951a,b), and approximated with the Yld2004-18p yield function. It indicates that all the results are consistent. Independently, uniaxial tension tests were carried out to measure the r value (width-to-thickness strain ratio in uniaxial tension), which can be related to the slope of the yield locus at the loading point. The results are listed in Table 1 and, for a given tensile direction, are found to be very close from each other for this parameter.

Note that the plane stress yield locus corresponding to Yld2000-2d would also lead to an excellent agreement with other yield surfaces in Fig. 6. In a different example on an Al-Li alloy sheet sample, Figs 7 and 8 shows that the flow stresses and r values as a function of the angle between rolling and tensile directions are generally well predicted with the plane stress Yld2000-2d and the general stress Yld2004-18p. Nevertheless, although the yield loci (not shown here) are very close from each other, this figure indicates that the fine details are better captured with Yld2004-18p, which is not surprising considering the higher number of anisotropy coefficients.

5.4 Identification

Many issues have been addressed in determining yield surfaces experimentally. For instance, the definition of yield has been the subject of dis-

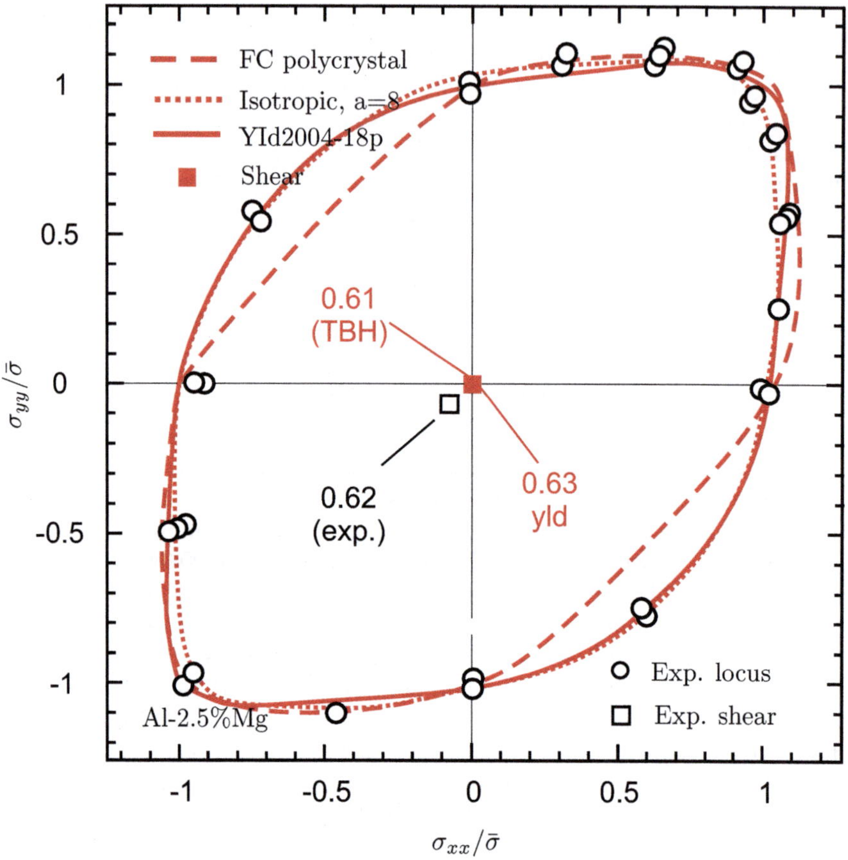

Figure 6. Yield locus for Al-2.5% binary alloy measured experimentally, calculated with crystal plasticity and approximated by Yld2004-18p (adapted from Barlat et al. 1997, with permission from Elsevier).

cussion (Paul, 1968; Hecker, 1976). Multiaxial experiments have been used to characterize a yield surface as, for instance described by Hecker (1976). In spite of many improvements over the last decades, multiaxial testing is still tedious, difficult to interpret and not suitable for quick material characterization for more practical applications. Therefore, other methods are necessary to identify constitutive coefficients for practical process simulations and a few of them are described below for sheet metals.

Anisotropic properties can be assessed by performing uniaxial tension tests in the $\boldsymbol{e}_x$ and $\boldsymbol{e}_y$ axes (rolling and transverse directions, respectively), and in direction at ξ degrees with respect to e_x. Practically, anisotropy is characterized by the yield stresses σ_0, σ_{45}, σ_{90}, the r values r_0, r_{45}, r_{90}, their respective average $\bar{q} = (q_0 + 2q_{45} + r_{90}/4)$ and variations $\Delta\bar{q} = (q_0 - 2q_{45} + r_{90}/2)$ where the subscript denotes the ξ value. Directional tension of wide specimens can be used to characterize plane strain tension anisotropy (Wagoner and Wang, 1979; Taha et al., 1995). This test does not produce a uniform state of stress within the specimen and generally leads to more experimental scatter than the uniaxial tension test.

The balanced biaxial yield stress (σ_b) is an important parameter to measure. This stress can be obtained by conducting a hydraulic bulge test (Young et al., 1981). In this test, a sheet blank is clamped between a die with a large circular opening and a holder. A pressure is gradually applied under the blank, which bulges in a quasi-spherical shape. The curvature and strains at the pole of the specimen are measured independently using mechanical or optical instruments. The stress is simply obtained from the membrane theory using the calculated thickness. This test is interesting not only because it gives information on the yield surface but also because it allows measurements of the hardening behavior up to strains of about twice those achieved in uniaxial tension. However, the yield point is not well defined in this test because of the low curvature of the specimen in the initial stage of deformation.

Another limitation of the bulge test is that, since the biaxial stress is not exactly balanced, measures of the corresponding strain state might lead to substantial errors. Thus, Barlat et al. (2003) proposed the disk compression test, which gives a measure of the flow anisotropy for a balanced biaxial stress state, assuming that hydrostatic pressure has no influence on plastic deformation. Pöhlandt et al. (2002) proposed to use a biaxial tensile testing machine to determine this coefficient.

Simple shear tests (Rauch and Schmitt, 1989) can be carried out to

Table 1. Measured and predicted r values (Lankford coefficients) for Al-2.5% Al-Mg binary alloy.

	r_0	r_{90}
Experimental	0.26	0.27
Crystal plasticity	0.17	0.26
Yld2004-18p	0.20	0.20
Yld2000-2d	0.20	0.20
Isotropic	1.00	1.00

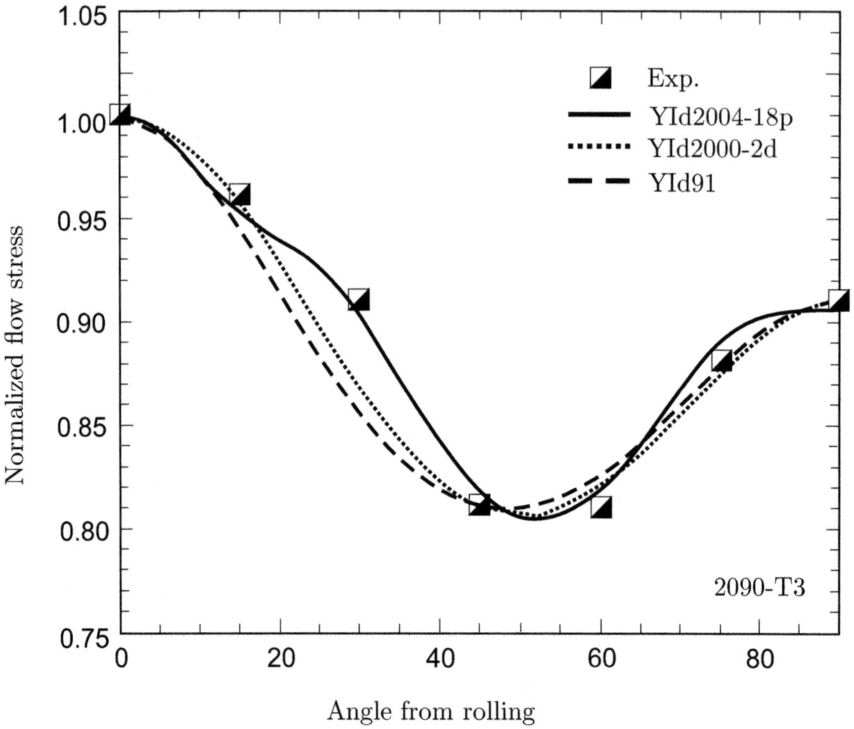

Figure 7. Flow stress anisotropy for Al-Li 2090-T3 sheet sample.

characterize the anisotropic behavior of the simple shear flow stress. A relatively simple device mounted on a standard tensile machine is needed for this test. A rectangular specimen is clamped with two grips, which move in opposite directions relative to each other.

From the tests, the identification of the constitutive parameters that best describe the material is not as straightforward as it looks. For instance, the yield stresses can be used as input data to calculate the anisotropic yield function coefficients. However, the yield stress from the bulge test is not very accurate. Moreover, any stress at yield is determined in the region of the stress-strain curve where the slope is the steepest, which might lead to additional inaccuracy. Finally, the yield stress is associated with a very small plastic strain and might not reflect the anisotropy of the material over a larger strain range. For these reasons, the flow stresses at equal amount of plastic work along different loading paths are more representative as input

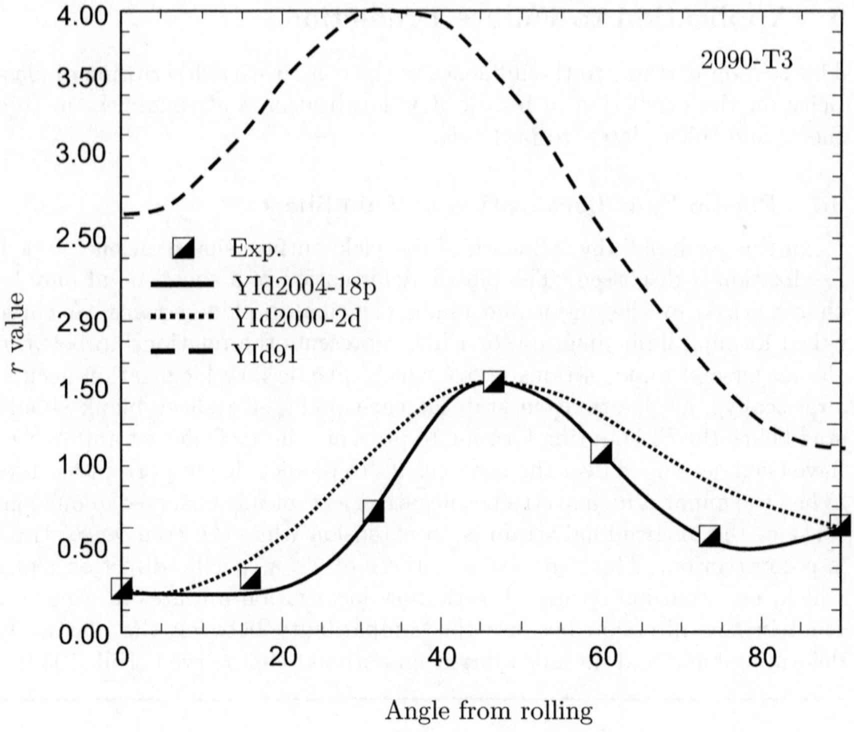

Figure 8. r value anisotropy for Al-Li 2090-T3 sheet sample.

data instead of the yield stress (Barlat et al., 2004).

Similar remarks hold for r values, which can be defined as instantaneous quantities at yield or as the standard slope of the width strain-thickness strain curve over a given deformation range in tension. In the one hand, the yield stresses and instantaneous r values at yield are more appropriate to define the coefficients of the yield function. On the other hand, stresses (called flow stresses) defined at a given amount of plastic work and standard r values characterize the average behavior of the material over a finite deformation range better. These values are likely to be more suitable and more descriptive of the average response over a certain strain range. In general, when the most appropriate test information is selected, the constitutive coefficients are determined using optimization methods.

6 Application to Failure Prediction

This section pertains to the influence of the constitutive description of plasticity on the prediction of plastic flow localization and toughness in thin sheets and thick plates, respectively.

6.1 Plastic Flow Localization in Thin Sheet

In this example, the influence of the yield surface shape on plastic flow localization is discussed. The plastic deformation of a sheet metal may be characterized by the major and minor principal in-plane strains. A curve called forming limit diagram, or FLD, represents the relationship between the major and minor strains above which, plastic flow localization occurs. Practically, the deformation state at each point of a sheet blank should stay below the FLD during forming to prevent failure. Different approaches have been developed over the last decades to predict this important feature. When the minor principal strain is negative, as typically observed in uniaxial tension, the longitudinal strain is an extension while the transverse strain is a contraction. Thus, in between, there exists a specific direction along which, no extension occurs. Plastic flow localization initiates in a narrow band in this direction because the compatibility between the plastically deforming band and its unloading neighborhood is preserved (Hill, 1950).

Exercise 18

Find the direction of no extension for an isotropic material subjected to uniaxial loading.

Hint: Write the plastic strain tensor increment and transform it with Equations (13) and (14) applied to strain components.

Answer of exercise 18

The direction of no extension is at 54.7^0 from the tension axis.

Exercise 19

Find the direction of no extension for an anisotropic material (characterized by a r value) subjected to uniaxial loading.

Hint: Similar as exercise 18. However, the plastic strain tensor increment is different. Note that $r = \mathrm{d}\varepsilon_2/\mathrm{d}\varepsilon_3$ for tension in direction $\boldsymbol{e}_1$.

Answer of exercise 19

The direction of no extension is at

$$\arctan\sqrt{1+\frac{1}{r}}$$

from the tension axis.

However, when both surface strains are positive, a direction of no extension in the plane of the sheet does not exist. As a result, plastic flow localization is triggered by another mechanism. A widely used approach is to assume that there is an imperfection (Marciniak and Kuczyński, 1967) in the form of a band perpendicular to the major principal strain direction that leads to a strain gradient as represented in Fig. 9. Because of the reduced cross-section, plastic deformation is higher in the band than in the homogeneous region and, eventually, plastic flow localization occurs. This is the case considered in this section. Additional details are discussed in Barlat (1989); Yoon and Barlat (2006). Other articles pertains to the influence of the yield surface on the forming limits such as Bassani et al. (1979); Barlat (1989); Barlat and Richmond (1987); Lian et al. (1989); Kuroda and Tvergaard (2000); Inal et al. (2005); Dasappa et al. (2012).

Quantities defined in the imperfection are identified with subscript i while those outside, the so-called homogenous region, are identified with the subscript h. The severity of the imperfection is characterized by the ratio of the thicknesses in the region i to that in the region h, i.e., $\delta = t_i/t_h \leq 1$. The material is assumed to be isotropic. Similar conclusions can be obtained for anisotropic materials but the mathematical formulation is significantly more complex. For the sake of simplicity, the rate sensitive stress-strain behavior of the sheet is represented by the following equation

$$\sigma_{\mathrm{r}} = K\left(\bar{\varepsilon}_0 + \bar{\varepsilon}\right)^n \left(\dot{\bar{\varepsilon}}/\dot{\bar{\varepsilon}}_0\right)^m \tag{57}$$

The homogeneous region of the material is subjected to a plane stress state with principal stresses σ_1 and σ_2. Figure 9 indicates that, in order to

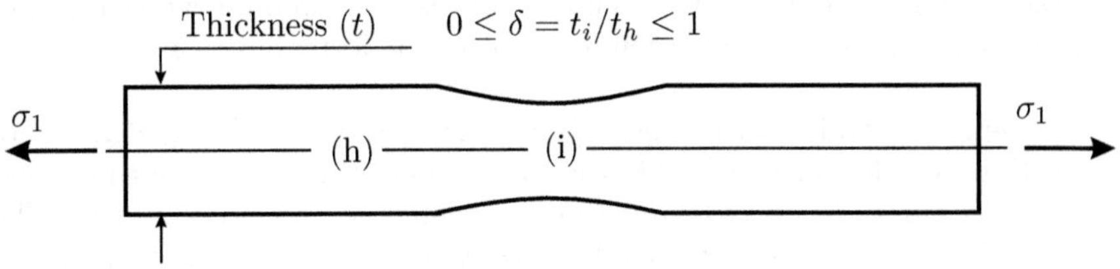

Figure 9. Neck developing in thin sheet.

transmit the force, the major principal stress σ_1 in the imperfection, i.e., σ_i, should be larger than that in the homogeneous region, σ_h. Based on equilibrium and compatibility, the governing equation for this model is (e.g., Barlat, 1989)

$$\left(\frac{\dot{\bar{\varepsilon}}_i}{\dot{\bar{\varepsilon}}_h}\right)^m \left(\frac{\bar{\varepsilon}_i}{\bar{\varepsilon}_h}\right)^n \left(\frac{\sigma_i/\sigma_r}{\sigma_h/\sigma_r}\right) = \frac{t_h}{t_i} = \frac{1}{\delta} \tag{58}$$

This equation means that three factors allow the stress to be larger in the imperfection than in the homogeneous region and, therefore, compensate for the decreasing value of δ. The first term in the left hand side of Eq. (58) is related to strain rate sensitivity. Since the strain rate is higher in the imperfection than in the homogeneous region, this term is greater than one. The second term is due to strain hardening. Again, this term is larger than one because the strain in the imperfection is higher than that in the homogeneous region. Finally, the third term is related to the yield surface shape. Indeed, the stress state corresponding to the homogenous region is represented by point P_h on the normalized yield locus in Fig. 10. The stress state, which corresponds to the imperfection, is represented by point P_i in this figure. In fact, in the Sowerby and Duncan (1971) interpretation of the localization process, Fig. 10, the stress state in the imperfection is evolving gradually towards plane strain, which is the plastic flow localization mode. As a result, the major stress in the imperfection increases as well because of this gradual change. The major stress in the imperfection is larger than that in the homogeneous region. Thus, the third term in the left hand side of Eq. (58) is larger or equal to one because of the yield surface shape effect. For instance, this term is always equal to one for Tresca and significantly larger than one for von Mises. Lian et al. (1989) discussed the yield surface shape effect in more details.

Equation (58) was integrated numerically for different hardening exponents (n), strain rate sensitivity parameters (m) and yield surface shapes (a) and the results are compared in Fig. 11. For a yield surface exponent of $a = 8$, this figure indicates that, as the strain hardening index changes from 0.2 to 0.3, or the strain rate sensitivity parameter increases from 0.0 to 0.02, the forming limit curve shifts upwards. It is well known that strain hardening and strain rate hardening improve the formability of sheet materials.

When the strain hardening exponent is $n = 0.2$ and the strain rate sensitivity coefficient is $m = 0.0$, the calculations were carried out for three different values of the yield function exponent, i.e., $a = 1, 2$ and 8. The first and second values correspond to Tresca and von Mises yield functions, respectively. Fig. 10 shows that this exponent has not effect of the forming limit for plane strain. Since this stress state is that corresponding to plastic

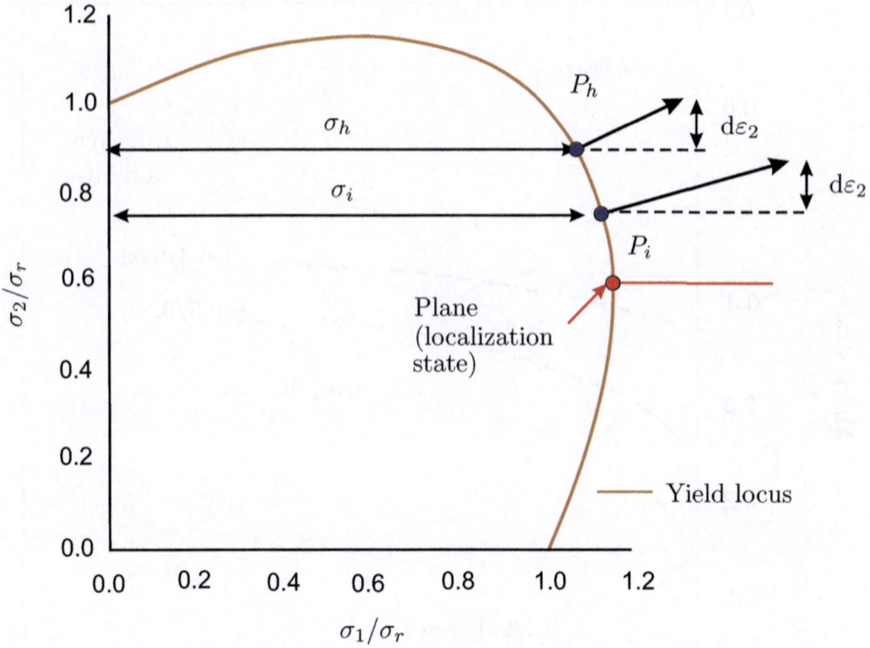

Figure 10. Sowerby and Duncan (1971) interpretation of necking in biaxially stretched sheet (yield locus is normalized by reference flow stress).

flow localization, i.e., the points P_h and P_i are superimposed in Fig. 9, there is not possible yield surface shape effect as the first term in the left hand side of Eq. (58) is always 1. For the Tresca material, $a = 1$, in balanced biaxial stretching, i.e., equal in-plane major stresses and plastic strains, the parameter

$$\frac{\sigma_i/\sigma_r}{\sigma_h/\sigma_r}$$

is equal to 1. In fact, this is true for any stretching case from plane strain to balanced biaxial stretching. Thus, the forming limit curve for a Tresca material is much lower than that calculated with the exponent $a = 8$.

In contrast, for a von Mises material, the yield surface shape for balanced biaxial stretching can be as high as about 1.15. This considerably counteracts the effect of the imperfection evolution represented by a decreasing value of δ. As a result, the forming limit curve for a von Mises material is much higher than that calculated with the exponent $a = 8$. These calculations indicate that the forming limit curves for isotropic von Mises and

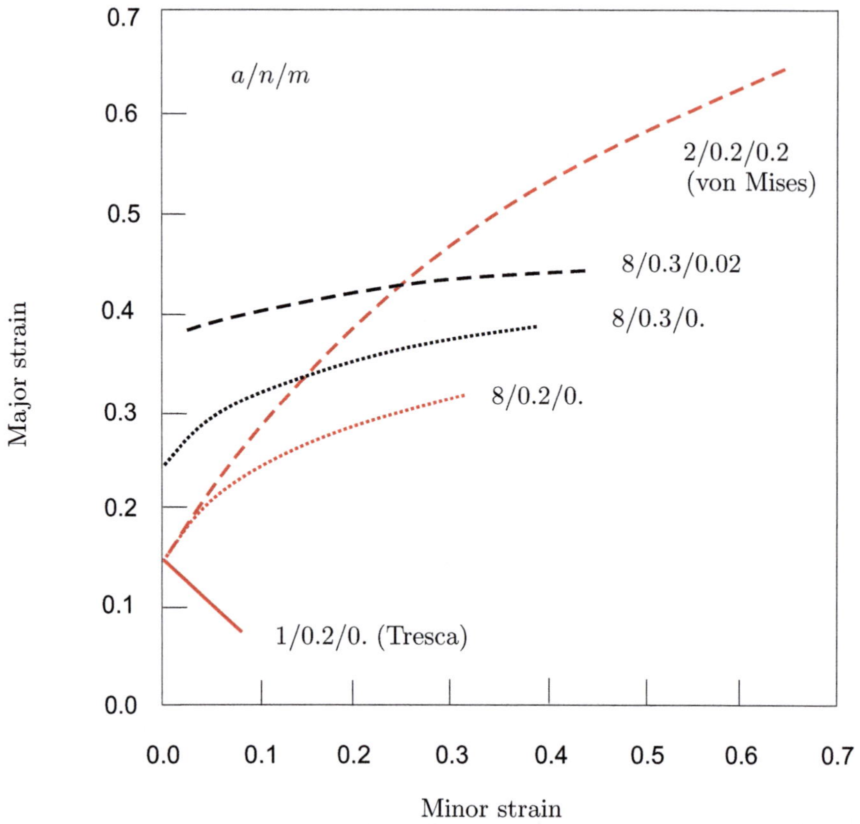

Figure 11. Forming limit diagram in the stretching range (both principal surface strain positive) for different yield surface shapes (a), hardening exponents (n) and strain rate sensitivity parameters (m).

Tresca materials are dramatically different. Although in many applications the use of either yield function is not significant, this example shows this is not the case for plastic flow localization. Therefore, as perfectly illustrated by this example, failure is strongly influenced by the details of the constitutive description.

6.2 Fracture Toughness in Thick Plate

This section discussed the work performed by Shuey et al. (2009) and Karabin et al. (2009). Plane strain fracture toughness K_{Ic} for a 7085-

T7X plate was investigated experimentally and numerically. This aerospace plate was aged at a given temperature after solution heat treatment to substantially enhance properties. In general, at a given temperature, the strength of a plate first increases as a function of the aging time, reach a maximum at the corresponding peak-aged time (τ_{PA}) and decreases. The time range after peak-age time is called over-aging. For aerospace plates, over-aging leads to better toughness and corrosion resistance compared to peak-aging. In this example, the 7085-T7X plate was subjected to different amounts of aging times from peak-age to $7.5 \times \tau_{PA}$. In depth investigations showed that the microstructure was about the same for all the aging times considered. The goal of this work was to predict the 25% difference in plane strain fracture toughness K_{Ic} between the materials aged at τ_{PA} and $5.5 \times \tau_{PA}$.

The main difference between all of these materials was the stress-strain curves represented in Fig. 12. For all the over-aging time considered, the uniform elongation is restricted to a range of about 6 to 7% because of the high strength of the various alloy conditions. These hardening curves look significantly different because of the stress range selected for the plot. However, they are not so different from each other when represented on a stress scale ranging from 0 to 600 MPa. The differences between these materials are more apparent by representing the rate of strain hardening as a function of strain in Fig. 13. Nevertheless, the rate of strain hardening appears to converge towards the same value after a strain of 0.06. The plane strain fracture toughness can be predicted using uniaxial tension test data as input. However, standard K_{Ic} models such as Hahn and Rosenfield (1968) do not lead to significant variations in plane strain fracture toughness for the materials aged at τ_{PA} and $5.5 \times \tau_{PA}$.

In order to predict fracture toughness, a standard test conducted on a compact tension (CT) specimen (Fig. 14a) was simulated with the finite element method. The mesh for half of the CT specimen is shown in Fig. 14b and a close-up view of the crack tip region in Fig. 14b. A blunted tip with a circular shape and a radius of about 6.5 μm was considered. The mesh size at the vicinity of the crack tip was about 1 μm. The FE simulations were performed with the Gurson (1977) model combined with the different hardening rates described in Figs. 12 and 13. Figure 15 illustrates the major strain distribution ahead of the crack tip for different stress intensity factors K_I. For low value of K_I, the major strain decreases as the distance from the crack tip increases. However, from a certain stress intensity factor, a strain peak develops at about 30 μm from the crack tip at a location with optimal strain and stress triaxiality, leading to a high rate of porosity growth. The plane strain fracture toughness K_{Ic} was defined as the stress intensity factor

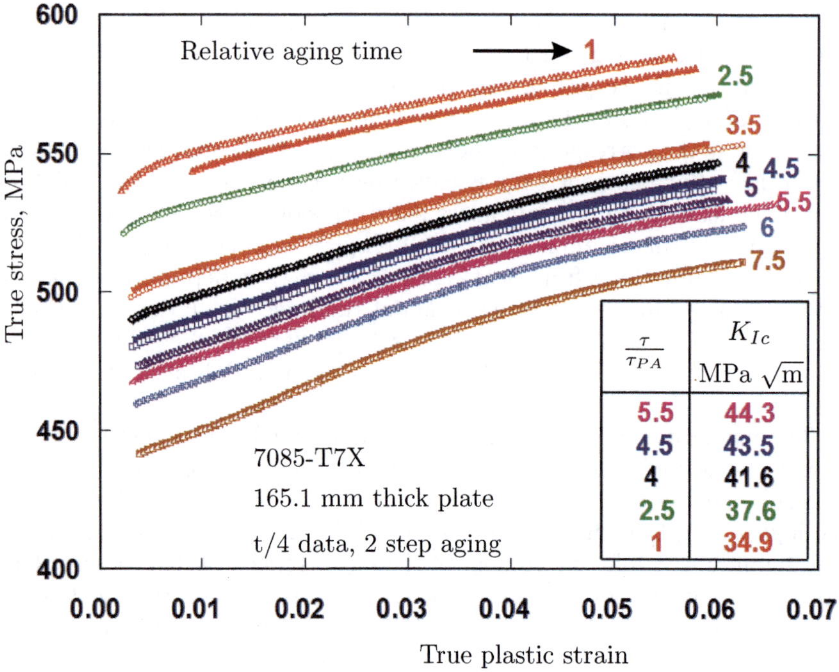

Figure 12. Duplicate stress-strain curves for 7085-T7X with different amounts of relative over-aging time (reprinted from Karabin et al. 2009, with permission from Springer).

K_I at which the strain peak starts to develop.

Based on this definition, the value of the predicted plane strain fracture toughness is represented as a function of the aging time in Fig. 16. With the Gurson model, the predicted values of K_{Ic} for τ_{PA} and $5.5\times\tau_{PA}$ are virtually the same. Again, it is worth repeating that the only varying factor in the simulations was the stress-stain curve. It is believed that, since the growth of the void volume fraction is not affected by strain hardening in the Gurson model (see Sect. 4.2), the corresponding predicted K_{Ic} values are the same. The simulations were also conducted with the modified version of the Gurson model proposed by Leblond et al. (1995), in which the void growth is lower for higher hardening rates. In this case, the predicted plane strain fracture toughness shown in Fig. 16 increases significantly with the aging time, indicating a strong influence of hardening on K_{Ic}, in good agreement with experimental results. Again, this example illustrates that a minor variation

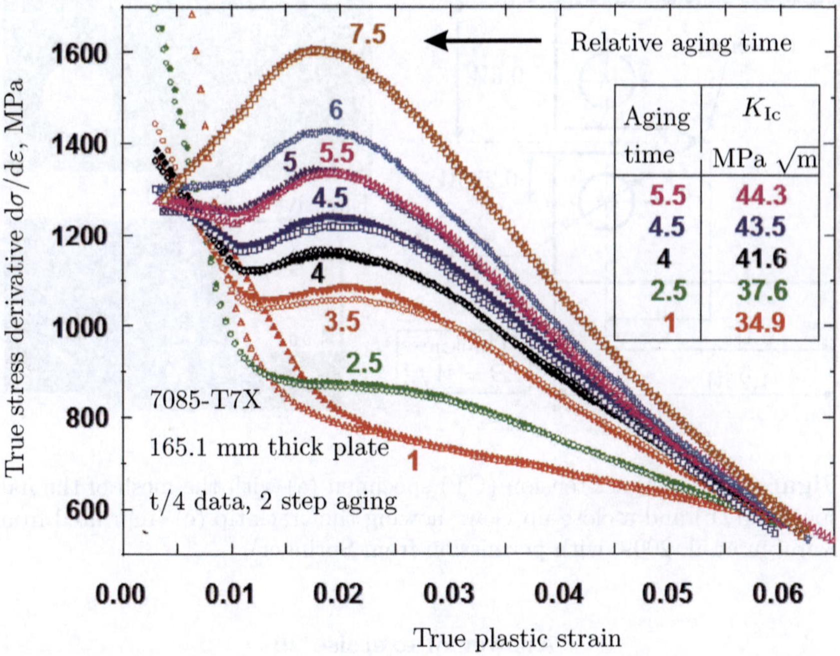

Figure 13. Duplicate rate of strain hardening for 7085-T7X with different amounts of over-aging time (reprinted from Karabin et al. 2009, with permission from Springer).

in the constitutive description of plasticity leads to a drastic difference in the failure properties of a material.

Exercise 20

Using the Irwin, Dugdale (in plane stress) and von Mises locus (in plane stress and plane strain) approximations, find the plastic zone size straight ahead of the crack tip. Compare with the results of Fig. 15.

Hint: Find these approximations in fracture mechanics books and estimate the values based on Fig. 12.

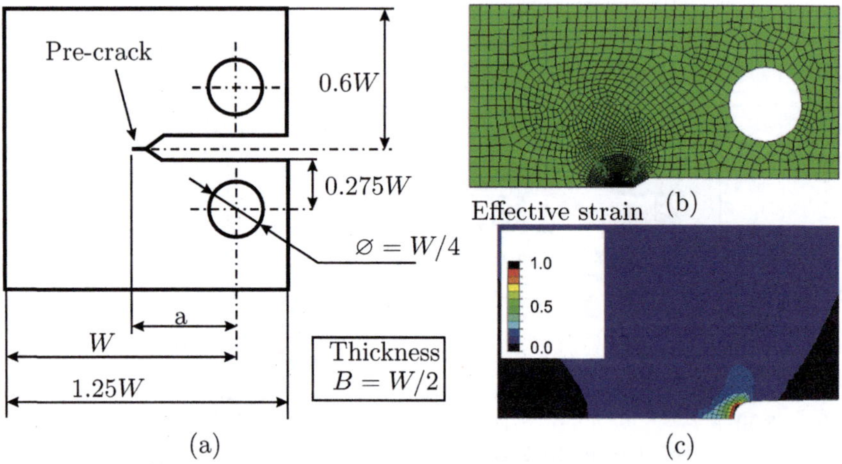

Figure 14. Compact tension (CT) specimen (a) with the mesh of the half specimen (b) and a close-up view showing the crack tip (c) (reprinted from Karabin et al. 2009, with permission from Springer).

Answer of exercise 20

For $K_I = K_{IC} \approx 35$ MPa$\sqrt{\text{m}}$, $\sigma_\text{y} = 550$ MPa and the Poisson's ratio $\nu \approx 0,3$:

- Irwin

$$d \approx \frac{1}{\pi}\left(\frac{K_I}{\sigma_\text{y}}\right)^2 \approx 1.3\,\text{mm}$$

- Dugdale

$$d \approx \frac{\pi}{8}\left(\frac{K_I}{\sigma_\text{y}}\right)^2 \approx 1.6\,\text{mm}$$

- von Mises locus (plane stress):

$$d \approx \frac{1}{2\pi}\left(\frac{K_I}{\sigma_\text{y}}\right)^2 \approx 0.6\,\text{mm}$$

- von Mises locus (plane strain):

$$d \approx \frac{(1-2\nu)^2}{2\pi}\left(\frac{K_I}{\sigma_\text{y}}\right)^2 \approx 0.1\,\text{mm}$$

The last result seems to be in better agreement with Fig. 15. This is reasonable because plane strain tension was considered in the simulations.

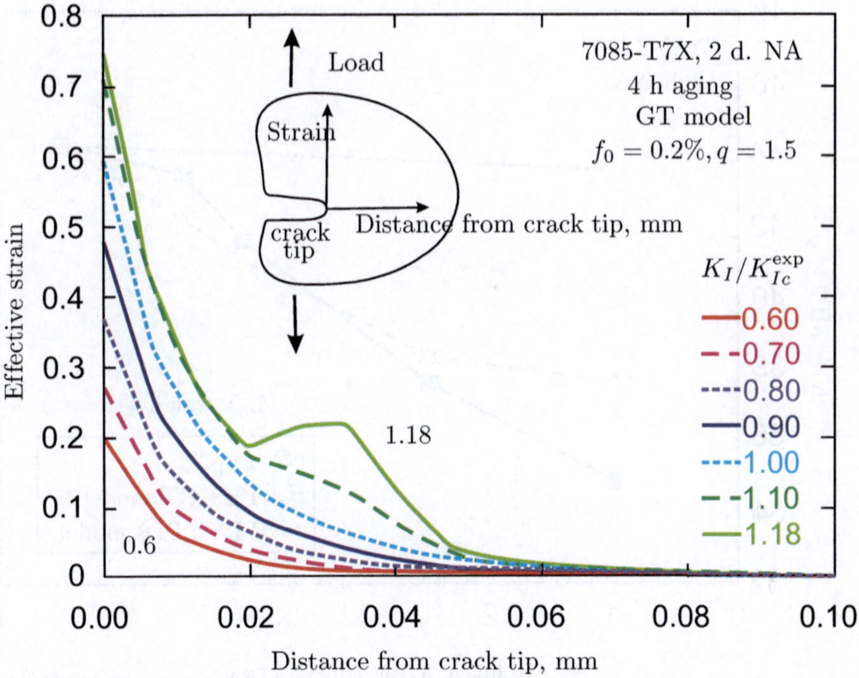

Figure 15. Effective plastic strain distribution ahead of the crack tip for different values of the stress intensity factor K_I (reprinted from Karabin et al. 2009, with permission from Springer).

Exercise 21

Using the Irwin and Dugdale (in plane stress) approximations, estimate the crack tip opening displacement (CTOD) for the different materials characterized in Fig. 12.

Hint: Find these approximations in fracture mechanics books and estimate the values as in exercise 20.

Answer of exercise 21

For $K_I = K_{IC} \approx 35$ MPa$\sqrt{\text{m}}$, $\sigma_\text{y} = 550$ MPa and Young modulus of $E \approx 70$ GPa for Al alloys:

- Irwin

$$\delta_t \approx \frac{4}{\pi} \frac{K_I^2}{E\sigma_\text{y}} \approx 40\,\mu\text{m}$$

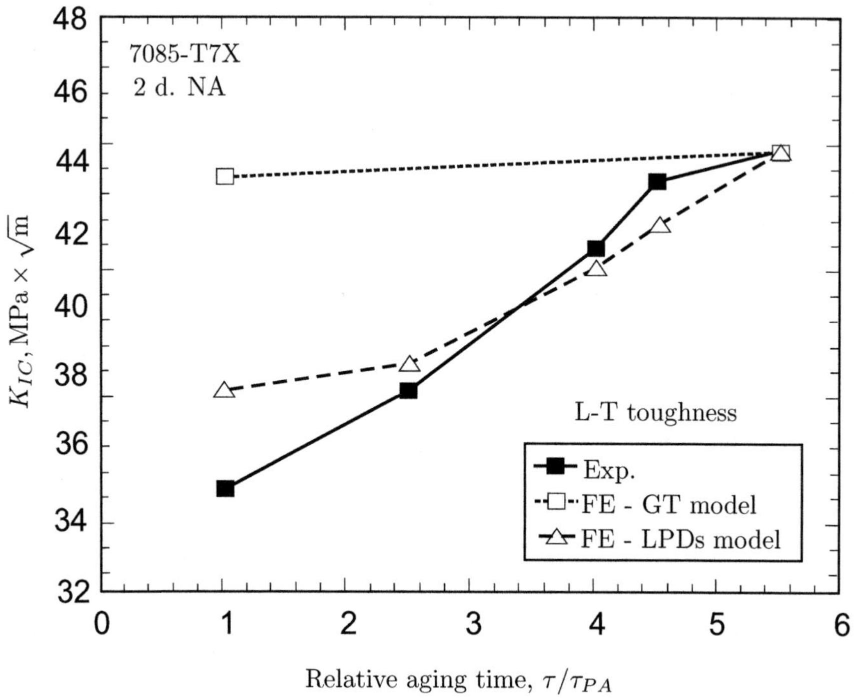

Figure 16. Experimental and predicted plane strain toughness K_{Ic} as a function of aging time. Predictions made with Gurson-Tvergaard (GT) and Leblond et al. (1995) (LPDs) models (reprinted from Karabin et al. 2009, with permission from Springer).

- Dugdale

$$\delta_t \approx \frac{K_I^2}{E\sigma_y} \approx 30\,\mu\text{m}$$

---

7 Conclusions

The scope of this discussion on modeling plasticity at low homologous temperature underlines the need of advanced constitutive models and test data specialized for the particular deformation mechanisms involved in the process of interest. Since microstructure controls the material behavior during plastic deformation, it should be embedded as much as possible in the constitutive model. However, there is a balance between this requirement, convenience for end-users and computation time. In that sense, the models described in this article are optimum in their application range for this balance at the current time. The applications examples at the end of this chapter clearly demonstrate that the details of the constitutive description significantly affect failure predictions.

Bibliography

H. Altenbach, A. Bolchoun, and V.A. Kolupaev. Phenomenological yield and failure criteria. In H. Altenbach and A. Öchsner, editors, *Plasticity of Pressure-Sensitive Materials*, pages 49–152. Springer, Berlin, Heidelberg, 2014.

H. Aretz. Applications of a new plane stress yield function to orthotropic steel and aluminium sheet alloys. *Modelling and Simulation in Materials Science and Engineering*, 12:491–509, 2004.

H. Aretz. A non-quadratic plane stress yield function for orthotropic sheet metals. *Journal of Materials Processing Technology*, 168:1–9, 2005.

D. Banabic, H. Aretz, D.S. Comsa, and L. Paraianu. An improved analytical description of orthotropy in metallic sheets. *International Journal of Plasticity*, 21(3):493–512, 2005.

F. Barlat. Forming limit diagram - prediction based on some microstructural aspects of materials. In R.H. Wagoner, K.S. Chan, and S.P. Keeler, editors, *Forming Limit Diagram - Concepts, Methods and Applications*, pages 275–301. TMS, Warrendale, PA, 1989.

F. Barlat and O. Richmond. Prediction of tricomponent plane stress yield surfaces and associated flow and failure behavior of strongly textured fcc sheets. *Mat. Sci. Eng.*, 95:15–29, 1987.

F. Barlat, D.J. Lege, and J.C. Brem. A six-component yield function for anisotropic materials. *International Journal of Plasticity*, 7(7):693–712, 1991.

F. Barlat, R.C. Becker, Y. Hayashida, Y. Maeda, M. Yanagawa, K. Chung, J.C. Brem, D.J. Lege, K. Matsui, S.J. Murtha, and S. Hattori. Yielding description of solution strengthened aluminum alloys. *International Journal of Plasticity*, 13(4):385–401, 1997.

F. Barlat, J.C. Brem, J.W. Yoon, K. Chung, R.E. Dick, D.J. Lege, F. Pourboghrat, S.-H. Choi, and E. Chu. Plane stress yield function for aluminum alloy sheets - Part I: Theory. *International Journal of Plasticity*, 19(9):1297–1319, 2003.

F. Barlat, O. Cazacu, M. Życzowski, D. Banabic, and J.W. Yoon. Yield surface plasticity and anisotropy. In D. Raabe, F. Roters, F. Barlat, and L.-Q. Chen, editors, *Continuum Scale Simulation of Engineering Materials - Fundamentals - Microstructures - Process Applications*, pages 145–177. Wiley-VCH Verlag GmbH, Berlin, 2004.

F. Barlat, H. Aretz, J.W. Yoon, M.E. Karabin, J.C. Brem, and R.E. Dick. Linear transformation-based anisotropic yield functions. *International Journal of Plasticity*, 21(5):1009–1039, 2005.

F. Barlat, J.W. Yoon, and O. Cazacu. On linear transformations of stress tensors for the description of plastic anisotropy. *International Journal of Plasticity*, 23(5):876–896, 2007.

J.L. Bassani, J.W. Hutchinson, and K.W. Neale. On the prediction of necking in anisotropic sheets. In H. Lippmann, editor, *Metal Forming Plasticity*, pages 1–13. Springer, Berlin, 1979.

J. Betten. Mathematical modeling of materials behaviour. *Math. Comp. Model.*, 11:702–708, 1988.

J.F.W. Bishop and R. Hill. A theory of the plastic distortion of a polycrystalline aggregate under combined stresses. *Phil. Mag.*, 42:414–427, 1951a.

J.F.W. Bishop and R. Hill. A theory of the plastic distortion of a polycrystalline aggregate under combined stresses. *Phil. Mag.*, 42:1298–1307, 1951b.

J.P. Boehler. On a rational formulation of isotropic and anisotropic hardening. In J.P. Boehler, editor, *Applications of Tensor Functions in Solid Mechanics*, pages 99–120. Springer, Wien, 1978.

J.P. Boehler and A. Sawczuk. Equilibre limite des sols anisotropes. *J. Mécanique*, 9:5–33, 1970.

F. Bron and J. Besson. A yield function for anisotropic materials. application to aluminum alloys. *International Journal of Plasticity*, 20(4-5): 937–963, 2004.

W. Burzyński. Über die Anstrengungshypothesen. *Schweiz. Bauz.*, 95:259–263, 1929.

O. Cazacu and F. Barlat. Generalization of Drucker's yield criterion to orthotropy. *Mathematics and Mechanics of Solids*, 6:613–630, 2001.

O. Cazacu and F. Barlat. Application of the theory of representation to describe yielding of anisotropic aluminum alloys. *Int. J. Engineering Science*, 41:1367–1385, 2003.

O. Cazacu and F. Barlat. A criterion for description of anisotropy and yield differential effects in pressure-insensitive metals. *International Journal of Plasticity*, 20(11):2027–2045, 2004.

O. Cazacu, B. Plunkett, and F. Barlat. Orthotropic yield criterion for hexagonal close packed metals. *International Journal of Plasticity*, 22 (7):1171–1194, 2006.

J.L. Chaboche. A review of some plasticity and viscoplasticity constitutive theories. *International Journal of Plasticity*, 24(10):1642–1693, 2008.

D.-S. Comsa and D. Banabic. Plane-stress yield criterion for highly-anisotropic sheet metals. In P. Hora, editor, *Proc. 7th Int. Conf. Works. on Numerical Simulation of 3D Sheet Metal Forming Processes*, pages 43–48. ETH Zurich, 2008.

P. Dasappa, K. Inal, and R. Mishra. The effects of anisotropic yield functions and their material parameters on prediction of forming limit diagrams. *Int. J. Solids Structures*, 49:3528–3550, 2012.

G.E. Dieter. *Mechanical Metallurgy.* McGraw Hill, London, 1988.

D.C. Drucker. Relation of experiments to mathematical theories of plasticity. *J. Appl. Mech.*, 16:349–357, 1949.

Z. Engel. Historical aspect of huber's work. In M. Pietrzyk, J. Kusiak, L. Sadok, and Z. Engel, editors, *Huber's Yield Criterion in Plasticity*, pages 1–7. Akademia Grniczo-Hutnicza, Kraków, 1994.

A.L. Gurson. Continuum theory of ductile fracture by void nucleation and growth - Part I: Yield criteria and flow rules for porous ductile media. *Trans. ASME. J. Eng. Materials and Technology*, 99:2–15, 1977.

G.T. Hahn and A.R. Rosenfield. Sources of fracture toughness: The relation between k and the ordinary tensile properties of metals. In *Applications Related Phenomena in Titanium Alloys*, number 432 in ASTM Special Technical Publication, pages 5–32. ASTM, Philadelphia, PA, 1968.

S.S. Hecker. Experimental studies of yield phenomena in biaxially loaded metals. In A. Stricklin and K.C. Saczalski, editors, *Constitutive Modeling in Viscoplasticity*, pages 1–33. ASME, New York, 1976.

A.V. Hershey. The plasticity of an isotropic aggregate of anisotropic face centred cubic crystals. *Trans. ASME. J. Appl. Mech.*, pages 241–249, 1954.

R. Hill. A theory of the yielding and plastic flow of anisotropic metals. *Proc. Roy. Soc. London*, A193:281–297, 1948.

R. Hill. *The Mathematical Theory of Plasticity.* Oxford University Press, Oxford, 1950.

R. Hill. Theoretical plasticity of textured aggregates. *Math. Proc. Cambridge Philos. Soc.*, 85:179–191, 1979.

R. Hill. Constitutive modelling of orthotropic plasticity in sheet metals. *J. Mech. Phys. Solids*, 38:405–417, 1990a.

R. Hill. A user-friendly theory of orthotropic plasticity in sheet metals. *Int. J. Mech. Sci.*, 35:19–25, 1990b.

J.H. Hollomon. Tensile deformation. *AIME Transactions*, 12:1–22, 1945.

W.F. Hosford. A generalized isotropic yield criterion. *Trans. ASME. J. Appl. Mech.*, 39:607–609, 1972.

W.F. Hosford and T.J. Allen. Twinning and directional slip as a cause for strength differential effect. *Met. Trans.*, 4:1424–1425, 1973.

Z.P. Huang and J. Wang. Nonlinear mechanics of solids containing isolated voids. *Applied Mechanics Review*, 59(4):210–229, 2006.

K. Inal, K.W. Neale, and Aboutajeddine. Forming limit comparisons for FCC and BCC sheets. *International Journal of Plasticity*, 21(6):1255–1266, 2005.

G.R. Johnson and W.H. Cook. A constitutive model and data for metals subjected to large strains, high strain rates and high temperatures. In *Proc. 7th International Symposium on Ballistics*, pages 541–547. The Hague, 1983.

M.E. Karabin, F. Barlat, and R.T. Shuey. Finite element modeling of plane strain toughness for 7085 aluminum alloy. *Metallurgical and Materials Transactions*, A40:354–364, 2009.

A.P. Karafillis and M.C. Boyce. A general anisotropic yield criterion using bounds and a transformation weighting tensor. *J. Mech. Phys. Solids*, 41:1859–1886, 1993.

U.F. Kocks, C.N. Tomé, and H.-R. Wenk. *Texture and Anisotropy*. University Press Cambridge, Cambridge, 1998.

A.S. Krausz and K. Krausz. *Unified Constitutive Laws of Plastic Deformation*. Academic Press, San Diego, 1996.

M. Kuroda and V. Tvergaard. Forming limit diagrams for anisotropic metal sheets with different yield criteria. *Int. J. Solids Structures*, 37:5037–5059, 2000.

R.A. Lebensohn and C.N. Tomé. A self-consistent anisotropic approach for the simulation of plastic deformation and texture development of polycrystals: Application to zirconium alloys. *Acta Metall. Mater.*, 41: 2611–2624, 1993.

J.B. Leblond, G. Perrin, and J. Devaux. An improved gurson-type model for hardenable ductile metals. *Eur. J. Mech. A, Solids*, 14:499–527, 1995.

J. Lian, F. Barlat, and B. Baudelet. Plastic behavior and stretchability of sheet metals. Part II: Effect of yield surface shape on sheet forming limit. *International Journal of Plasticity*, 5(2):131–148, 1989.

S.I. Liu. On representations of anisotropic invariants. *Int. J. Eng. Science*, 20:1099–1109, 1982.

R.W. Logan and W.F. Hosford. Upper-bound anisotropic yield locus calculations assuming pencil glide. *Int. J. Mech. Science*, 22:419–430, 1980.

Z. Marciniak and K. Kuczyński. Limit strains in the processes of stretch-forming sheet metal. *Int. J. Mech. Science*, 9:609–620, 1967.

D.L. McDowell. A perspective on trends in multiscale plasticity. *International Journal of Plasticity*, 26(9):1280–1309, 2010.

M. Meyers and K. Chawla. *Mechanical Behavior of Materials.* Cambridge University Press, Cambridge, 2009.

B. Paul. Macroscopic criteria for plastic flow and brittle fracture. In H. Liebowitz, editor, *Fracture*, volume 1, pages 313–495. Academic Press, San Diego, CA, 1968.

B. Plunkett, O. Cazacu, and F. Barlat. Orthotropic yield criteria for description of the anisotropy in tension and compression of sheet metals. *International Journal of Plasticity*, 24(5):847–866, 2008.

K. Pöhlandt, D. Banabic, and K. Lange. Equi-biaxial anisotropy coefficient used to descibe the plastic behavior of sheet metals. In M. Pietrzyk, editor, *Proceedings of the 5th ESAFORM Conference on Metal Forming, Krakow, Poland, April 2002*, pages 723–727, 2002.

D. Raabe, F. Roters, F. Barlat, and L.Q. Chen. *Continuum Scale Simulations of Engineering Materials - Fundamentals - Microstructures - Process Applications.* Wiley-VCH, Berlin, 2004.

E.F. Rauch and J.-H. Schmitt. Dislocation substructures in mild steel deformed in simple shear. *Mat. Sci. Eng.*, A113:441–448, 1989.

R.T. Rockafellar. *Convex Analysis.* Princeton University Press, Princton (NJ), 1972.

E. Schmid and W. Boas. *Kristallplastizität.* Springer, Berlin, 1935.

R.T. Shuey, F. Barlat, M.E. Karabin, and D.J. Chakrabarti. Experimental and analytical investigations on plane strain toughness for 7085 aluminum alloy. *Metallurgical and Materials Transactions*, A40:365–376, 2009.

Z. Sobotka. Theorie des plastischen fliessens von anisotropen krpern. *ZAMM*, 49:25–32, 1969.

R. Sowerby and J.L. Duncan. Failure in sheet metal in biaxial tension. *Int. J. Mech. Sci.*, 13:217–229, 1971.

W.A. Spitzig and O. Richmond. The effect of pressure on the flow stress of metals. *Acta Metal.*, 32:457–463, 1984.

T.B. Stoughton. A non-associated flow rule for sheet metal forming. *International Journal of Plasticity*, 18(5-6):687–714, 2002.

T.B. Stoughton and J.W. Yoon. Anisotropic hardening and non-associated flow in proportional loading of sheet metals. *International Journal of Plasticity*, 25(9):1777–1817, 2009.

H.W. Swift. Plastic instability under plane stress. *J. Mech. Phys. Solids*, 1: 1–18, 1952.

F. Taha, A. Graf, and W.F. Hosford. Plane-strain tension tests on aluminum alloy sheet. *Trans. ASME. J. Eng. Mater. Technol.*, 117:168–171, 1995.

G.I. Taylor. Plastic strains in metals. *J. Inst. Metals*, 62:307–324, 1938.

G.I. Taylor and H. Quinney. The plastic distortion of metals. *Phil. Trans. Roy Soc. London*, A230:323–362, 1931.

N. Tozawa. Plastic deformation behavior under the conditions of combined stress. In D.P. Koistinen and N.M. Wang, editors, *Mechanics of Sheet Metal Forming*, pages 81–110. Plenum Press, New York, 1978.

N. Tozawa and M. Nakamura. A biaxial compression testing method for thin sheets. *Plasticity and Processing*, 13:538–541, 1972.

Y. Tozawa and M. Nakamura. Methods for testing in-line compression of sheets. *Sosei to Kato*, 8:444–448, 1967.

H. Tresca. Mémoire sur l'écoulement des corps solides soumis à de fortes pressions. *Compte Rendus Acad. Sci. Paris*, 59:754–758, 1864.

V. Tvergaard. On localization in ductile materials containing spherical voids. *Int. J. Fracture*, 18:237–252, 1982.

E. Voce. A practical strain-hardening function. *Metallurgica*, 51:219–226, 1955.

R. von Mises. Mechanik der plastischen körper im plastisch deformablen zustand. *Göttinger Nachrichten Math. Phys. Klasse*, 1:582–592, 1913.

R. von Mises. Mechanik der plastischen formnderung von kristallen. *ZAMM*, 8:161–185, 1928.

R.H. Wagoner and N.-M. Wang. An experimental and analytical investigation of in-plane deformation of 2036-T4 aluminum sheet. *Int. J. Mech. Sci.*, 21:255–264, 1979.

C.C. Wang. A new representation theorem for isotropic functions, Part I and II. *Arch. Rat. Mech. Anal.*, 36:166–223, 1970.

J.W. Yoon and F. Barlat. *ASM Handbook*, volume 14B - Metalworking: Sheet Forming, chapter Modeling and simulation of the forming of aluminum sheet alloys, pages 792–826. ASM International, Materials Park (OH), 2006.

R.F. Young, J.E. Bird, and J.L. Duncan. An automated hydraulic bulge tester. *J. Appl. Metalwork*, 2:11–18, 1981.

M.H. Yu. Advances in strength theories for materials under complex stress state in the 20th century. *Appl. Mech. Rev.*, 55:198–218, 2002.

M. Życzkowski. *Combined Loadings in the Theory of Plasticity.* Polish Scientific Publisher, Warsaw, 1981.

Failure and Damage in Cellular Materials

Liviu Marsavina [*] and Dan M. Constantinescu [†]

[*] Universitatea Politehnica Timişoara, Timişoara, Romania
[†] Universitatea Politehnica Bucureşti, Bucureşti, Romania

Abstract This chapter presents the main aspects on failure and damage of cellular materials. Tensile, compression and fracture mechanics properties of plastic foams are presented and the main influence factors are investigated: density, temperature, loading speed and loading direction. Particularly for fracture toughness the mixed mode loading and size effects are discussed. The potential of digital image correlation as a tool to observe the damage of polyurethane foams is also highlighted.

1 Introduction

Foam materials have a cellular structure and hence behave in a complex manner, especially under conditions of progressive crush. This crush behavior is dependent on the geometry of the microstructure and on the characteristics of the parent material. Foam materials are often used as cores in sandwich construction, and in this application the material can be subjected to multi-axial stresses prior to and during crush. Well-known advantages of cellular materials are their excellent ability for energy adsorption, good damping behavior, sound absorption, excellent heat insulation and a high specific stiffness combined with a low weight. The combination of these properties opens a wide field of potential applications, i.e. as core materials in sandwich panels. A good knowledge of the behavior of different grades of foams is important for being able to design high performance sandwich composites adapted to the special needs of a particular application (Gibson and Ashby, 1997; Mills, 2007).

The properties of cellular materials are influenced by the properties of solid material (polymers, metals, ceramics), by the cellular structure topology (open or closed cells) and relative density ρ/ρ_s, with ρ density of cellular material and ρ_s the density of the solid material (Ashby, 2005).

Polyurethane (PU) foam is an engineering material for energy absorption and has been widely used in many applications such as packaging and cushioning. The mechanical testing of rigid PU foams under compression

H. Altenbach, T. Sadowski (Eds.), *Failure and Damage Analysis of Advanced Materials*,
CISM International Centre for Mechanical Sciences
DOI 10.1007/978-3-7091-1835-1_3

in the rise and transverse direction gives different deformation responses in each direction which are attributed to the anisotropy in the internal cellular structure.

There are two approaches to the modeling of the constitutive behavior of foam materials. The first is continuum modeling. A number of theories have been presented, namely the critical state theory, which is used in standard finite element codes such as ABAQUS, and enhancements have been developed to take account of specific foam behavior (Warren and Kraynik, 1997; Akay and Hanna, 1990; Mines et al., 1994; Mines and Jones, 1995; Li and Mines, 2002). To calibrate such a numerical model based on material behavior separate traction, compression, shear and hydrostatic tests are needed to provide the appropriate mechanical properties of the studied foam (Li et al., 2000). These models, such as that of Gibson and Ashby (1997), are based on the assembly of geometric symmetric cells (rectangular prism, cubic, etc.) and relate analytically elasticity and yield stress to the foam relative density. The second approach is micro-modeling, in which the actual cellular structure is modeled (Jin et al., 2007; Chen et al., 1998; Ren and Silberschmidt, 2008; Gibson, 1985). This approach has the advantage of differentiating between micro-mechanical failure modes, but it is computationally demanding for complete sandwich structures with progressive crush. The architecture is determined by the cell wall thickness, the size distribution, the shape of the cells and the structure of the foam which is simulated by an assembly of walls and struts. Complex modeling approaches based on finite element method try to describe as finely as possible the foam microstructure. The continuum approach has been well proven, and can be used with standard finite element codes, being computationally efficient for modeling the progressive crush of foam. However, the approach assumes smooth stress gradients in the material, which implies that the foam consists of strain-hardening cells.

Strain rate and temperature effects on the crush behavior of foams were studied by Li et al. (2000). Following, Mines (2007) studies strain rate effects on Divinycell PVC foam, Rohacell PMI foam and Alporas aluminum foam. His impact tests used standard static test rigs, with the higher rate of loading being achieved using a high rate servo hydraulic machine which can achieve crosshead speeds of up to 10 m/s. As he mentions in the conclusions: "The conduct of impact materials tests requires careful design and data analysis, in order to filter out inertial and structural effects and hence to measure true material properties". Such a statement has to be considered.

Saint-Michel et al. (2006) have evaluated the mechanical properties of studied foams in a quite wide relative density range (from 0.3 to 0.85) and present the microstructural characterization and the mechanical behavior of such materials. Their experimental results are then compared in the linear domain to the theoretical approaches of Gibson and Ashby (1997) and Christensen and Lo (1979, 1986). The modeling is then extended to the description of the mechanical behavior in the non-linear domain. Viot et al. (2005) carried out tests on polypropylene foams under high strain rate compression tests on a flywheel for higher strain rates and the material behavior has been determined as a function of two parameters, density and strain rate. The sample compression was filmed with a high speed camera monitored by the flywheel software as to obtain displacement and strain fields during tests. A modified split Hopkinson pressure bar procedure was used to conduct dynamic compressive experiments by Chen et al. (2002), and later by Song et al. (2005), based on experimental results, developed a phenomenological model which describes the compressive and failure behaviors at various strain rates and environmental temperatures for syntactic foams. To complete our brief overview we have to mention that Gong et al. (2005a) and Gong et al. (2005b) have performed more thorough research on understanding the responses of open cell foams to uniaxial compression in the rise and transverse directions. They also characterized the cell and ligament morphology of PU foams with various cell sizes and experimentally studied the mechanical properties of these foams. The Kelvin cell model was used to describe the initial elastic response of the foams under uniaxial compression. The nonlinear aspects of the compressive response and crushing of open cell foams were also studied based on this anisotropic cell model.

Other complexities in the constitutive behavior of foams also occur. The post-collapse behavior is influenced by the air pressure enclosed in the closed cell foam which is compressed. Properties for polymeric foams are viscoelastic and hence time dependent. Recovery after loading is also time dependent, and matters are further complicated if foam damage has occurred.

We initially started testing different grades of foams as: PVC foam, Coremat, extruded polystyrene, polyurethane foam with density 200 kg/m^3, polyurethane foam with density 40 kg/m^3, expanded polystyrene (Apostol et al., 2007; Marsavina et al., 2008b). Initially we tested the Coremat core in traction, and polyurethane foams with densities of 40 kg/m^3 and 200 kg/m^3 in traction, compression, and three-point bending. For the bending of the 200 kg/m^3 foam we have also impregnated it with polyester and epoxy resins on the upper and lower faces of the specimens and studied the influence of such a layer on the behavior of the foam (Marsavina et al., 2009).

Present results concentrate on the mechanical testing of three densities of polyurethane foams of 35 kg/m^3, 93 kg/m^3 and 200 kg/m^3. It is studied the influence of the speed of loading from 2 mm/min up to 6 m/s and of the temperature at three levels which are considered as: -60 °C, 23 °C and 80 °C. The mechanical testing presented here is mainly dedicated to the compressive response of these foams as to study their densification behavior on one hand, and the recovery of the foams after unloading on the other hand. In these tests initial strain rates started from a value of 0.0014 s^{-1} to a maximum value of 545 s^{-1}. Specimens were tested in the *rise direction* (notated as *direction 3 – out of plane*) of the foam and in one *in-plane direction (direction 1)*. Differentiating the foam properties according to the testing direction is an issue of practical interest and significance. Other important results were obtained by studying the influence of the temperatures which are encountered in engineering applications.

2 Behavior of Cellular Materials in Tension and Compression

The PU foams cells morphology and dimensions for the three densities were studied before testing through optical microscopy (OM) and scanning electron microscopy (SEM). An Olympus optical microscope, model BX 51, having a maximum magnification factor of 200, made possible the measurement of the cells dimensions (length, width, and cell wall thickness). For the SEM analyses the specimens were covered with a very thin layer of gold (as foam is non-conductive from electrical point of view) and kept in vacuum for 14 hours. It is important to mention than in the following figures the rise (out of plane) direction of the foam is always in vertical position, as to be able to notice the orientation of the cells for each density. Some of the already damaged cells were destroyed during vacuuming and in Fig. 1 several empty cells can be noticed. In Fig. 1a) a SEM image of the foam with the density of 35 kg/m^3 is shown; the closed cells have many "wrinkles", damaged areas, and microcracks. The cells are having the maximum length of 683 μm and the minimum length of 130 μm, respectively; wall thickness is in between 22.4 and 30 μm. When the density is 93 kg/m^3 (Fig. 1b) cells sizes are becoming almost equal on the main directions being in between 541 μm and 180 μm, having a wall thickness quite similar to the ones before, from 19 μm to 35.4 μm. Cells surface has a neat aspect. Finally, for the 200 kg/m^3 density foam (Figs. 1c) and d) main sizes of the cells are in the interval 472 μm to 110 μm and wall thickness in between 20.7 and 35 μm. To summarize, the wall thickness is in average of 26-27 μm for all the three densities and maximum cells length decreases from 683 μm for 35 kg/m^3, to 541 μm for 93 kg/m^3, and to 472 μm for 200 kg/m^3. A more evident elongation of the cells on the rise direction is noticed for the cells with densities of 35 kg/m^3 and 200 kg/m^3.

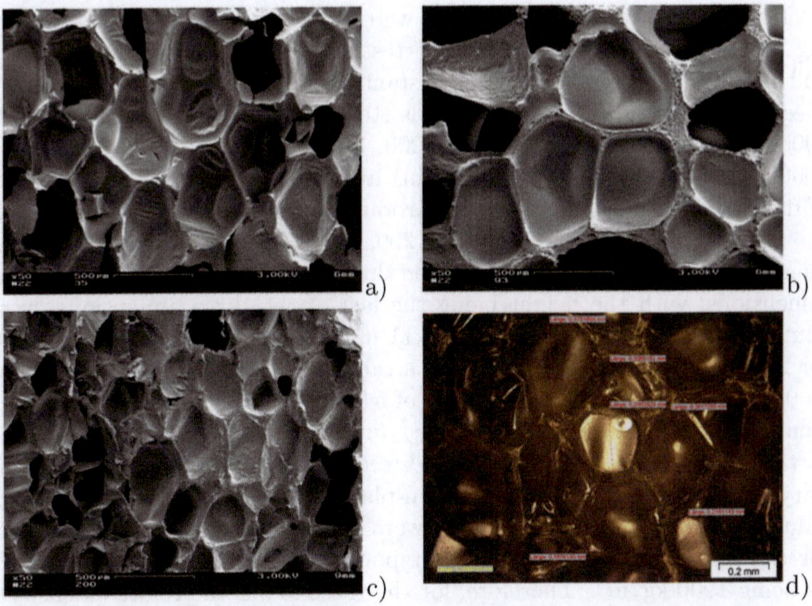

Figure 1. SEM images of the cell morphology for the PU foams with densities: a) 35 kg/m^3; b) 93 kg/m^3; c 200 kg/m^3; d OM image for 200 kg/m^3

2.1 Experimental Determination of Foam Properties in Tension and Compression

All three grades of foams were tested in traction and in compression. Traction tests were done on a Zwick-Roell testing machine, model Z250, having a 10 kN force cell at testing speeds of 2, 54, 200 and 500 mm/min. The machine is equipped with an environmental chamber which can work in between -80 °C and 120 °C. The selected temperatures for our tests were -60 °C, 23 °C and 80 °C, as to cover a range of temperatures possible in engineering applications, from aerospace at low temperatures to automotive when an extreme hot environment may appear. A micro extensometer used during tests has independently moving levers and is suitable for materials with small rigidity as not introducing a bending loading; it can be used for low and high temperatures with special protected extensions made from ceramic material. Specimens of type 1B were cut according to standard ISO 527-1[1].

[1] ISO 527-1: 2012 Plastics - Determination of tensile properties - Part 1: General principles

Tension and compressions tests were done on a hydraulic MTS testing machine specially conceived for testing polymers by Apostol (2011), and presented by Apostol and Constantinescu (2012). Maximum testing speed is 6 m/s and our testing speeds started from 2 mm/min going up to 40000 mm/min (2, 6, 18, 54, 125, 200, 350, 500, 1000, 2000, 3500, 6000, 10000, 20000, 30000, 40000 mm/min) and then 1, 3, and 6 m/s. Foams of densities 35 and 93 kg/m^3 were produced by a Romanian company and the 200 kg/m^3 foam is Divinycell H 200, produced by DIAB. As the specimens were cut from PU plates of given thickness the approximate specimens dimensions, with the height being the last of the three dimensions, were: 25x25x24 mm for 35 kg/m^3, 15x15x11 mm for 93 kg/m^3, 12x12x11.9 mm for 200 kg/m^3. Therefore the initial strain rate started from a value as low as 0.00139 s^{-1} to a maximum value of 545.45 s^{-1}. For the tested compression specimens the *rise (out of plane) direction* of the foam was notated as *direction 3* and *one of the in-plane directions* as *direction 1*; some preliminary tests showed that on both the in-plane directions practically the same values of the mechanical properties were obtained. The solid density (both for rigid and flexible PU foams) is reported by Gibson and Ashby (1997) as being 1200 kg/m^3. Therefore, for the three foams the relative density is approximately: 0.03, 0.08, and 0.17.

For the testing we used a specially designed MTS Composite testing machine, capable of reaching 8 m/s with the help of a three stage valve, valve that is being used for speeds higher than 0.7 m/s. This machine is also equipped with a piezoelectric quartz-crystal load cell washer. In our tests we haven't exceeded 6 m/s for safety reasons, due to the small height of our specimens. The machine is controlled by creating a command line program that carries on the task required for testing. The acquisition has been done by using a fast measurement buffer of 1024 values at a rate of 35 kHz for the beginning of the test and by using the highest possible acquisition rate provided by the machine in a normal manner at 5 kHz till the end of the test. In order to be sure of the generated results, a comparison has been done by using a SIGMA oscilloscope manufactured by Nicolet Technologies capable of measurements up to 500 kHz. By using the piezoelectric load cell washer together with the oscilloscope the obtained data have been compared to the data obtained by using only the machine, and latter on we took the decision to use only the results provided by the testing machine as they proved to be correct. On the other hand this method is simpler, and by using the crosshead movement to measure displacement and calculate strain, the conventional characteristic curve is generated.

For each testing case (density, temperature, speed) five specimens were tested and the representative one was selected; if a test gave suspicious results it was disregarded. The volume of obtained data is significant and only few of them are presented hereby.

Specimens are compressed up to when specimen height becomes 1.5-2 mm (maximum strain reaching a little bit more than 90 %) and data were recorded with specific frequency of data acquisition depending on the loading speed as to obtain a convenient volume of data, not in excess; for the recovery of the foams the same speed of unloading was chosen as 0.6 mm/min, always sampling data at the same frequency of 0.5 Hz which was found to be sufficient for all loading speeds at the three temperatures of testing regardless the speed of testing and density. Foam recovery strain values are established having as a reference the moment when unloading starts and in the following figures is named *recovered strain.*

2.2 Effect of Density, Forming Direction, and Speed of Loading

Tensile Testing Conventional stress-strain characteristic curves were established only for the loading speeds of 2, 54, 200, and 500 mm/min for the three studied densities of 35, 93, and 200 kg/m^3. The temperatures of testing were, as mentioned before, -60 °C, 23 °C and 80 °C. Tests are done on *direction 1* as specimens were cut from polyurethane plates. Each curve is an average of three tests. The corresponding diagrams are shown in Figs. 2, 3 and 4. As one can notice the low negative temperature makes the behavior of the foams more fragile than at 23 °C, but not as much as expected. The curves for the higher two speeds of loading at 23 °C (dashed lines) are coming above the curves resulting at -60 °C (doted lines) for the lower speed of loading 2 mm/min for the foam with density 35 kg/m^3. For -60 °C the increase of speed of loading doesn't change much the characteristic curves, especially for the higher densities of 93 kg/m^3 and 200 kg/m^3. On the other hand the temperature of 80 °C increases significantly the ductile behavior of foams, with greater emphasis on the 93 and 200 kg/m^3 densities.

The compared mechanical properties are: modulus of elasticity, maximum stress at yielding - notated as maximum stress, and elongation at failure. Obtained results are given in Tables 1, 2, and 3 for the three increasing densities. The increase of speed of loading doesn't increase significantly the values of the moduli of elasticity for each temperature level and density, regardless the temperature, but these values are significantly different at different temperatures. For each of the densities the moduli of elasticity decrease about 2 times or even more - mainly for the 200 kg/m^3 foam - when the temperature is increased from -60 °C to 80 °C (see corresponding columns in Tables 1-3). Maximum stress (at yielding) is about the same for the 35 kg/m^3 density foam regardless the temperature level, and - in

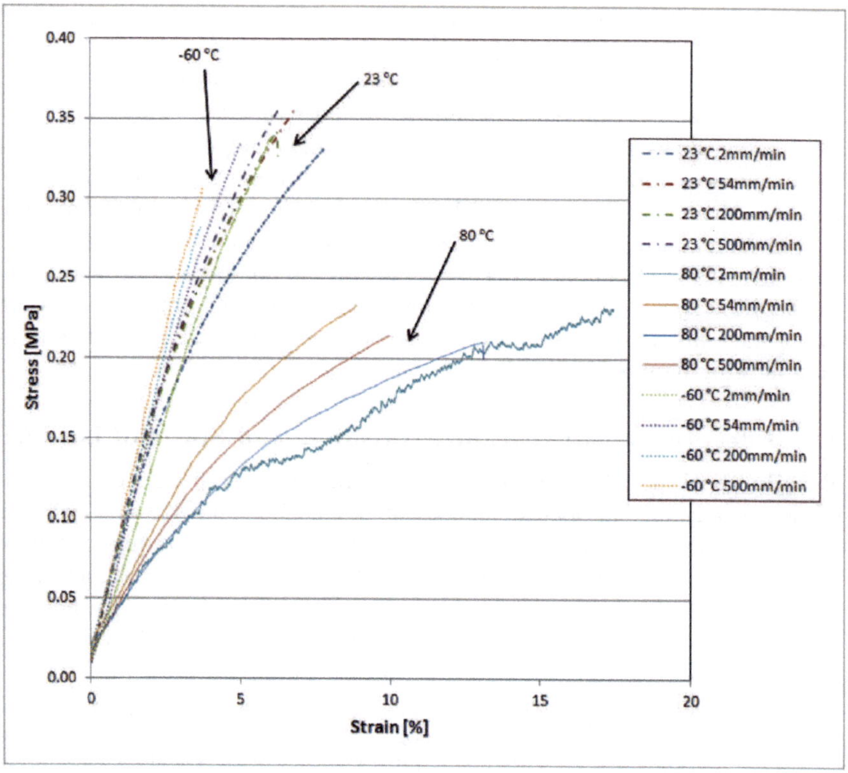

Figure 2. Tensile stress-strain characteristic curves for foam with density 35 kg/m^3

average - is reduced to half for the densities 93 kg/m^3 and 200 kg/m^3 when temperature is increased from -60 °C to 80 °C. Elongation at failure increases about 3 times for the density 35 kg/m^3, about 9 times (even more for 500 mm/min) for density 93 kg/m^3, and more than 10 times for the highest density of 200 kg/m^3 when temperature increases. At the temperature of 23 °C elongation at failure is about the same for 35 and 93 kg/m^3 densities, decreasing with the increase of testing speed; for 200 kg/m^3 elongation is greater. For -60 °C the decrease is 2-3 times as compared to 23 °C for the 93 and 200 kg/m^3 densities, and less than once for the 35 kg/m^3 density. Finally, for 80 °C compared to 23 °C, elongation at failure doubles for the density 35 kg/m^3 (Table 1), and is 4-5 times greater for the other

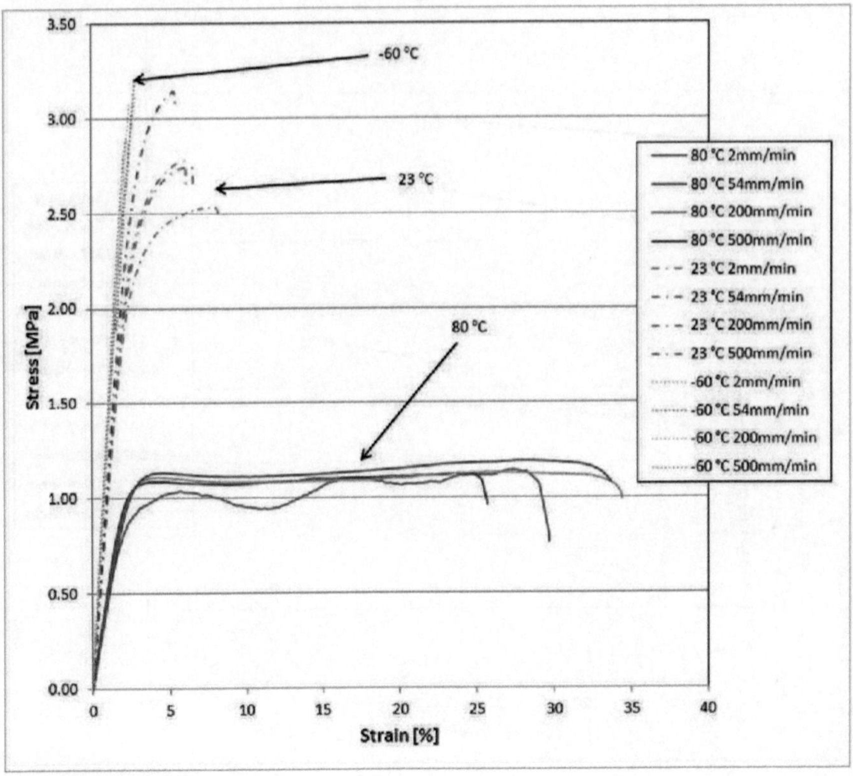

Figure 3. Tensile stress-strain characteristic curves for foam with density 93 kg/m^3

two densities of 93 and 200 kg/m^3 (Tables 2, 3).

Clearly, the temperature influence on the discussed mechanical properties is significant for each foam densities. For each temperature of testing and density of the foam the increase of the speed of loading in the mentioned interval (2 to 500 mm/min) reduces somehow the elongation at failure, but keeps about the same modulus of elasticity and maximum stress.

Compression Testing Tests in compression were done for the three foam densities and the three levels of temperature at the testing speeds: 2, 6, 18, 54, 125, 200, 350, 500, 1000, 2000, 3500, 6000, 10000, 20000, 30000, 40000 mm/min. Speeds of 1 m/s (60000 mm/min), 3 m/s (180000 mm/min), and 6 m/s (360000 mm/min), were also considered. For each testing case

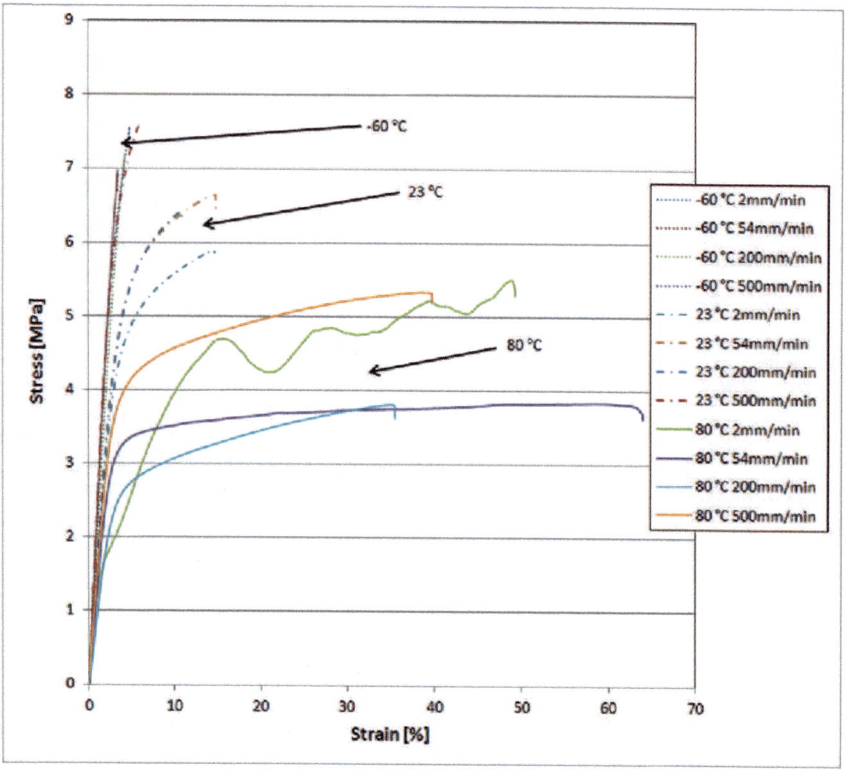

Figure 4. Tensile stress-strain characteristic curves for foam with density 200 kg/m^3

(density, temperature, speed) five specimens were tested and an average value will be presented in the following discussion; if a test gave suspicious results it was disregarded. The volume of obtained data is significant and only a part, considered as relevant, is presented hereby.

The direction of testing, notated as *direction 3* or *rise direction* and *direction 1* or *in-plane direction*, was considered as an additional parameter to influence the mechanical properties for each density and temperature.

Compression is produced till the specimen height becomes 1.5-2 mm (strain becoming about 90 %), followed by unloading and foam recovery controlled with 0.6 mm/min which was found to be sufficient for all loading speeds at the three temperatures of testing regardless the speed of testing, temperature and density. Foam recovery values are established having as a

Table 1. Traction mechanical properties for the foam of density 35 kg/m^3 at different temperatures

Speed of testing [mm/min]	Modulus of elasticity [MPa]			Maximum stress [MPa]			Elongation at failure [%]		
	-60 °C	23 °C	80 °C	-60 °C	23 °C	80 °C	-60 °C	23 °C	80 °C
2	7.09	8.67	3.94	0.30	0.33	0.21	5.52	7.37	12.52
54	7.98	8.24	3.95	0.29	0.36	0.22	4.48	7.10	12.18
200	7.66	8.55	3.65	0.26	0.34	0.13	3.85	5.75	13.00
500	8.38	8.13	3.54	0.27	0.32	0.23	3.58	5.20	11.00

Table 2. Traction mechanical properties for the foam of density 93 kg/m^3 at different temperatures

Speed of testing [mm/min]	Modulus of elasticity [MPa]			Maximum stress [MPa]			Elongation at failure [%]		
	-60 °C	23 °C	80 °C	-60 °C	23 °C	80 °C	-60 °C	23 °C	80 °C
2	169.01	105.84	59.58	2.83	2.48	1.16	2.88	8.07	29.00
54	137.61	117.46	67.74	2.72	2.82	1.19	2.63	6.50	26.67
200	149.65	107.61	73.84	2.66	2.86	1.40	2.41	6.03	23.67
500	143.22	112.73	53.78	2.49	3.08	1.17	2.23	5.03	29.00

Table 3. Traction mechanical properties for the foam of density 200 kg/m^3 at different temperatures

Speed of testing [mm/min]	Modulus of elasticity [MPa]			Maximum stress [MPa]			Elongation at failure [%]		
	-60 °C	23 °C	80 °C	-60 °C	23 °C	80 °C	-60 °C	23 °C	80 °C
2	279.49	185.42	121.20	7.22	6.06	4.82	3.70	13.00	48.00
54	259.94	190.41	139.66	7.85	6.68	4.25	3.93	12.33	57.00
200	258.22	176.08	96.70	6.66	6.21	3.66	3.80	9.95	41.33
500	259.51	237.43	168.97	7.17	7.49	5.17	3.40	6.90	36.67

reference the moment when unloading starts and in the following tables are notated as non-dimensional.

- Foam with density of 35 kg/m^3

  In Fig. 5 are presented the characteristic curves obtained at 23 °C on direction 1 for all speeds of testing up to 6 m/s. When yielding starts a hardening behavior is noticed in all curves. For speeds starting from 10000 mm/min, although stress-strain values are filtered, the curves show "peaks and valleys" as local instabilities are to be clearly seen up to strains of 40 %, while cells walls are damaged in an unstable manner. On direction 3 (Fig. 6) a plateau at yielding is obtained and local different types of damages which probably appear are to be

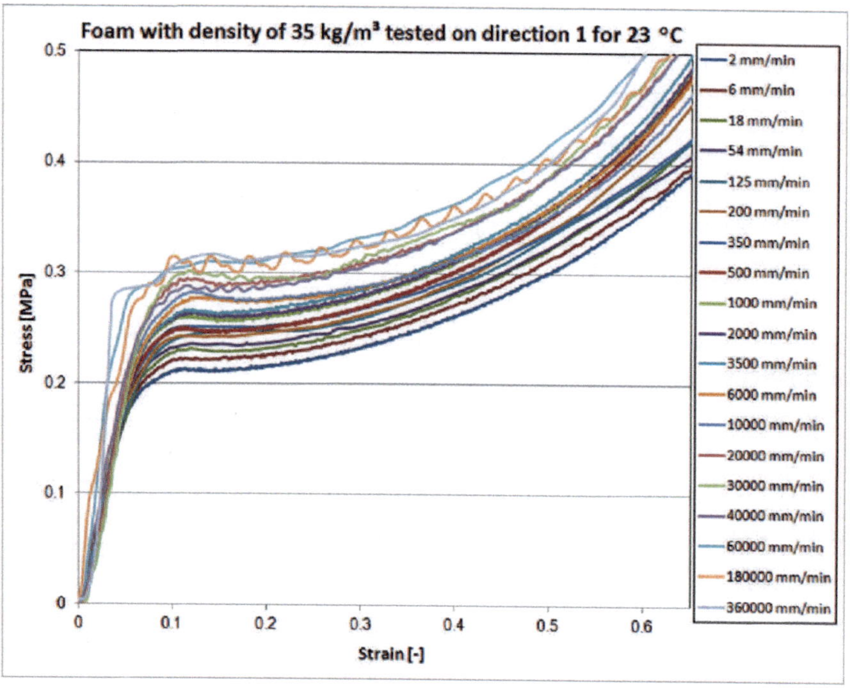

Figure 5. Stress-strain diagrams in compression on direction 1 at 23 °C (35 kg/m^3)

noticed. Clearly phenomena are difficult to be quantified as besides various local failure mechanisms one should also consider the rapid loading influence. At 6 m/s the obtained curve keeps the general trend but stress is first decreasing on the yielding plateau and then is increasing with the onset of densification.

When comparing the mechanical properties obtained in compression as modulus of elasticity, maximum stress, foam recovery for all testing speeds in between the two directions it is to be clearly seen in Table 4 that the direction of testing influences the results. The modulus of elasticity is greater on direction 3 than on direction 1, and from a speed of 500 mm/min it is two times greater; only at the higher speeds of 1 m/s, 3 m/s and 6 m/s it is less than twice bigger. Maximum stress (when yielding starts) is increasing slowly with the speed of loading being greater on direction 3 than 1. The

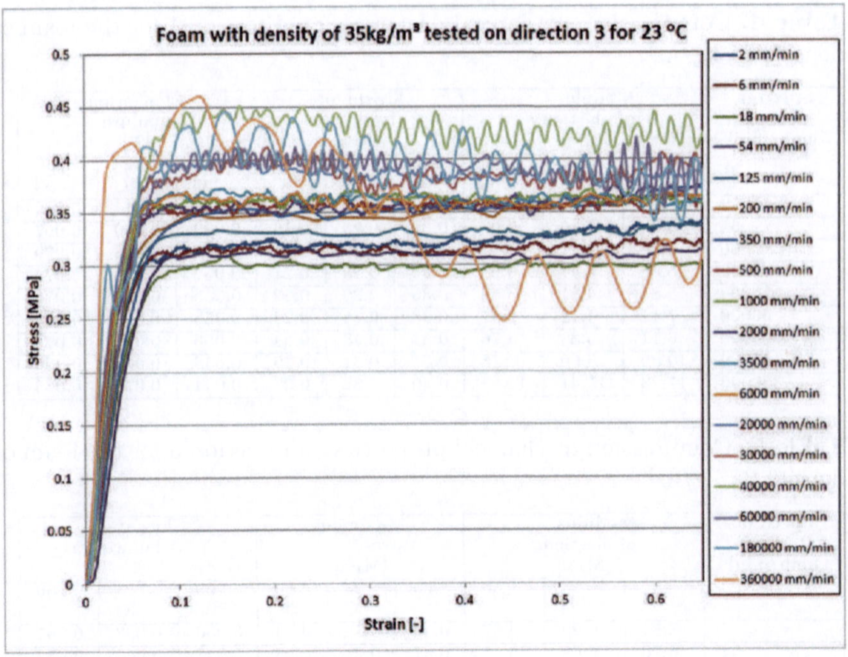

Figure 6. Stress-strain diagrams in compression on direction 3 at 23 °C (35 kg/m^3)

foam recovers better on the in plane direction (direction 1) than on rise direction (direction 3) and generally decreases with the increase of speed of loading, showing that higher speeds produce irreversible damage processes.

For selected speeds of loading the influence of the temperature on the properties for direction 1 is given in Table 5. The increase of temperature at each speed of testing reduces the modulus of elasticity and the maximum stress. For each temperature the modulus increases with the testing speed up to 1 m/s, followed by significant increases for 3 m/s and 6 m/s. Maximum stress decreases slightly with the increase of temperature for each testing speed, but increases at each temperature level with the increase of speed. Foam recovery varies in an opposite way: increases with the increase of temperature for each testing speed, but somehow decreases at each temperature level with the increase of speed. On direction 3 the trends of variation discussed

Table 4. Compression mechanical properties on direction 1 for the foam of density 35 kg/m^3

Speed of testing [mm/min]	Modulus of elasticity [MPa]			Maximum stress [MPa]			Elongation at failure [%]		
	-60 °C	23 °C	80 °C	-60 °C	23 °C	80 °C	-60 °C	23 °C	80 °C
2	4.21	4.07	2.75	0.29	0.21	0.16	0.140	0.084	0.210
54	5.04	4.30	3.08	0.32	0.25	0.19	0.092	0.080	0.097
500	5.26	4.21	3.02	0.31	0.26	0.21	0.065	0.073	0.079
6000	5.88	4.47	3.08	0.34	0.28	0.21	0.057	0.069	0.078
20000	5.87	4.81	3.50	0.36	0.29	0.22	0.055	0.082	0.076
40000	7.11	4.11	3.61	0.38	0.28	0.24	0.055	0.078	0.074
60000	6.11	5.38	3.76	0.33	0.28	0.24	0.056	0.078	0.074
180000	10.89	8.60	7.18	0.37	0.31	0.29	0.046	0.078	0.066
360000	17.23	15.41	12.97	0.40	0.32	0.37	0.044	0.072	0.063

Table 5. Compression mechanical properties on direction 3 for the foam of density 35 kg/m^3

Speed of testing [mm/min]	Modulus of elasticity [MPa]			Maximum stress [MPa]			Elongation at failure [%]		
	-60 °C	23 °C	80 °C	-60 °C	23 °C	80 °C	-60 °C	23 °C	80 °C
2	8.35	6.29	4.67	0.40	0.29	0.24	0.056	0.068	0.061
54	8.70	6.60	6.45	0.37	0.31	0.28	0.048	0.062	0.054
500	9.03	9.31	7.06	0.41	0.36	0.32	0.041	0.044	0.049
6000	10.47	9.25	8.17	0.44	0.37	0.34	0.043	0.049	0.049
20000	10.48	9.06	9.21	0.45	0.40	0.35	0.048	0.044	0.053
40000	10.92	10.92	8.56	0.47	0.44	0.37	0.051	0.050	0.052
60000	10.57	9.74	7.68	0.48	0.41	0.34	0.034	0.051	0.053
180000	17.32	14.60	14.90	0.50	0.43	0.41	0.031	0.046	0.045
360000	22.28	21.00	17.71	0.66	0.46	0.49	0.033	0.041	0.049

above are same as on direction 1 for all three properties (Table 5). The modulus of elasticity is significantly greater on direction 3 than on direction 1. Maximum stress is greater and foam recovery is smaller for all testing speeds and all test temperatures (values compared for the same parameters) on direction 3 than on direction 1.

It is also interesting to analyze the response of the foam with density of 35 kg/m^3 on direction 1 by observing the characteristic curves at -60 °C and 80 °C, as compared to the ones presented already in Fig. 7 for 23 °C. In Fig. 7 at -60 °C, for the selected speeds, yielding is produced on a plateau and the fragile wall damage is to be seen in the local ups and downs. For the lower speeds of testing at 2, 54, and 500 mm/min the characteristic curves move one by one to the right, as the onset of densification is produced later. For the higher selected speeds of 6000, 20000, and 40000 mm/min densification starts

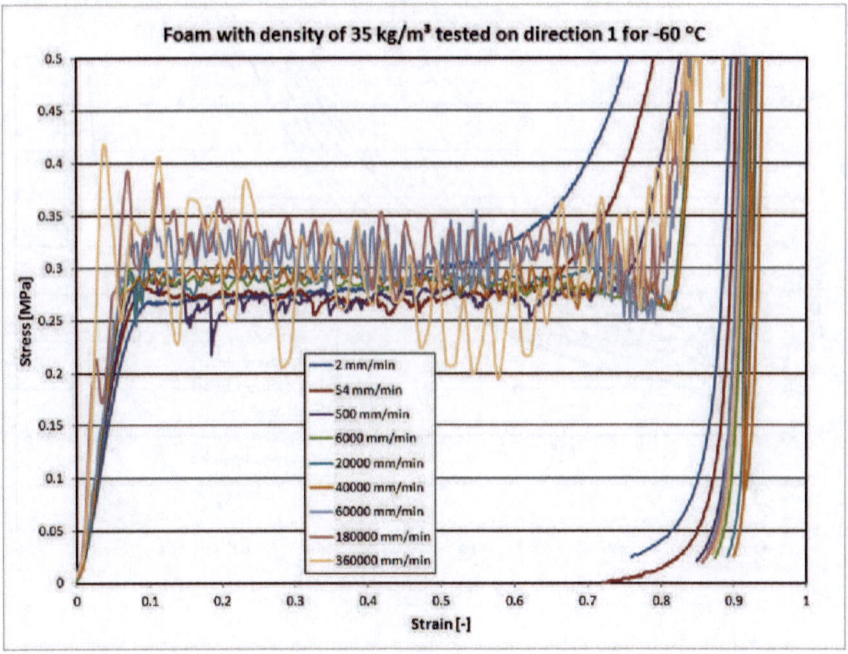

Figure 7. Selected stress-strain diagrams in compression on direction 1 at -60 °C (35 kg/m^3)

at about the same moment and the three curves superpose at a strain of approximately 80 %. At 1, 3 and 6 m/s densification starts also at about 80 % strain (Fig. 7). For the same foam and a temperature of testing of 80 °C, yielding is produced with hardening as the loading speed is increased (Fig. 8). Curves have a smooth variation; only from 40000 mm/min starts to appear an evident influence of the local instabilities, test being done with an initial strain rate of 56/s.

On direction 3 at -60 °C, on the plateau region, the cells of the foam are breaking in a fragile manner, but curves stay together up to 40000 mm/min (Fig. 9). From 60000 mm/min up to 360000 mm/min (1m/s to 6 m/s) the behavior of the 35 kg/m^3 density foam is quite difficult to be predicted for such compressive tests. Although the presented curves are averaged from the experimentally obtained ones, in Fig. 9 it is to be seen that for higher speeds the stress-strain diagrams go initially above and then (from 60 %) bellow the curves

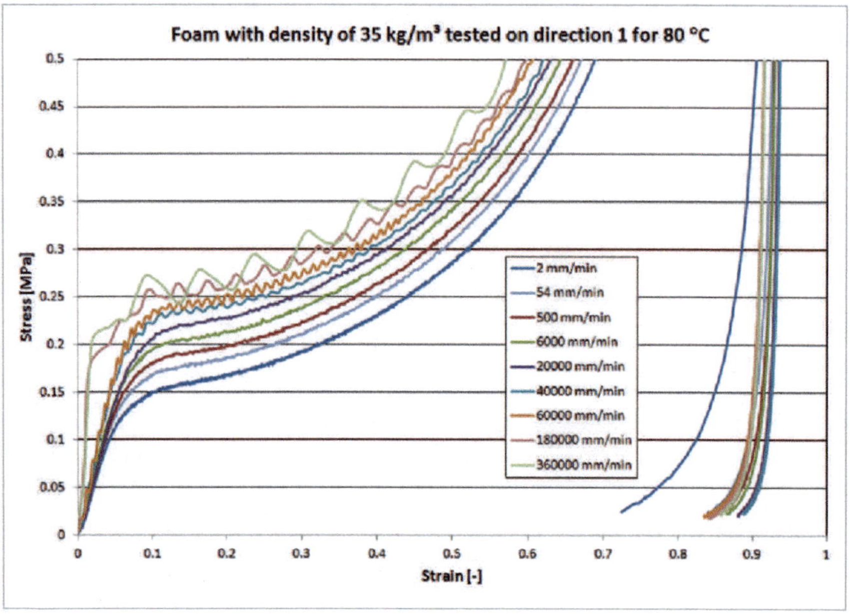

Figure 8. Selected stress-strain diagrams in compression on direction 1 at 80 °C (35 kg/m^3)

which result at lower speeds. For the temperature of 80 °C (Fig. 10) the yielding plateau shifts up in a more evident way as before with the increase of speed of testing (see also Table 5 for maximum stress values), and from 1 m/s testing speed the curves show a random variation, especially above a strain of 50 %. Again, it is difficult to quantify correctly tests done at speeds of m/s.

- Foam with density of 93 kg/m^3
  At 23 °C we have represented in Fig. 11 for direction 1 and in Fig. 12 for direction 3 all the stress-strain curves which we have obtained experimentally for 19 speeds of testing.

  Both figures are overcrowded with curves but show the general trends: hardening in the yielding region and more irregular variations when testing speed is towards the highest values. When yielding is produced the maximum stress drops more on direction 3 than on direction 1. It is interesting to notice that on direction 3 the 19 curves

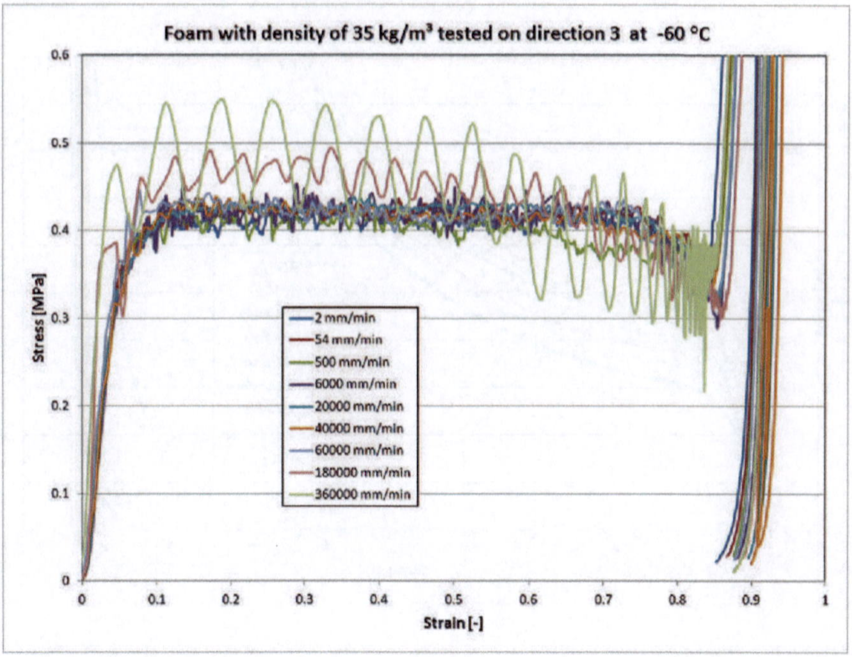

Figure 9. Selected stress-strain diagrams in compression on direction 3 at -60 °C (35 kg/m^3)

gather in three groups as first 6 - 2 to 200 mm/min, then 10 - 350 to 40000 mm/min, and finally 3 speeds - from 1 to 6 m/s.

Table 6 compares, for selected speeds of testing, the variation of the foam properties at the three temperatures on direction 1. Table 7 does the same thing, but on direction 3.

At all temperatures the moduli are generally greater on direction 1 than on direction 3, mainly when speed of testing increases; exception is the temperature of 80 °C for which at speeds starting from 1 m/s the values are almost the same on both directions. Maximum stress decreases on both directions with the increase of temperature, but increases with the speed of loading, as happens in fact for the foam of density 35 kg/m^3. As values, they are about 5-8 times greater than for the lower density foam. Foam recovery increases with the increase of temperature but is about the same on each direction at the temperatures of -60 °C and 23 °C for speed of 1 to 6 m/s; values

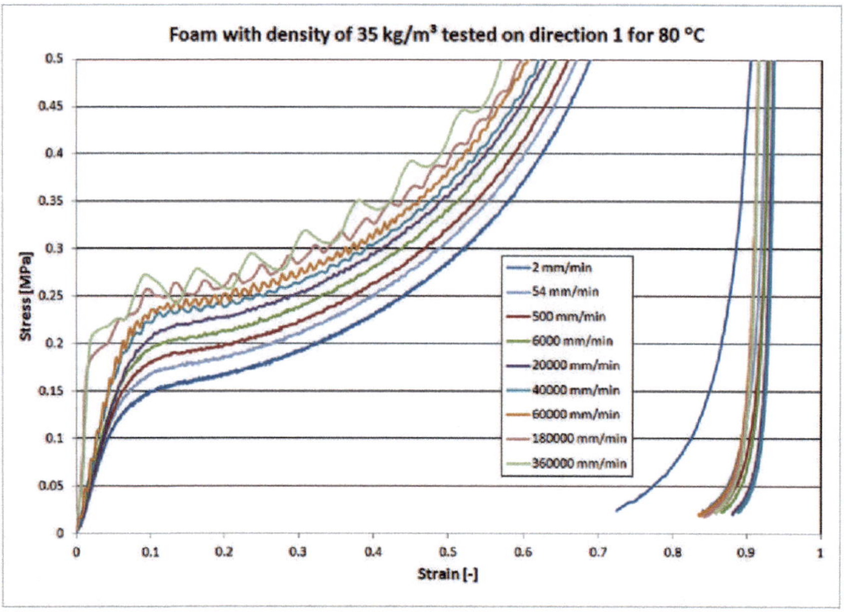

Figure 10. Selected stress-strain diagrams in compression on direction 3 at 80 °C (35 kg/m^3)

are 8-9 % at -60 °C and 25-30 % at 80 °C.

As seen, recovery is greater at 80 °C but, again, doesn't change with the increase of the last three speeds. For all temperatures recovery is in fact greater on the rise direction (direction 3) than on the transverse one (direction 1).

Figures 13 and 14 present selected stress-strain curves on direction 1 for -60 °C and 80 °C, respectively. Only parts of the curves containing the yielding region for strains up to 60 % are shown. The issues of onset of densification and densification region are not discussed here. On direction 1 yielding is in between 2-3 MPa at -60 °C (Fig. 13) and between 1-2 MPa for 80 °C (Fig. 14) for all speeds of testing. On direction 3 (Figs. 15 and 16) yielding is mostly produced around a value of 2.5 MPa for -60 °C, and again between 1-2 MPa for 80 °C. This confirms that for this foam in-plane and rise directions produce a not too much different response with the variation of speed of loading and temperature.

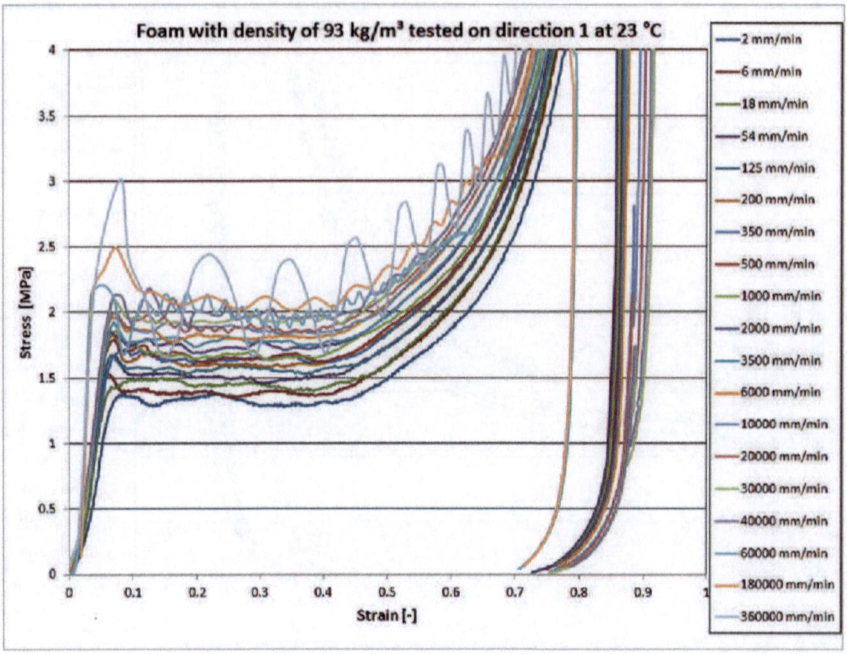

Figure 11. Stress-strain diagrams in compression on direction 1 at 23 °C (93 kg/m^3)

- Foam with density of 200 kg/m^3
  For the highest density of the tested foams when tests are done at 23 °C on direction 1 we show in Fig. 17 all 19 obtained curves by changing the testing speed; they go up as speed is increased with a clear difference when speed becomes of the order of m/s. Yielding is produced with hardening. For the same types of curves obtained after testing on direction 3 (Fig. 18) yielding stress is increasing and the "trembling" of the curves indicate different mechanisms of failure on the rise direction. It is interesting to notice that the difference between the maximum and minimum yielding stress is greater on direction 3 than on direction 1. In Table 10 is registered only the maximum stress at yielding which is greater on direction 3 than on direction 1.

  The modulus of elasticity is also greater on direction 3 than 1. It is almost constant on direction 1 for speeds between 125 mm/min and 20000 mm/min and then increases with almost 50 % when speed

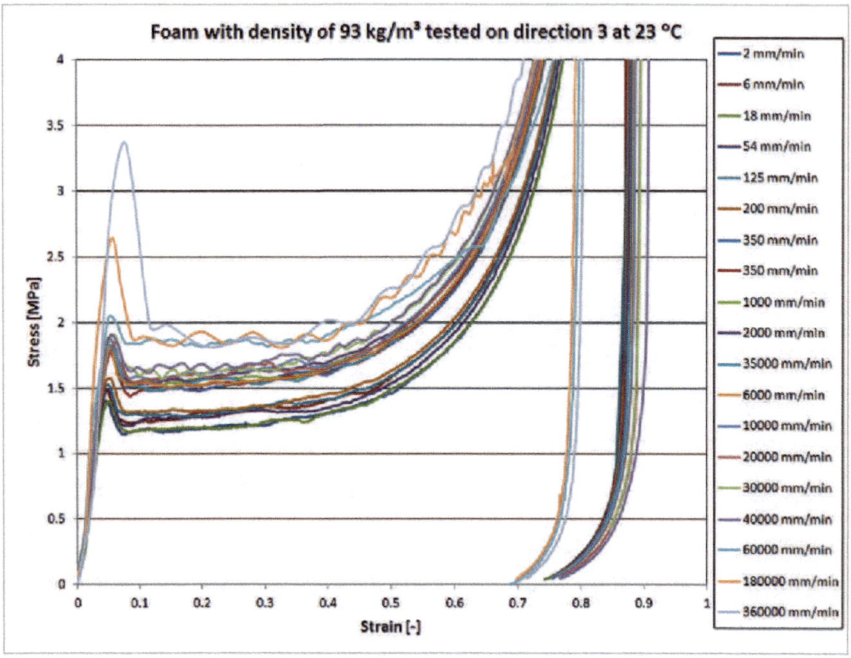

Figure 12. Stress-strain diagrams in compression on direction 3 at 23 °C (93 kg/m^3)

reaches 6 m/s. On direction 3 the modulus increases constantly from 125 mm/min and doubles in value when speed of testing becomes 6 m/s.

Foam recovery decreases slightly with the increase of speed of loading on both directions, being a little bit greater on direction 1 as 13-11 %, compared to 11-10 % on direction 3.

For the three temperatures of testing the mechanical properties obtained on directions 1 and 3 are recorded in Table 8, respectively Table 9. On both directions the modulus of elasticity decreases with the increase on temperature at each speed of testing, but increases with the increase of the speed at each temperature. Let's say at 6 m/s it is on direction 3 compared with direction 1 greater with: 37.4 % at -60 °C, 60.7 % at 23 °C, and with 7.1 % at 89 °C. Generally, the smaller differences for the moduli are obtained at 80 °C. In fact at 23 °C and 80 °C the corresponding values at each speed of testing are

Table 6. Compression mechanical properties on direction 1 for the foam of density 93 kg/m^3

Speed of testing [mm/min]	Modulus of elasticity [MPa]			Maximum stress [MPa]			Elongation at failure [%]		
	-60 °C	23 °C	80 °C	-60 °C	23 °C	80 °C	-60 °C	23 °C	80 °C
2	58.45	30.95	48.93	2.38	1.42	1.19	0.075	0.110	0.181
54	57.48	38.81	53.90	2.48	1.64	1.41	0.076	0.105	0.181
500	60.35	40.57	54.50	2.60	1.80	1.55	0.083	0.107	0.177
6000	63.52	43.56	62.60	2.70	2.00	1.77	0.079	0.120	0.192
20000	70.13	45.93	63.32	2.80	2.06	1.86	0.072	0.139	0.211
40000	89.34	45.48	60.79	2.82	2.12	1.93	0.064	0.142	0.220
60000	70.90	60.42	39.64	3.13	2.32	1.52	0.086	0.079	0.250
180000	93.67	79.23	55.26	3.83	2.51	2.56	0.082	0.075	0.256
360000	107.67	90.14	61.30	4.63	3.02	2.81	0.079	0.072	0.249

Table 7. Compression mechanical properties on direction 3 for the foam of density 93 kg/m^3

Speed of testing [mm/min]	Modulus of elasticity [MPa]			Maximum stress [MPa]			Elongation at failure [%]		
	-60 °C	23 °C	80 °C	-60 °C	23 °C	80 °C	-60 °C	23 °C	80 °C
2	41.72	41.07	32.09	2.11	1.39	1.00	0.116	0.131	0.206
54	47.39	41.75	36.12	2.31	1.50	1.20	0.108	0.126	0.210
500	50.52	45.95	38.28	2.41	1.64	1.33	0.105	0.125	0.207
6000	52.77	45.03	41.14	2.48	1.80	1.51	0.105	0.135	0.218
20000	55.68	46.62	42.93	2.51	1.82	1.63	0.110	0.141	0.245
40000	57.39	48.95	44.37	2.48	1.95	1.63	0.111	0.149	0.245
60000	66.35	53.54	38.97	3.01	2.04	1.45	0.093	0.087	0.295
180000	84.02	69.43	50.55	3.46	2.64	1.93	0.090	0.088	0.287
360000	92.98	83.92	68.04	4.15	3.36	2.91	0.081	0.086	0.266

not so much different on directions 1 and 3, especially on direction 1.

As mentioned (Figs. 17 and 18) maximum stress is greater on direction 3 than on direction 1, and decreases with the increase of temperature at each speed (Tables 8 and 9).

Foam recovery is increasing with the increase of temperature at each speed and is greater on direction 1 than on direction 3 (Tables 8 and 9). At temperatures of -60 °C and 23 °C the recovery is about the same on both directions, but is greater at 80 °C.

As seen in Figs. 19 and 20 (-60 °C and 80 °C) for tests done on direction 1 for some selected speeds, yielding shows some hardening for both temperatures and is produced in between 7-9 MPa at -60 °C, and 2-4 MPa at 80 °C. For the same temperatures on direction 3 (Figs. 21 and 22) yielding is in between 9-12 MPa at -60 °C, and 3-5 MPa at 80 °C. So, on one hand the increase of temperature reduces

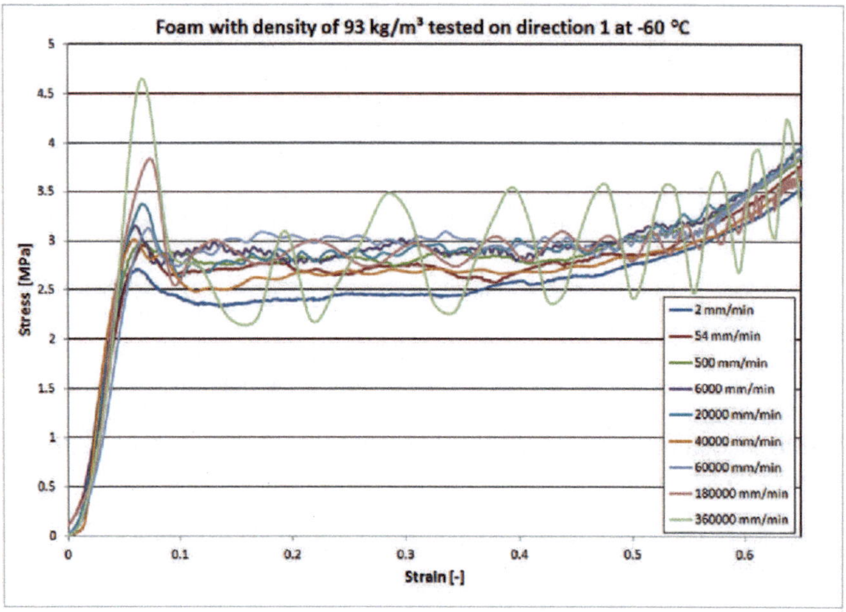

Figure 13. Selected stress-strain diagrams in compression on direction 1 at -60 °C (93 kg/m^3)

the yielding plateau average value, and on the other hand on direction 3 these values are greater than on direction 1. On direction 3 at -60 °C the foam behaves in a "fragile" manner as the walls of the cells break suddenly, especially when speed of testing is increased - the curves show, as seen, many fluctuations.

3 Fracture Toughness of Cellular Materials Under Static and Dynamic Loading

Rigid foams show a linear elastic behavior in tension, and a brittle fracture, so the Linear Elastic Fracture Mechanics (LEFM) applies for these materials. The design based on LEFM needs to determine the fracture parameters: Stress Intensity Factors (SIFs) K_i [MPa m$^{0.5}$] or Energy Release Rate (ERR) G_i [N/m], with $i =$ I, II, III the mode of fracture, which depend

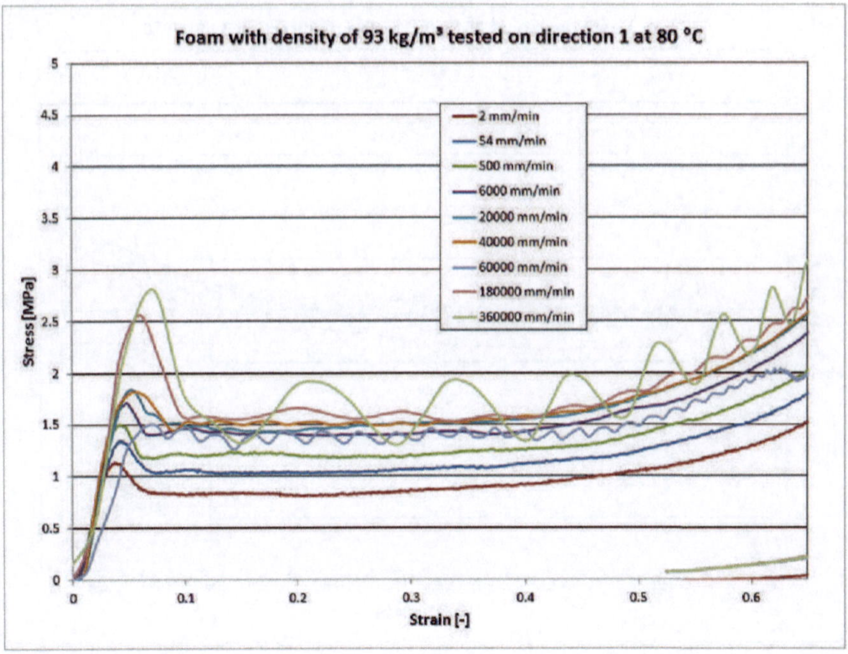

Figure 14. Selected stress-strain diagrams in compression on direction 1 at 80 °C (93 kg/m^3)

on load, crack dimensions, geometry of cracked body and Young's modulus for the ERR, and to compare it with the fracture toughness of the materials expressed by K_{Ic} or G_{Ic}, which represent an important material property. A crack in a linear elastic material loaded in mode I will propagate unstable when:

$$K_{\mathrm{I}} = K_{\mathrm{Ic}} \quad \text{or} \quad G_{\mathrm{I}} = G_{\mathrm{Ic}} \tag{1}$$

The determination of the fracture toughness for cellular materials becomes an important task for many researches. There are two approaches: experimental investigations and micromechanical modeling.

3.1 Experimental Determination of Fracture Toughness

Testing methods of polymeric foams are reviewed by Landrock (1995); Brown (1999, 2002) and Ward and Sweeney (2004). However, there are no standard prescriptions for fracture toughness determination of cellular mate-

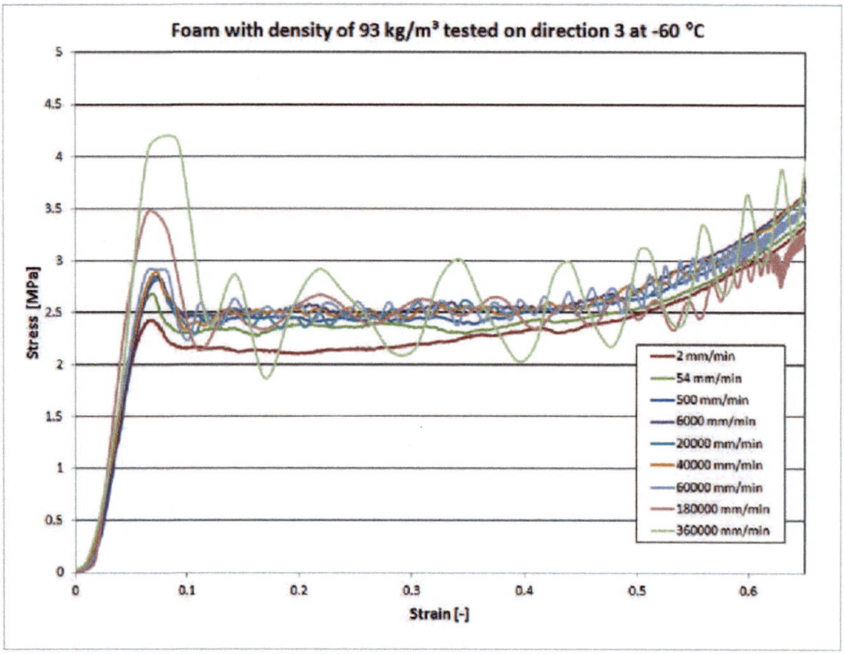

Figure 15. Selected stress-strain diagrams in compression on direction 3 at -60 °C (93 kg/m^3)

rials, and the plane strain fracture toughness of plastic materials methodology ASTM D5045-99[2] is often used. Single Edge Notched Bending (SENB) specimens (Fig. 23) and Compact Tension (CT) specimens (Fig. 24) are recommended because they exhibit a predominantly bending stress state, which allows smaller specimen sizes to achieve plane strain conditions. If the material is supplied in the form of a sheet, the specimen thickness, B, should be identical with the sheet thickness. The plain strain condition could be achieved only if specimen thickness B is big enough, and also the ligament in the crack area $(W-a)$ is sufficient to avoid excessive plasticity. The introduction of a crack in the specimen is possible by machining a sharp notch. Subsequently, one can initiate a natural crack by inserting a fresh razor blade and tapping. If a natural crack cannot be successfully ini-

[2]ASTM D5045-99 Standard Test Methods for Plane-Strain Fracture Toughness and Strain Energy Release Rate of Plastic Materials

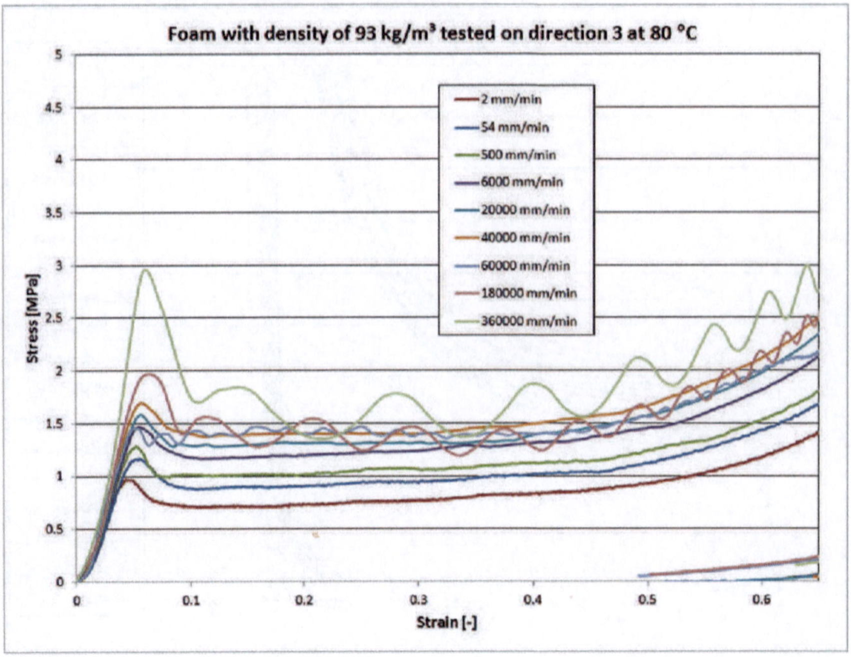

Figure 16. Selected stress-strain diagrams in compression on direction 3 at 80 °C (93 kg/m^3)

tiated by tapping, a sufficiently sharp crack can alternatively be generated by sliding or sawing a new razor blade across the notch root.

Most of the tests for determining the mode I fracture toughness of cellular materials were carried out using SENB specimens. SENB specimens were adopted also for mixed mode, but under a Four Point Bending load, as done by Hallström and Grenestedt (1997). However, other types of specimens were used to investigate the mixed mode fracture toughness, like: Double Cantilever Beam (DCB), Fig. 25 (Siriruk et al., 2011), Asymmetric Semi-Circular Bend (ASCB), Fig. 26 (Marsavina et al., 2014), Compact Tension Shear (CTS), Fig. 27 (Noury et al., 1998), End Notch Flexure (ENF) specimen, Fig. 28 (Burman, 1998).

The determination of fracture toughness requires the recording of load versus crack opening displacement curve P-u during testing. The displacement measurement can be performed using the machine's position trans-

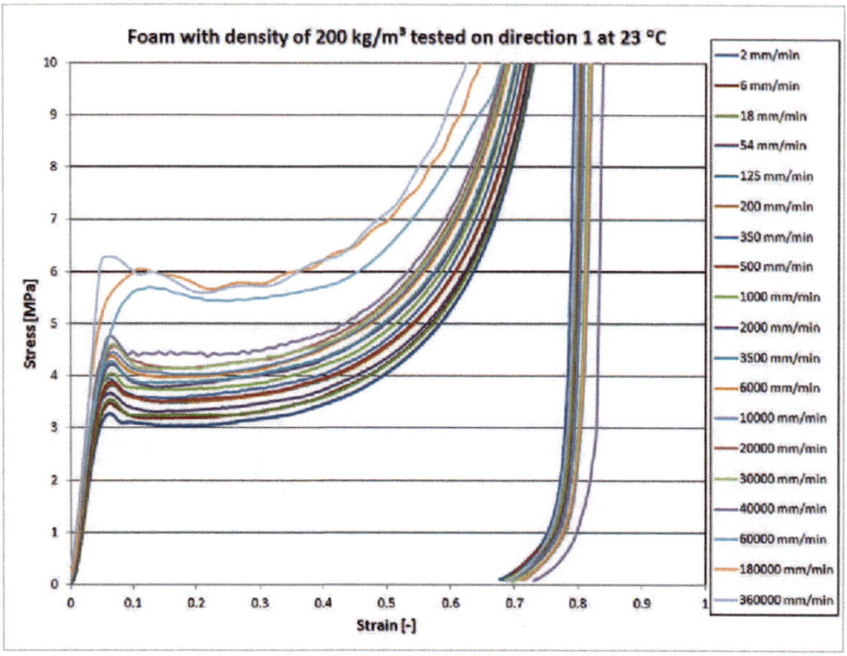

Figure 17. Stress-strain diagrams in compression on direction 1 at 23 °C (200 kg/m^3)

ducer. The load-displacement data must be corrected for system compliance, loading-pin penetration and sample compression by performing a calibration of the testing system. If an internal displacement transducer is not available, or has insufficient precision, then an externally applied displacement-measuring device may be used as illustrated in Fig. 29 for the SENB configuration. For CT specimens, a clip gage can be mounted across the loading pins, Fig. 30. For both the SENB and CT specimens, the displacement should be taken at the loading point.

In the ideal case this is a linear diagram with an abrupt drop of load at the initiation of crack growth. In this case fracture toughness can be calculated from the maximum load. For the case when the P-u curve does not exhibit a load drop the value of stress intensity factors K_Q is obtained following the ASTM D5045-99 procedure, Fig. 31:

- drawing a straight line (AB) to determine the initial compliance $C = \tan\theta$, which represents the slope of line (AB);

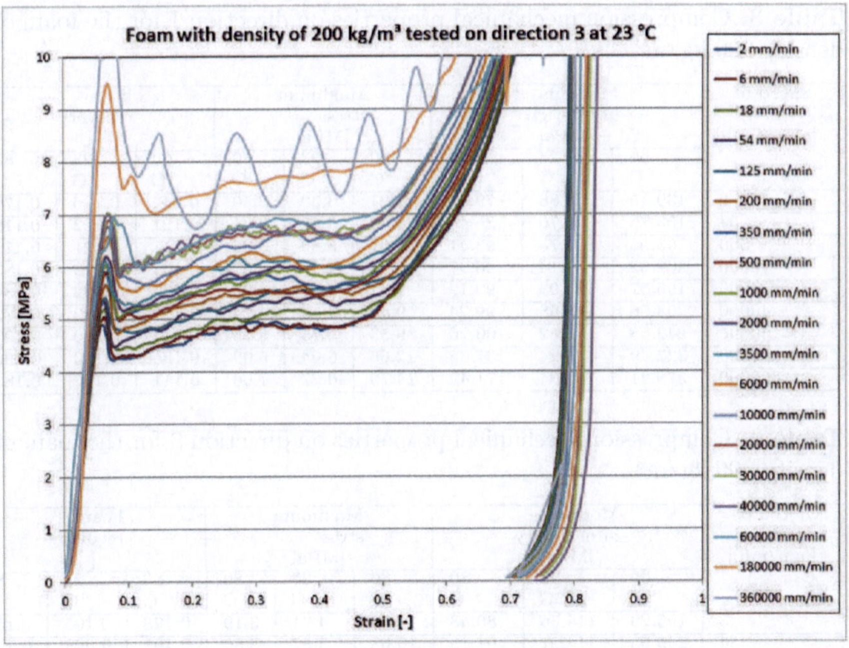

Figure 18. Stress-strain diagrams in compression on direction 3 at 23 °C (200 kg/m^3)

- draw a second line (AB') with the compliance 5% greater than that of line (AB). If the maximum force that the specimen was able to sustain, $P_{\max}$, falls within lines (AB) and (AB'), use $P_{\max}$ to calculate K_Q. If $P_{\max}$ falls outside line (AB) and line (AB'), then use the intersection of line (AB') and the load curve as P_Q;

- if $P_{\max}/P_Q < 1.1$ use P_Q in the calculation of K_Q. However, if $P_{\max}/P_Q > 1.1$, the test is invalid.

The calculation value of the stress intensity factor could be obtained:

- for CT specimen

$$K_Q = P_Q \frac{f\left(\frac{a}{W}\right)}{B\, W^{\frac{1}{2}}} \tag{2}$$

Table 8. Compression mechanical properties on direction 1 for the foam of density 200 kg/m^3

Speed of testing [mm/min]	Modulus of elasticity [MPa]			Maximum stress [MPa]			Foam recovery [-]		
	-60 °C	23 °C	80 °C	-60 °C	23 °C	80 °C	-60 °C	23 °C	80 °C
2	139.15	86.88	79.87	7.21	3.30	2.50	0.144	0.126	0.179
54	153.76	94.76	85.46	8.19	3.67	2.91	0.121	0.122	0.176
500	163.95	105.75	93.41	8.67	3.89	3.26	0.111	0.121	0.179
6000	208.32	109.14	96.07	9.05	4.35	3.69	0.104	0.119	0.183
20000	196.57	105.93	99.12	9.41	4.57	3.99	0.099	0.122	0.184
40000	208.58	115.58	99.24	9.52	4.69	4.24	0.101	0.122	0.182
60000	242.58	116.52	100.26	9.53	5.65	3.75	0.112	0.110	0.193
180000	322.49	133.24	137.35	12.66	6.04	5.09	0.105	0.110	0.190
360000	379.21	154.34	171.49	14.79	6.27	7.00	0.111	0.110	0.183

Table 9. Compression mechanical properties on direction 3 for the foam of density 200 kg/m^3

Speed of testing [mm/min]	Modulus of elasticity [MPa]			Maximum stress [MPa]			Foam recovery [-]		
	-60 °C	23 °C	80 °C	-60 °C	23 °C	80 °C	-60 °C	23 °C	80 °C
2	196.29	114.56	80.43	9.58	4.91	3.19	0.123	0.105	0.154
54	209.53	114.27	91.19	10.68	5.37	3.80	0.101	0.098	0.156
500	219.74	126.78	91.54	11.17	5.92	4.26	0.086	0.091	0.157
6000	224.95	134.74	99.93	11.80	6.57	4.86	0.075	0.089	0.159
20000	226.03	134.07	109.67	12.32	6.83	5.07	0.077	0.091	0.157
40000	225.81	149.94	119.80	13.49	7.00	5.28	0.081	0.095	0.160
60000	265.79	161.63	109.11	12.83	7.27	4.50	0.089	0.100	0.161
180000	414.33	229.47	153.45	15.94	9.49	6.75	0.084	0.102	0.160
360000	520.91	248.01	183.68	18.10	11.28	8.87	0.082	0.099	0.169

where for 0.2 < a/W<0.8:

$$f\left(\frac{a}{W}\right) = \left(2+\frac{a}{W}\right)\times \times \frac{\left[0.886+4.64\frac{a}{W}-13.32\left(\frac{a}{W}\right)^2+14.72\left(\frac{a}{W}\right)^3-5.6\left(\frac{a}{W}\right)^4\right]}{\left[1-\left(\frac{a}{W}\right)\right]^{\frac{3}{2}}} \quad (3)$$

where P_{Q} - load [N], B - specimen thickness [mm], W - specimen width [mm], a - crack length [mm].

- for SENB specimen with $L/W = 4$

$$K_{\mathrm{Q}} = P_{\mathrm{Q}} \frac{f\left(\frac{a}{W}\right)}{B\, W^{\frac{1}{2}}} \quad (4)$$

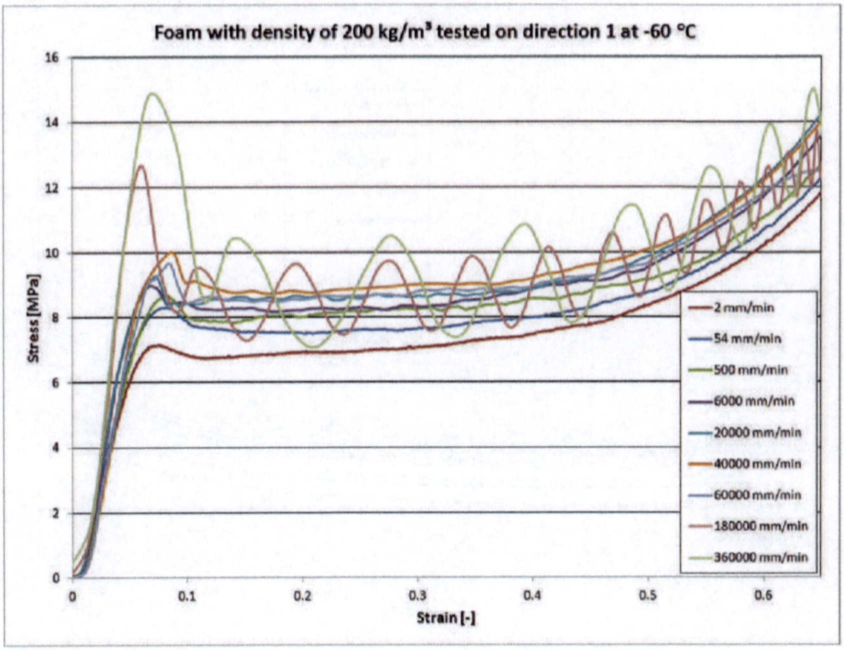

Figure 19. Selected stress-strain diagrams in compression on direction 1 at -60 °C (200 kg/m^3)

where $(0 < a/W < 1)$

$$f\left(\frac{a}{W}\right) = 6\left(\frac{a}{W}\right)^{\frac{1}{2}} \times$$
$$\times \frac{\left[1.99 - \left(\frac{a}{W}\right)\left(1 - \frac{a}{W}\right)\left(2.15 - 3.93\left(\frac{a}{W}\right) + 2.7\left(\frac{a}{W}\right)^2\right)\right]}{\left(1 + 2\frac{a}{W}\right)\left(1 - \frac{a}{W}\right)^{\frac{3}{2}}} \quad (5)$$

and P_{Q} - load [N], B - specimen thickness [mm], W - specimen width [mm], a - crack length [mm].

After K_{Q} calculation the validity is checked as:

$$a, B, (W - a) \geq 2.5\left(\frac{K_{\mathrm{Q}}}{\sigma_{\mathrm{y}}}\right)^2 \quad (6)$$

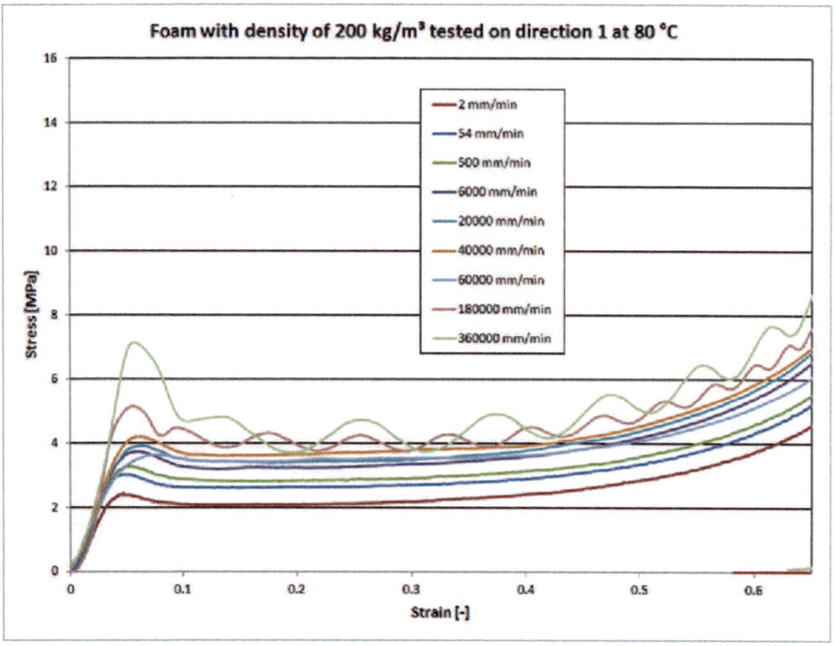

Figure 20. Selected stress-strain diagrams in compression on direction 1 at 80 °C (200 kg/m^3)

If condition (6) is fulfilled then the fracture toughness is equal with K_{Q}:

$$K_{\mathrm{Ic}} = K_{\mathrm{Q}} \tag{7}$$

Otherwise the test is not a valid, and should be repeated on larger specimens (increasing the dimensions in increments of 1.5 times).

The determination of G_{Ic} requires an accurate integration of the load versus loading point displacement curve, which requires an accurate displacement determination using a displacement transducer. In principle, G_{Ic} can be obtained using

$$G_{\mathrm{Ic}} = p\frac{K_{\mathrm{Ic}}^2}{E} \tag{8}$$

with $p = 1$ for plane stress and $p = 1 - \nu^2$ for plane strain, E Young's modulus and ν Poisson's ratio. It should be mentioned that for plastic foams E must be obtained at the same time and at same temperature conditions

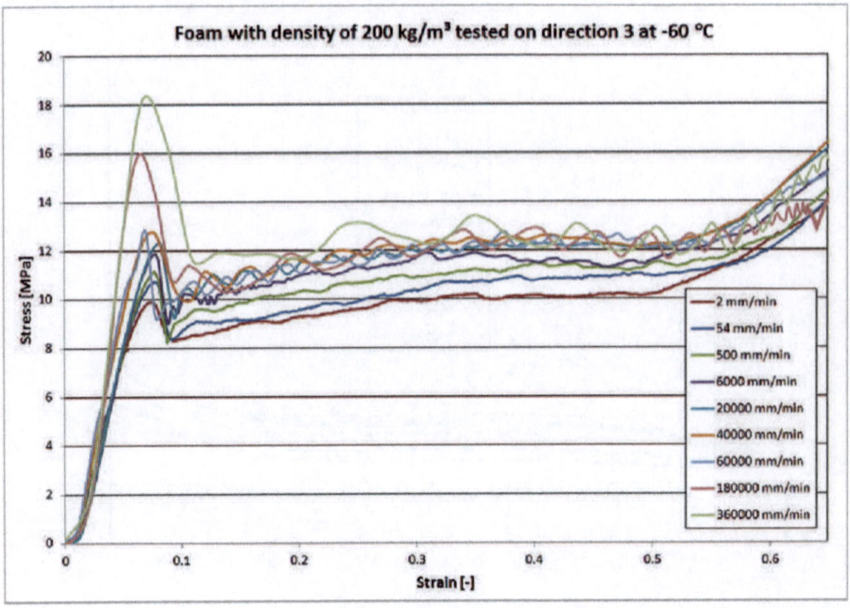

Figure 21. Selected stress-strain diagrams in compression on direction 3 at -60 °C (200 kg/m^3)

for the fracture test because of viscoelastic effects. Many uncertainties are introduced by this procedure and it is considered preferable to determine G_{Ic} directly from the energy derived from integration of the load versus displacement curve up to the same load point as used for K_{Ic} according with ASTM 5045-99.

The compliance method is also used for determination of G_{Ic} (Fowlkes, 1974)

$$G_{\mathrm{Ic}} = \left[\frac{P_{\mathrm{c}}^2}{2B} \frac{\mathrm{d}C}{\mathrm{d}a} \right]_{a=a_{\mathrm{c}}} \tag{9}$$

with P_{c} - load at crack initiation, a_{c} - crack length at crack initiation, $C = (u/P)$ compliance of the specimen.

To employ this method it is necessary to:

- determine the compliance curve versus crack length $C = f(a)$, Fig. 32;
- measure the load at crack initiation P_{c} and the corresponding crack length a_{c} from load - displacement curve recorded during fracture test;

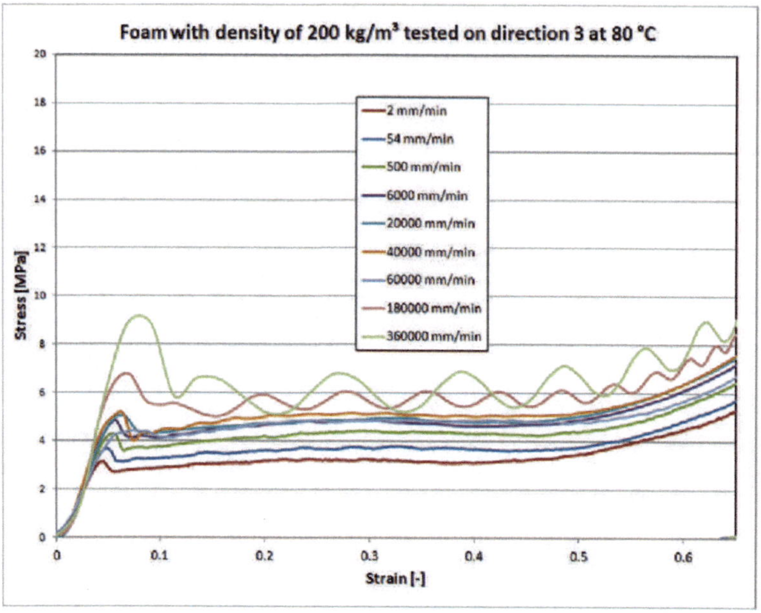

Figure 22. Selected stress-strain diagrams in compression on direction 3 at 80 °C (200 kg/m^3)

- calculate the G_{Ic} by applying relation (9).

3.2 Effect of Density, Forming Direction, Loading Speed

McIntyre and Anderson (1979), investigating the fracture toughness of polyurethane foams of different densities using SENB specimens, found that the K_{Ic} is independent of crack length and a linear correlation with density exists up to 200 kg/m^3 density. Above this value the correlation became non-linear. Also, a linear relationship between K_{Ic} and density (in the range 90 – 235 kg/m^3) was obtained by Danielsson (1996) for PVC foams using SENB specimens with two different span-to-depth ratios and centre cracked sheet specimens.

Figure 33 presents the effect of relative density on mode I fracture toughness results available in literature for different types of plastic foams using SENB specimens.

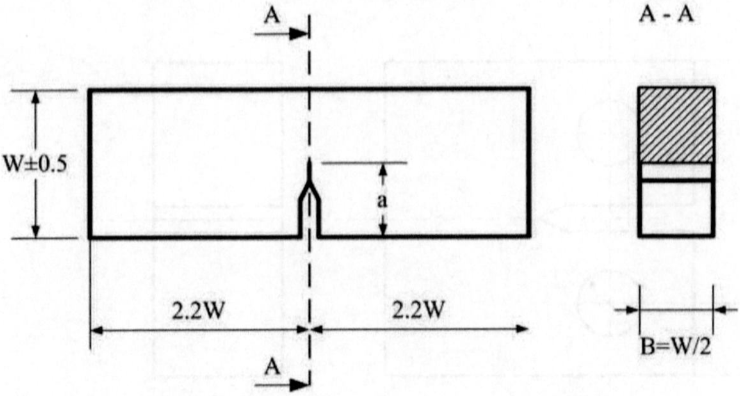

Figure 23. Single Edge Notched Bending specimen

Viana and Carlsson (2002) presented results for mode I fracture toughness of different densities ($\rho = 36, 80, 100, 200$ and 400 kg/m^3) PVC foams. SENB specimens and the standard procedure described by ASTM D5045 were used. Also, Kabir et al. (2006) employed the same procedure for determining the fracture toughness of PVC and PUR foams. Specimens with dimensions width $B = 8.5 - 25.4$ mm, height $W = 2B$, span $L = 4W$ and crack length $a = 0.5W$ were considered. The influence of density, specimen size, loading rate and cell orientation was investigated. Density has a significant effect on fracture toughness, which increases more than 7 times when the foam density increases 3.5 times.

Marsavina et al. (2012) presented the influence of density (from 40 to 300 kg/m^3) on fracture toughness for PUR foams and showed that the fracture toughness increases 22 times when the density increases 7.5 times.

Poapongsakorn and Carlsson (2013) determined the fracture toughness of closed cell PVC foams of densities from 45 to 200 kg/m^3 using SENB specimens loaded in Three Point Bending (TPB), respectively in Four Point Bending (FPB). The results show that the values of K_{Ic} obtained on FPB loading are 1.8-1.93 times higher than those obtained for TPB loading at the same density.

Comparing two different closed cell foams PVC foam R260 and PUR240 foam with almost the same density, showed a considerable difference in fracture toughness, as mentioned by Kabir et al. (2006)

$$K_{\mathrm{Ic}}^{\mathrm{PUR240}} = 0.30\ \mathrm{MPa}\sqrt{\mathrm{m}} \quad < K_{\mathrm{Ic}}^{\mathrm{R260}} = 0.72\ \mathrm{MPa}\sqrt{\mathrm{m}}$$

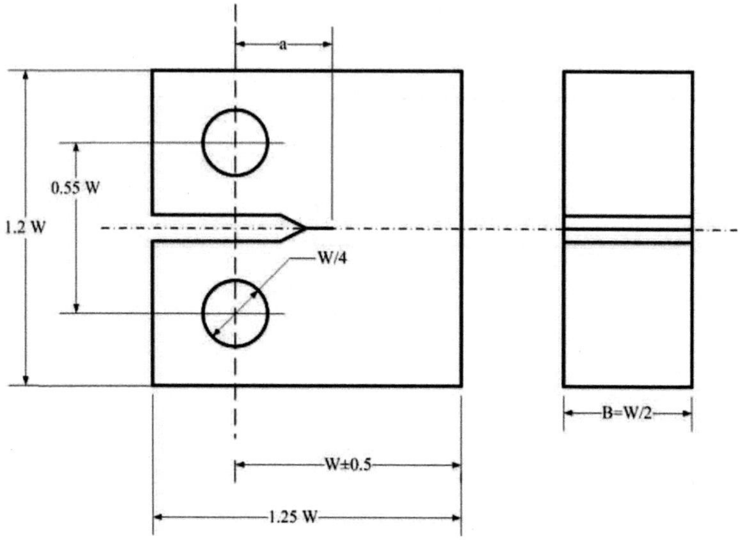

Figure 24. Compact Tension specimen

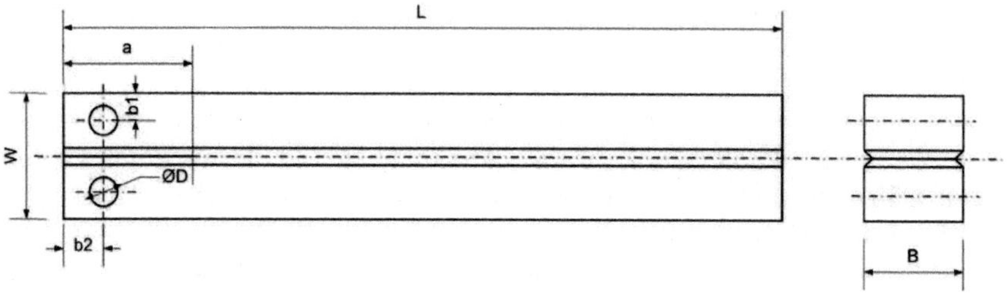

Figure 25. Double Cantilever Beam specimen

which could be explained by higher fracture toughness of solid PVC polymer (2.45 MPa $m^{0.5}$) comparing with fracture toughness of PUR solid (0.3-0.4 MPa $m^{0.5}$).

The influences of density and loading speed for PVC foams were provided by Kabir et al. (2006) and for PUR foam by Marsavina and Linul (2010), Fig. 34, where in parenthesis is indicated the density in kg/m^3. For the PVC foam the fracture toughness increases with loading speed and this increase is lower for the high density foam: from 22% for 75 kg/m^3 density

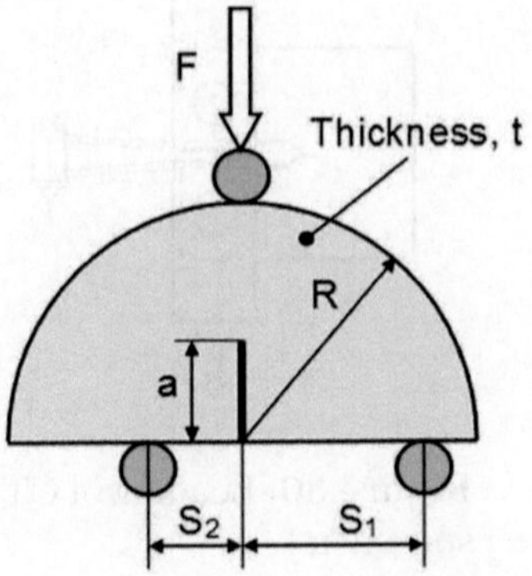

Figure 26. Asymmetric Semi-Circular Bend specimen

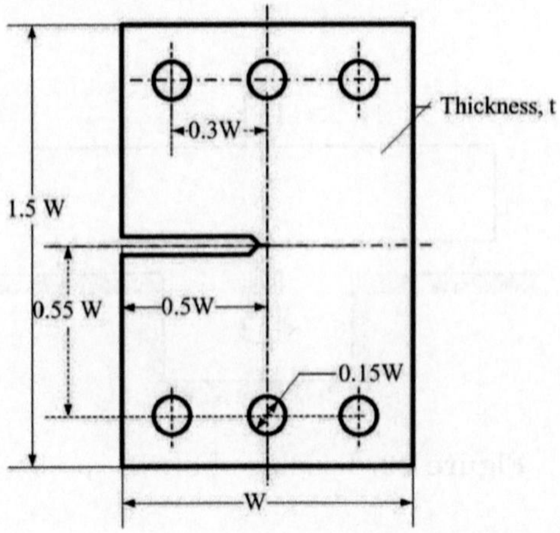

Figure 27. Compact Tension Shear specimen

to 18% for 260 kg/m^3 density for 100 times increase of loading speed. In contrary for PUR foam the fracture toughness decreases with 11% when loading speed increases 100 times.

Experimental results of Kabir et al. (2006) show that the fracture toughness increases with 18% when the loading rate is increased with two orders

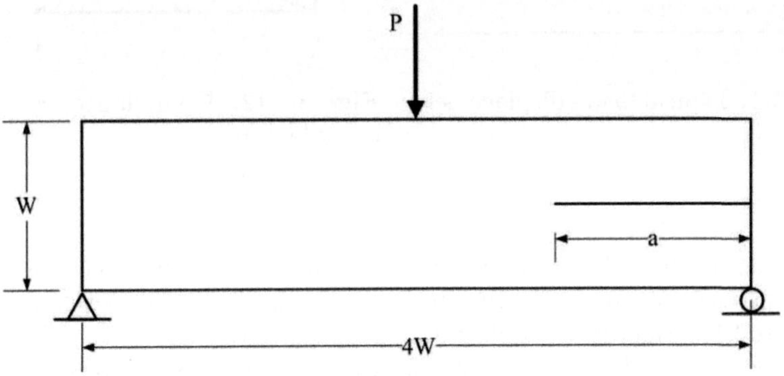

Figure 28. End Notch Flexure specimen

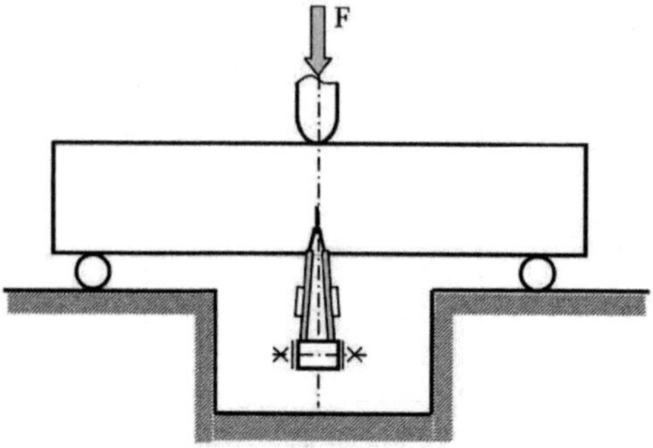

Figure 29. Loading of SENB specimen

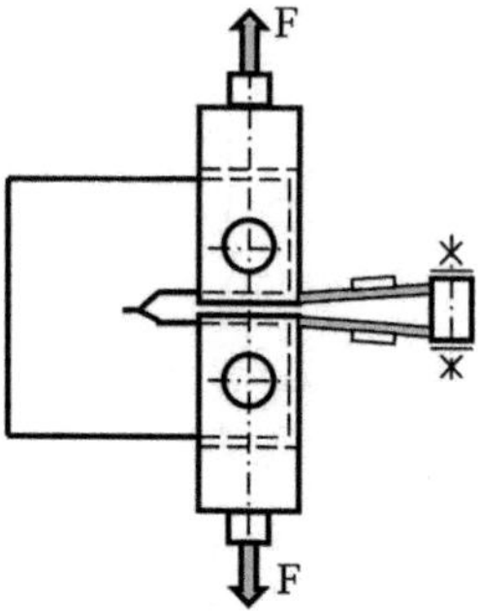

Figure 30. Loading of CT specimen

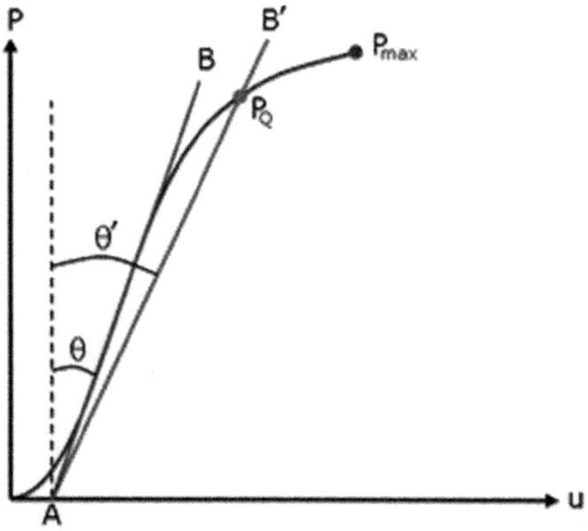

Figure 31. Typical load - displacement curve

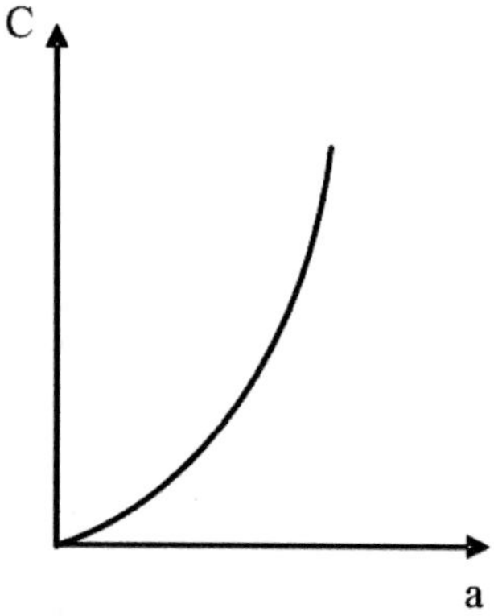

Figure 32. Compliance vs. crack length curve

of magnitude, for flow direction, while for rise direction loading speed appears to have no influence. Fracture toughness values in rise direction are higher up to 21%, than those in flow direction, Fig. 35.

Less experimental data are available for fracture toughness expressed by the critical energy release rate G_{Ic}. Fig. 36 presents some data from literature. Saenz et al. (2011) determined the G_{Ic} for PVC and PES foams using SENB and DCB specimens. The PVC foams have a linear elastic

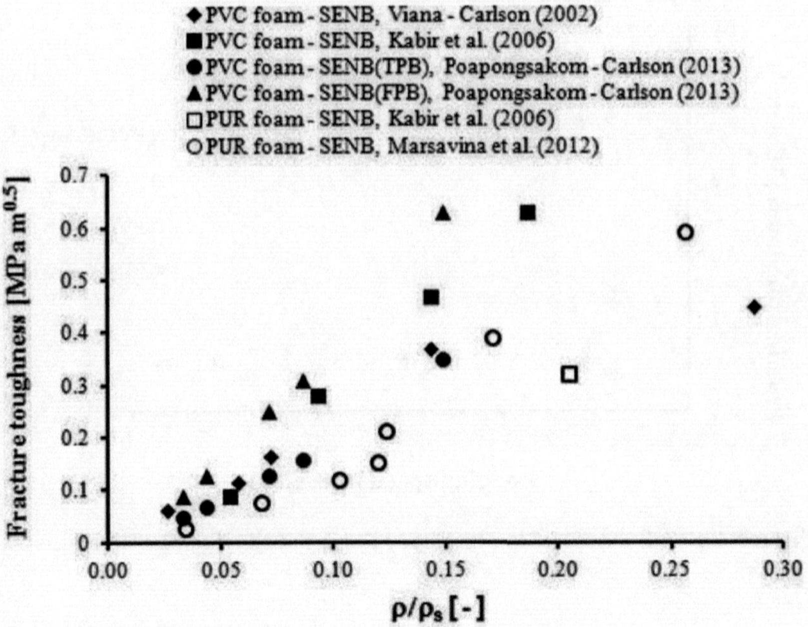

Figure 33. Fracture toughness versus relative density for different plastic foams

behavior, while the thermoplastic PES foams displayed much more ductility and substantially larger toughness values at a comparable foam density. Fowlkes (1974) presents some results obtained for PUR foam of density 87.8 kg/m^3 using different types of specimens. The fracture toughness of PUR foam is compared with PVC and PES foams. Poapongsakorn and Carlsson (2013) investigated the effect of loading configuration and cell size on fracture toughness of closed cell PVC foams. The fracture toughness obtained on SENB specimens loaded in three point bending (TPB) is approximately three times lower than those in four point bending (FPB). The explanation is that for the TPB configuration most of the deformation, with contribution of the foam indentation, occurs in the ligament of the specimen in the cracked region.

3.3 Effect of Mixed Mode Loading

For the general mixed mode case a fracture criterion should provide:

- the angle of crack initiation θ_c,

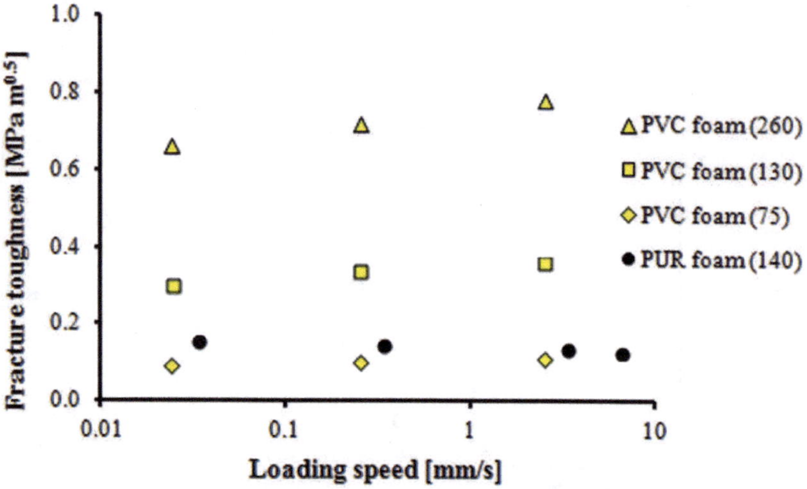

Figure 34. Influences of loading speed and density for plastic foams

- a critical combination of stress intensity factors: K_{I} and K_{II}, and fracture toughness K_{Ic} should exist

$$F\left(K_{\mathrm{I}}, K_{\mathrm{II}}, K_{\mathrm{Ic}}\right) = 0 \tag{10}$$

The singular stress field around a crack under mixed mode loading can be written in polar coordinates (r, θ) as

$$\begin{aligned}
\sigma_{rr} &= \frac{1}{4\sqrt{2\pi r}}\left[K_{\mathrm{I}}\left(5\cos\frac{\theta}{2} - \cos\frac{3\theta}{2}\right) + K_{\mathrm{II}}\left(-5\sin\frac{\theta}{2} + 3\sin\frac{3\theta}{2}\right)\right], \\
\sigma_{\theta\theta} &= \frac{1}{4\sqrt{2\pi r}}\left[K_{\mathrm{I}}\left(3\cos\frac{\theta}{2} + \cos\frac{3\theta}{2}\right) + K_{\mathrm{II}}\left(-3\sin\frac{\theta}{2} - 3\sin\frac{3\theta}{2}\right)\right], \\
\tau_{r\theta} &= \frac{1}{4\sqrt{2\pi r}}\left[K_{\mathrm{I}}\left(\sin\frac{\theta}{2} + \sin\frac{3\theta}{2}\right) + K_{\mathrm{II}}\left(\cos\frac{\theta}{2} + 3\cos\frac{\theta}{2}\right)\right]
\end{aligned} \tag{11}$$

Maximum Circumferential Tensile Stress (MTS) Criterion According to Erdogan and Sih (1963) criterion, based on $\sigma_{\theta\theta}$ stress from Eq. (11), the crack growth starts radially from the crack tip at an angle $\theta = \theta_{\mathrm{c}}$ perpendicular to the maximum circumferential tensile stress $\sigma_{\theta\theta\max}$. The crack propagation becomes unstable when $\sigma_{\theta\theta,\max}$ reaches a critical value

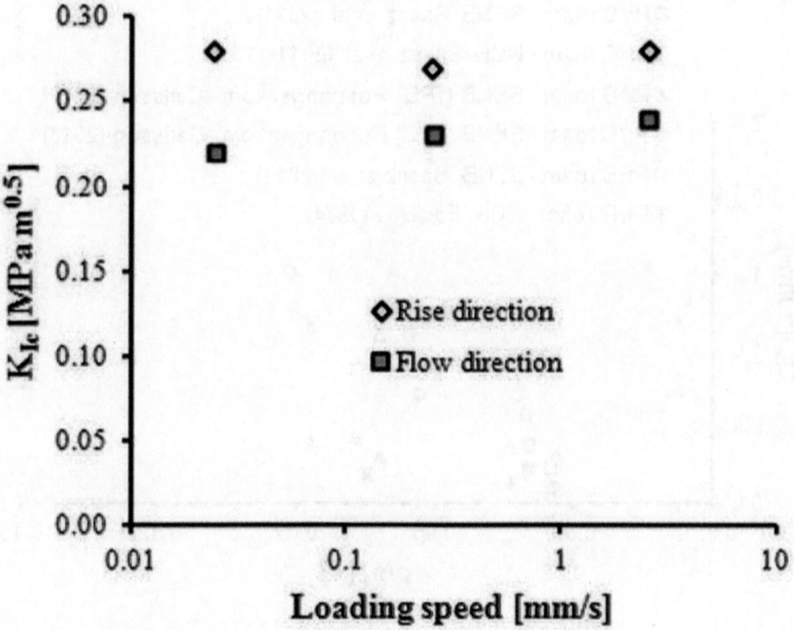

Figure 35. Influence of loading speed on fracture toughness for H130 PVC foam, after Kabir et al. (2006)

$\sigma_{\rm cr}$, which is a material parameter. The crack propagation angle can be found from

$$\left.\frac{\partial \sigma_{\theta\theta}}{\partial \theta}\right|_{\theta=\theta_{\rm c}} = 0 \quad \text{and} \quad \left.\frac{\partial^2 \sigma_{\theta\theta}}{\partial \theta^2}\right|_{\theta=\theta_{\rm c}} < 0 \tag{12}$$

leading to

$$K_{\rm I} \sin\theta_{\rm c} + K_{\rm II}\,(3\cos\theta_{\rm c} - 1) = 0 \tag{13}$$

Solving Eq. (13) the crack propagation angle results

$$\theta_{\rm c} = -\arccos\left(\frac{3K_{\rm II}^2 + K_{\rm I}\sqrt{K_{\rm I}^2 + 8\,K_{\rm II}^2}}{K_{\rm I}^2 + 9\,K_{\rm II}^2}\right), \tag{14}$$

and the fracture criterion, Eq. (10) could be expressed:

$$\cos\frac{\theta_{\rm c}}{2}\left(K_{\rm I}\cos^2\frac{\theta_{\rm c}}{2} - \frac{3}{2}K_{\rm II}\sin\theta_{\rm c}\right) = K_{\rm Ic} \tag{15}$$

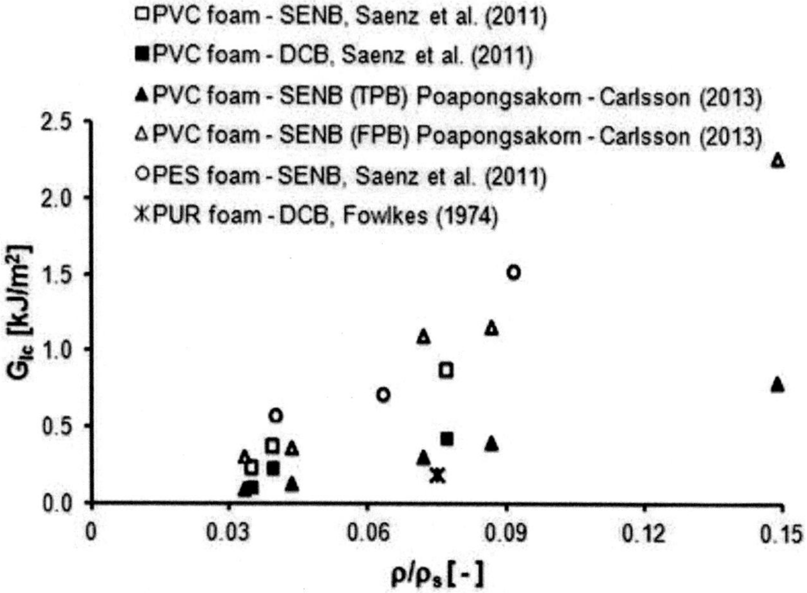

Figure 36. Fracture toughness G_{Ic} versus relative density for different plastic foams

Minimum Strain Energy Density (SED) Criterion Sih (1974) postulated fracture occurs in the direction where the strain energy density is minimum, at a critical distance r_0. For mixed mode loading the strain energy density W, can be expressed

$$W = \frac{1}{2}\sigma_{ij}\varepsilon_{ij} = \frac{S}{r_0}, \tag{16}$$

where S represents the strain energy density factor

$$\begin{aligned} S &= \frac{1}{16\pi\mu}\left\{(1+\cos\theta)\,(\kappa-\cos\theta)\,K_{\mathrm{I}}^2 + 2\sin\theta\,(2\cos\theta-\kappa+1)\,K_{\mathrm{I}}K_{\mathrm{II}}\right. \\ &+ \left.[(\kappa+1)\,(1-\cos\theta)+(1+\cos\theta)\,(3\cos\theta-1)]\,K_{\mathrm{II}}^2\right\}, \end{aligned} \tag{17}$$

with μ the shear modulus, $\kappa = 3-4\nu$ for plane strain, and $\kappa = (3-\nu)/(1+\nu)$ for plane stress.

According to this criterion the crack propagates when

$$\left.\frac{\partial S}{\partial \theta}\right|_{\theta=\theta_c} = 0 \quad \text{and} \quad \left.\frac{\partial^2 S}{\partial \theta^2}\right|_{\theta=\theta_c} > 0, \tag{18}$$

respectively the fracture initiates when S reaches a critical value S_{cr}

$$S_{\mathrm{cr}} = \frac{\kappa - 1}{8\pi\mu} K_{\mathrm{Ic}}^2. \tag{19}$$

Maximum Energy Release Rate Criterion (Gmax) Hussain et al. (1974) investigates the infinitesimal kink of a crack at an angle θ, and expressed the energy release rate in terms of the stress intensity factors of the initial crack

$$G(\theta) = \frac{1}{E'} \left(K_{\mathrm{I}}^2(\theta) + K_{\mathrm{II}}^2(\theta) \right), \tag{20}$$

where

$$\left\{ \begin{array}{c} K_{\mathrm{I}}(\theta) \\ K_{\mathrm{II}}(\theta) \end{array} \right\} = \left(\frac{4}{3 + \cos^2(\theta)} \right) \left(\frac{1 - \frac{\theta}{\pi}}{1 + \frac{\theta}{\pi}} \right)^{\frac{\theta}{2\pi}} \left\{ \begin{array}{c} K_{\mathrm{I}} \cos(\theta) + \frac{3}{2} K_{\mathrm{II}} \sin(\theta) \\ K_{\mathrm{II}} \cos(\theta) - \frac{1}{2} K_{\mathrm{I}} \sin(\theta) \end{array} \right\} \tag{21}$$

resulting

$$\begin{aligned} G(\theta) = \frac{4}{E'} \left[\frac{1}{3 + \cos^2(\theta)} \right]^2 \left(\frac{1 - \frac{\theta}{\pi}}{1 + \frac{\theta}{\pi}} \right)^{\frac{\theta}{\pi}} \times \\ \times \left\{ \left[1 + 3\cos^2(\theta) \right] K_{\mathrm{I}}^2 + 8 \sin(\theta) \cos(\theta) K_{\mathrm{I}} K_{\mathrm{II}} + \left[9 - 5\cos^2(\theta) \right] K_{\mathrm{II}}^2 \right\} \end{aligned} \tag{22}$$

The angle of crack propagation θ_{c} is found by maximizing $G(\theta)$

$$\left. \frac{\partial G}{\partial \theta} \right|_{\theta = \theta_{\mathrm{c}}} = 0 \quad \text{and} \quad \left. \frac{\partial^2 G}{\partial \theta^2} \right|_{\theta = \theta_{\mathrm{c}}} < 0, \tag{23}$$

Crack extension occurs when $G(\theta_{\mathrm{c}})$ reaches the critical value G_{Ic}

$$G(\theta_{\mathrm{c}}) = \frac{K_{\mathrm{Ic}}^2}{E} \tag{24}$$

resulting

$$\begin{aligned} & 4 \left(\frac{1}{3 + \cos^2(\theta_{\mathrm{c}})} \right)^2 \left(\frac{1 - \frac{\theta_{\mathrm{c}}}{\pi}}{1 + \frac{\theta_{\mathrm{c}}}{\pi}} \right)^{\frac{\theta_{\mathrm{c}}}{\pi}} \times \\ & \left[\left(1 + 3\cos^2(\theta_{\mathrm{c}}) \right) \left(\frac{K_{\mathrm{I}}}{K_{\mathrm{Ic}}} \right)^2 + 8 \sin(\theta_{\mathrm{c}}) \cos(\theta_{\mathrm{c}}) \left(\frac{K_{\mathrm{I}} K_{\mathrm{II}}}{K_{\mathrm{Ic}}^2} \right) \right. \\ & \left. + \left(9 - 5\cos^2(\theta_{\mathrm{c}}) \right) \left(\frac{K_{\mathrm{II}}}{K_{\mathrm{Ic}}} \right)^2 \right] = 1 \end{aligned} \tag{25}$$

Equivalent Stress Intensity Factor (ESIF) Criterion Richard (1985); Richard et al. (2005) proposed a generalized fracture criterion, based on the equivalent stress intensity factor K_{eq}, which is defined similar to the maximum principal stress σ_{eq}, as

$$K_{eq} = \frac{K_I}{2} + \frac{1}{2}\sqrt{K_I^2 + 4\left(\alpha K_{II}\right)^2} \leq K_{Ic} \tag{26}$$

with $\alpha = K_{Ic}/K_{IIc}$. Crack starts to propagate when K_{eq} reaches the fracture toughness of the material K_{Ic}. For the crack initiation angle Richard (1985) proposed the expression:

$$\theta_c = \mp\left(155.5^0 \frac{|K_{II}|}{|K_I| + |K_{II}|}\right) - 83.4^0 \left(\frac{|K_{II}|}{|K_I| + |K_{II}|}\right)^2, \tag{27}$$

which represents a correlation with a considerable number of experiments.

Only few studies present the mixed mode fracture of polymeric foams, and most of them investigate PVC foams. Burman (1998) presented fracture toughness results for two commercial foams, Rohacell WF51 (density 52 kg/m^3) and Divinycell H100 (density 100 kg/m^3). The mode I fracture toughness K_{Ic} was obtained on SENB specimens and has values 0.08 MPa m$^{0.5}$ for WF51, respectively 0.21 MPa m$^{0.5}$ for H100. He also determined the Mode II fracture toughness using End-Notch Flexure (ENF) specimen with values of 0.13 MPa m$^{0.5}$ for WF51, respectively 0.21 MPa m$^{0.5}$ for H100.

Hallström and Grenestedt (1997) investigated mixed mode fracture of cracks and wedge shaped notches in expanded PVC foams. Different types of specimens made of Divinycell H100 were investigated and the non singular T-stress was considered in formulation of fracture criteria. It was concluded that for predominantly mode II the use of T-stress improved the fracture predictions. Three different densities of PVC foams were investigated by Noury et al. (1998) using a Compact Tensile Specimen with Arcan fixtures to produce mixed mode conditions. The ratio between mode II and mode I fracture toughness K_{IIc}/K_{Ic} was found to be between 0.4 and 0.65 depending on foam density. For mixed mode loading the Richard fracture criterion Richard (1985) gives better predictions of fracture limit and crack initiation angle.

Marsavina et al. (2014) presented mixed mode fracture results obtained on three types of PUR foams of densities 100, 145 and 300 kg/m^3. ASCB specimens were used (R = 40 mm, a = 20 mm, S_1 = 30 mm), which have the advantage that all types of mixed modes could be obtained only by changing the position of the S_2 support ($S_2 = S_1$ = 30 mm for pure mode

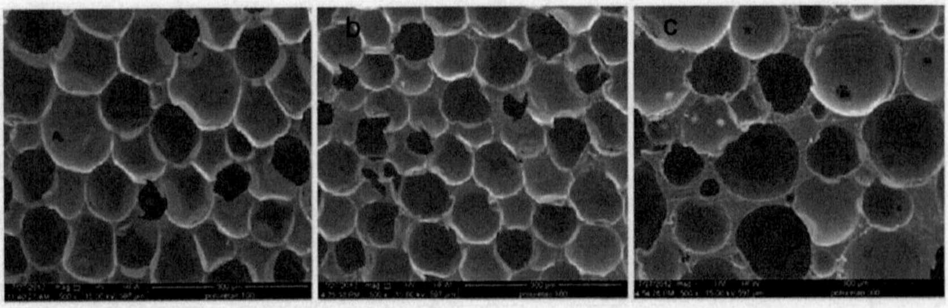

Figure 37. SEM microstructures for investigated PUR foams (magnification 500x): a) density 100 kg/m^3, b) density 145 kg/m^3, c) density 300 kg/m^3

I, $S_2 = 2.66$ mm for pure mode II). The mixed mode stress intensity factors $K_{\rm I}$ and $K_{\rm II}$ were obtained according with Ayatollahi et al. (2011):

$$K_i = \frac{F}{2Rt}\sqrt{\pi a}\, Y_i\,(a/R,\ S_1/R,\ S_2/R)\ , \quad i = \text{I, II}, \tag{28}$$

where the non-dimensional SIFs $Y_i\ (a/R, S_1/R, S_2/R)$ were determined by polynomial fitting after finite element analysis, and for $a/R = 0.5$ and $S_1/R = 0.75$ have expressions:

$$\begin{aligned} Y_{\rm I}\,(S_2/R) &= 6.235\,(S_2/R)^3 - 15.069\,(S_2/R)^2 + 17.229\,(S_2/R) - 1.0612 \\ Y_{\rm II}\,(S_2/R) &= 1.884\,(S_2/R)^5 - 7.309\,(S_2/R)^4 + 5.037\,(S_2/R)^3 \\ &+ 2.78\,(S_2/R)^2 - 5.075\,(S_2/R) + 1.983 \end{aligned}$$

Tests were performed on a Zwick/Roell of 5 kN testing machine at room temperature with a loading rate of 2 mm/min. Support rollers of diameter 20 mm were used for the bending fixture. For each position of support S_2 four specimens were tested. Brittle fracture was observed for all tested specimens with an abrupt drop of the load to zero after reaching the maximum load. The linear elastic behavior was confirmed during the tests when no cushioning and plastic deformations remain after finishing the test.

Figure 37 shows the microstructure of the three investigated PUR foams. Figures 38 a) - c) present the mixed mode fracture toughness results together with the fracture limit curves predicted by the four fracture criteria mentioned above. It can be observed that for low density foams (100 and 145 kg/m^3 densities) the experimental results fall between the Gmax and ESIF criteria. While for the foam with 300 kg/m^3 the experimental data are more scattered and close to SED and ESIF criteria. Based on these

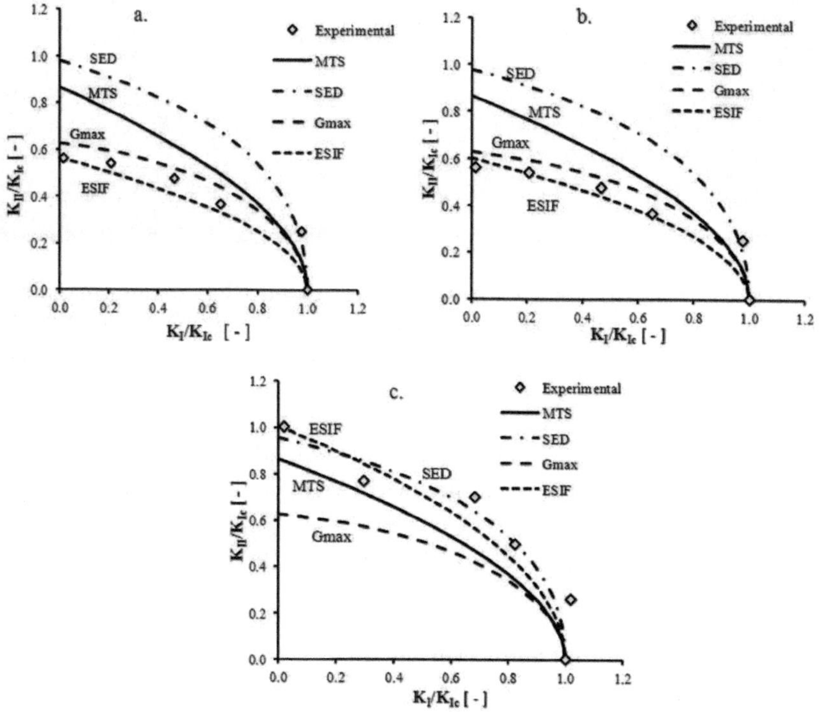

Figure 38. Comparison between fracture limit curves and experimental data for PUR foams of different densities: a) density 100 kg/m^3, b) density 145 kg/m^3, c) density 300 kg/m^3

results it could be concluded that for rigid PUR foams the Richard's ESIF criterion (Richard, 1985) is more reliable to predict mixed mode fracture. This could be also explained by the fact that it takes into account the ratio between mode I and mode II fracture toughness $\alpha = K_{\text{Ic}}/K_{\text{IIc}}$. The ratio a decreases with increasing relative density, and the same tendency was found by Noury et al. (1998) for PVC foams. Crack propagation angle was measured on each specimen. Figure 39 presents the mean values of the crack propagation angle θ_c measured on the specimens versus applied mixed mode loading $Me = \arctan(K_{\text{II}}/K_{\text{I}})$, side by side with the predicted crack propagation angles by theoretical criteria. It could be observed that for predominantly mode I loadings Me<45^0 the measured values are in good agreement with the predicted ones. For predominantly mode II loading

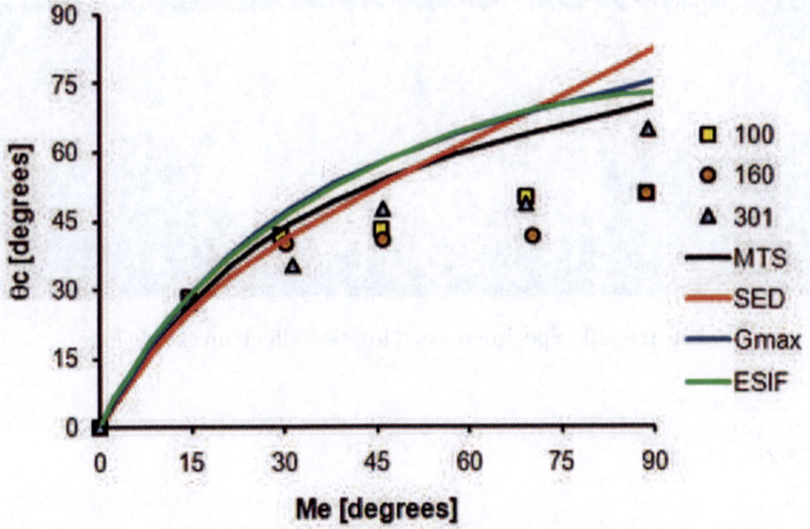

Figure 39. Comparison between predicted crack propagation angle and experimental data for PUR foams of different densities

($Me > 45^0$) the experimental crack propagation angles differ from the predicted values. It can be also observed that the foam with density 300 kg/m^3, with a microstructure close to a porous solid gives closer propagation angles to theoretical predictions, developed for brittle solid materials.

3.4 Size Effect

Using the asymptotic matching, Bažant (2002) derived a relative simple law bridging using classical plasticity and fracture mechanics approaches. The size effect is defined as the dependence of the nominal stress $\sigma_N = (3P_{max}S)/(2BW^2)$ as a function of the characteristic size of the specimen, considered the width of the specimen W. The size effect is best highlighted in a plot of Log(σ_N) versus Log W. If the failure of the foam obeys linear elastic fracture mechanics (LEFM), the logarithmic size effect plot would have to be a straight line with the slope -1/2. A ductile behavior following the strength of material approach with no size effect would be a horizontal line $\sigma_N = \sigma_f$, with σ_f being fracture or yield stress. The obtained experimental results are asymptotic to these approaches having the form (Bažant,

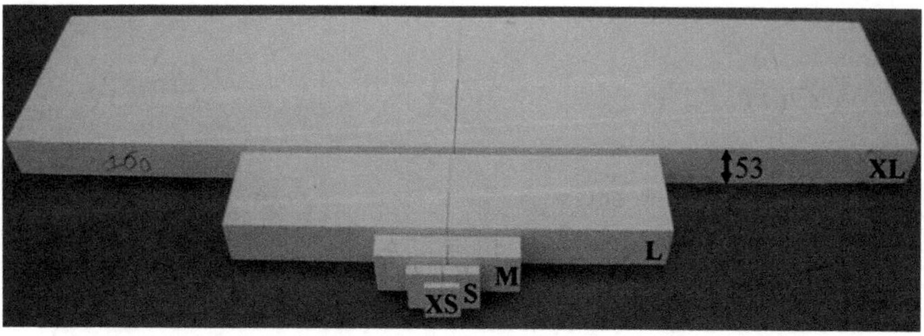

Figure 40. Specimen sizes for size effect investigations

2002)

$$\sigma_{\mathrm{N}} = \frac{\sigma_{\mathrm{N0}}}{\sqrt{1 + \frac{W}{W_0}}} \tag{29}$$

where σ_{N0} and W_0 are fitting parameters.

This law was experimentally validated for different types of materials, including PVC foams, by Bažant et al. (2003). They presented results for Divinycell H100 foam loaded in tension, obtaining the fitting parameters $W_0 = 3.18$ mm and $\sigma_{N0} = 2.48$ MPa and fitted the fracture toughness results obtained by Zenkert and Backlund (1989) for Divinicell H200 PVC foam using SENB specimens.

The size effect was investigated using SENB specimens made from three different densities of PUR foams (100, 145 and 300 kg/m^3) (Marsavina et al., 2014). Five different types of specimens, named Extra Small, Small, Medium, Large and Extra Large, Fig. 40, were cut from the same plate and had the plate thickness B of approximately 53 mm for densities 100 and 145, respectively 25 mm for 300 kg/m^3 density. The employed specimens were geometrically similar in two dimensions with length-to-width ratio 4. The mean values of the specimen dimensions and the obtained experimental results are shown in Table 5 (Marsavina et al., 2014). It should be mentioned that the plane strain condition (6) was not satisfied for all specimen sizes.

Fig. 41 presents the experimental data and the fitting curve using Eq. (29), and in Table 10 are shown the fitting parameters. The results envisage that for all specimen sizes Linear Elastic Fracture Mechanics describes better their behaviour, Fig. 41. The fitting parameters σ_{N0} and W_0 increase with increasing foam density.

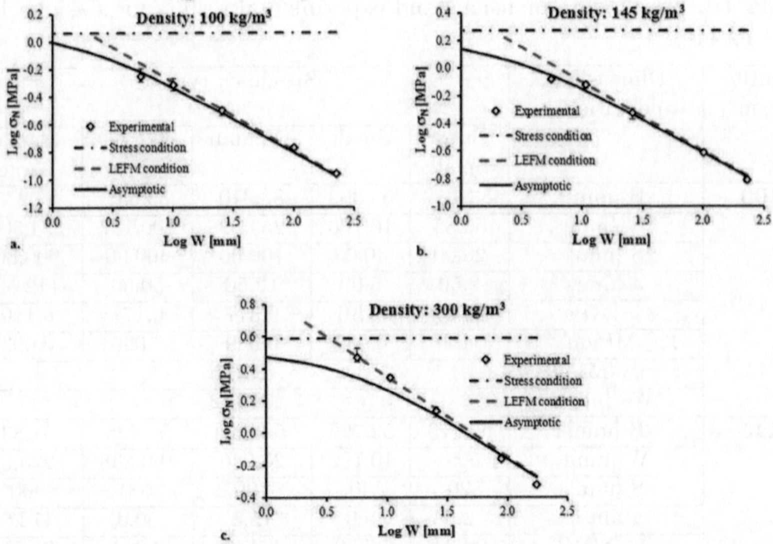

Figure 41. The size effect results expressed in terms of Logσ_N versus Log W for PUR foams: a) density 100 kg/m^3, b) density 145 kg/m^3, c) density 300 kg/m^3

3.5 Dynamic Fracture Toughness

Less investigations were carried out for the determination of the dynamic fracture toughness for plastic foams. Kabir et al. (2006) investigated the Mode I dynamic fracture toughness of a 260 kg/m^3 density PVC foam and obtained a maximum value of 2.74 MPa m$^{0.5}$, which is approximately 3.75 times higher than the static fracture toughness of the same foam. Mills and Kang (1994) used a falling mass on a compact tension specimen made by Polystyrene (PS) in order to determine the dynamic fracture toughness, and Mills (2007) also proposed a correlation between the dynamic fracture toughness and foam density. Effect of impregnation layer on fracture toughness of PUR rigid foam of 200 kg/m^3 density was investigated by Marsavina and Sadowski (2008).

Marsavina et al. (2008a) presented a comparison between static and dynamic fracture toughness of PUR foams. Instrumented impact tests were carried on according with EN ISO 179-2:2000[3]. The dynamic fracture tests

[3]EN ISO 179-2:2000 Plastics - Determination of Charpy impact properties. Part 2: Instrumented impact test

Table 10. Specimen dimensions and experimental results for size effect of PUR foams

Density [kg/m^3]	Dimensions and variables	Specimen type				
		Extra Small	Small	Medium	Large	Extra Large
100	B [mm]	53.125	53.302	53.310	53.687	53.275
	W [mm]	5.385	10.106	25.452	100.187	224.500
	S [mm]	20.00	40.00	100.00	400.00	900.00
	a [mm]	2.50	5.00	12.50	50.00	112.50
	σ_N [MPa]	0.567	0.491	0.317	0.171	0.110
	K_{Ic}[MPam$^{0.5}$]	0.071	0.087	0.089	0.096	0.093
	σ_{N0} [MPa]	**1.275**				
	W_0 [mm]	**1.542**				
145	B [mm]	52.173	52.296	52.232	51.725	51.83
	W [mm]	5.55	10.792	25.946	100.838	226.6
	S [mm]	20	40	100	400	900
	a [mm]	2.5	5.0	12.5	50.0	112.5
	σ_N [MPa]	0.834	0.759	0.475	0.244	0.155
	K_{Ic} [MPa m$^{0.5}$]	0.105	0.135	0.133	0.137	0.131
	σ_{N0} [MPa]	**1.651**				
	W_0 [mm]	**2.232**				
300	B [mm]	25.37	25.33	25.312	25.27	25.295
	W [mm]	5.655	10.584	25.57	87.9725	173.65
	S [mm]	20	40	100	350	700
	a [mm]	2.5	5	12.5	43.75	87.5
	σ_N [MPa]	2.957	2.211	1.367	0.688	0.476
	K_{Ic} [MPa m$^{0.5}$]	0.375	0.392	0.383	0.361	0.354
	σ_{N0} [MPa]	**3.297**				
	W_0 [mm]	**4.545**				

were carried on with an instrumented hammer KB Prüftechnik (Germany) having a mass of 2.04 kg, length 0.386 m, drop height 0.742 m, drop angle 157.32^0, hammer energy 7.5 J, and impact velocity 3.815 m/s. The tub has a built-in electronic sensor which allows recording the load with 1 MHz frequency. A four-channel data acquisition A/D card[4] was used for recording the load in time, and then load - displacement curve was determined. A check for energy loses due to friction was performed prior to testing and it was found that the frictional loss was 0.059 J which represents 0.4 % of the nominal energy of the pendulum 14.847 J, which fulfils the standard

[4]AdLink NuDAQ PCI-9812

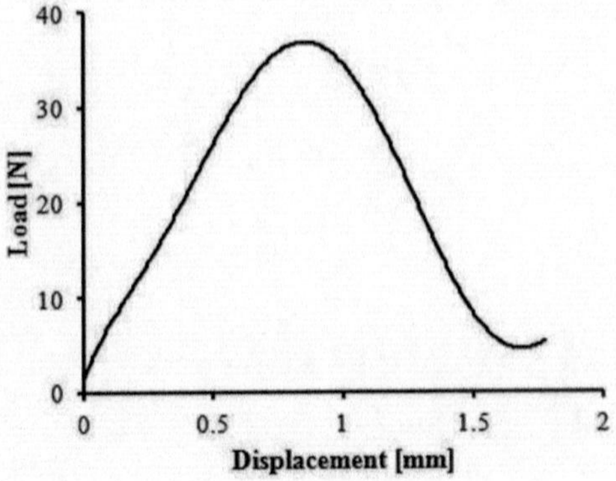

Figure 42. Typical load - displacement curve from instrumented impact test of PUR foam of 100 kg/m^3 density

requirement that the energy loss due to friction should be less than 1%. Tests were performed at room temperature.

The dynamic fracture toughness K_{Id} was determined on SENB specimens made from PUR foams of three densities (100, 145 and 300 kg/m^3) following the same procedure employed for static tests using Eqs. (2) - (3). A typical load - displacement curve obtained at impact test is shown in Fig. 42 for 100 kg/m^3 foam density.

The dynamic fracture toughness results in plane and out of plane are shown in Fig. 43. Both static and dynamic fracture toughness increases with density, Fig. 44. Minimum value of $K_{\mathrm{Id}} = 0.19$ MPa m$^{0.5}$ corresponds to 100 kg/m^3 density, while the maximum value $K_{\mathrm{Id}} = 1.05$ MPa m$^{0.5}$ was obtained for the foam with 300 kg/m^3 density. Fig. 44 presents the dynamic fracture toughness versus static fracture toughness for the PUR foam of different densities. A linear interpolation indicates that the dynamic fracture toughness is 2.87 times higher that the static one.

3.6 Micromechanical Models for Predicting Fracture Toughness

Micromechanical models relates the fracture toughness of the foam K_{Ic} on the fracture strength of the solid material from cell walls σ_{fs}, the relative density ρ/ρ_{s}, and a geometric characteristic of the cell structure: usually cell length l or strut thickness h, Fig. 45. Gibson and Ashby (1997) assumed that the crack tip is located at half-edge length and considered an elastic mode I stress field at the crack tip, Fig. 46. They start from the stress singularity at the tip of a crack of length $2a$ and normal to remote loading

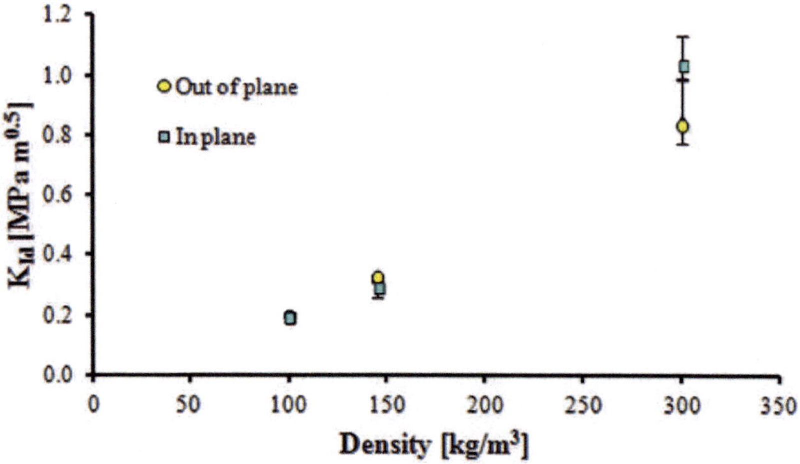

Figure 43. Dynamic fracture toughness versus density

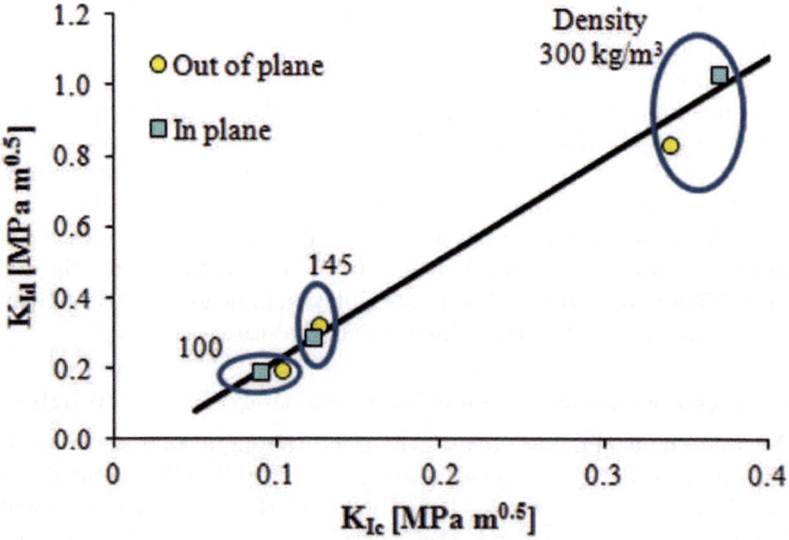

Figure 44. Dynamic fracture toughness versus static fracture toughness

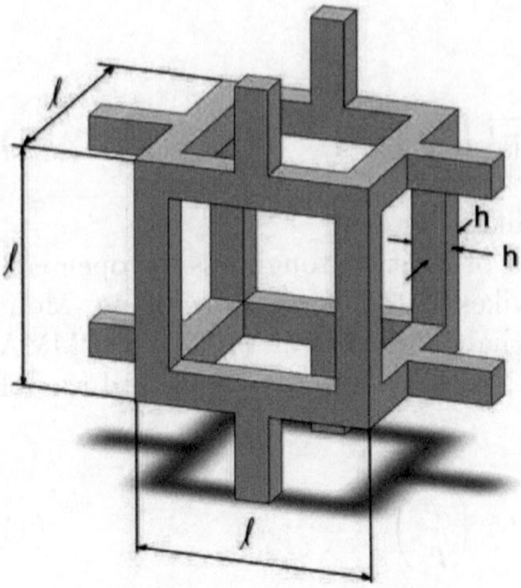

Figure 45. Cell geometry

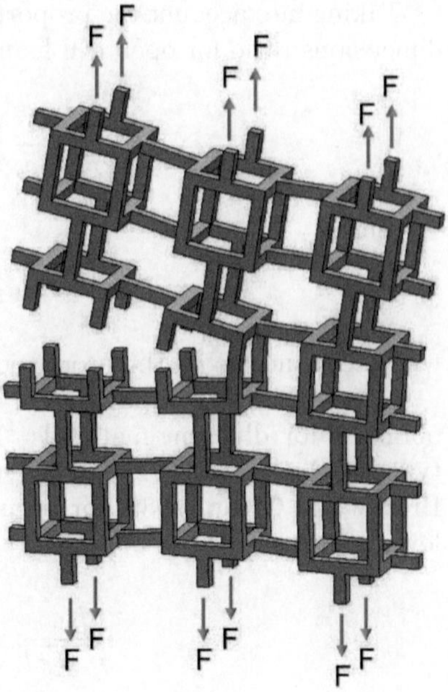

Figure 46. Cracked open cell structure

σ in an elastic continuum solid, at distance r (on a direction $\theta = 0$) from crack tip. They considered only the singular term in the Irwin's stress field solution, and the bending of struts

$$\sigma_{yy} = \frac{K_{\mathrm{I}}}{\sqrt{2\pi r}} = \frac{\beta\,\sigma\sqrt{\pi\,a}}{\sqrt{2\,\pi\,r}}, \tag{30}$$

with β a constant depending on the geometry of the cracked body.

The first unbroken edge is placed at distance $l/2$ in front of the crack and is subjected to a force F

$$F = l\int_0^l \sigma_{yy}\,\mathrm{d}r = \sqrt{2}\,\beta\,\sigma\,l^2\sqrt{\frac{a}{l}}. \tag{31}$$

The edges fail when the stress produces by bending moment $M = Fl/2$ reaches the ultimate tensile strength σ_{fs}

$$\sigma_{fs} = \frac{\frac{Fl}{2}}{\frac{h^3}{6}} = \frac{3Fl}{h^3} = \frac{3\sqrt{2}\,\beta\,K_{\mathrm{I}}\,l^{\frac{5}{2}}}{\sqrt{\pi}h^3} \Rightarrow K_{\mathrm{Ic}} = \frac{\sqrt{\pi\,l}}{3\sqrt{2}\,\beta}\sigma_{\mathrm{fs}}\left(\frac{h}{l}\right)^3. \tag{32}$$

Taking into account the proportionality between relative density and cell dimensions ratio for open cell foam, as given by Gibson and Ashby (1997)

$$\left(\frac{\rho}{\rho_s}\right) \propto \left(\frac{h}{l}\right)^2, \tag{33}$$

it results

$$K_{\mathrm{Ic}} = C\sigma_{\mathrm{fs}}\sqrt{\pi\, l}\left(\frac{\rho}{\rho_s}\right)^{3/2}, \tag{34}$$

where C contains all the proportionality constants.

Comparing with experimental data of fracture toughness for open cells obtained for different materials: Fowlkes (1974) for polyurethane, McIntyre and Anderson (1979) for polyurethane, Maiti et al. (1984) for PMMA, Brezny and Green (1989) for ceramic foams, results C = 0.65, and the following correlation was obtained

$$\frac{K_{\mathrm{Ic}}}{\sigma_{\mathrm{fs}}\sqrt{\pi\, l}} = 0.65\left(\frac{\rho}{\rho_s}\right)^{3/2} \tag{35}$$

Experimental results on reticulated carbon foams compared to the ones obtained with relation (34) show that this is valid only for $h/l > 10$, as presented by Huang and Gibson (1991).

A similar correlation was proposed by Green (1985) for ceramic foams, considering elastic deformation in shell theory of hollow sphere model for foam cells

$$\frac{K_{\mathrm{Ic}}}{\sigma_{\mathrm{fs}}\sqrt{\pi\, l}} = 0.28\left(\frac{\rho}{\rho_s}\right)^{1.3} \tag{36}$$

However, the model described by Eq. (34) doesn't take into account the tension of cell walls. Linul and Marsavina (2011) investigating a 2D solid representative volume of polymeric foam with a square structure (Fig. 47) showed that in the first unbroken strut the stress field contains both components: tension and bending, Fig. 48.

Choi and Sankar (2005) relate the stress intensity factors to the stress field in the crack tip ligament of the foam taking into account both the tension and bending of a cell wall. They used an effective crack length by multiplying the cell length l with a non-dimensional factor α, and found the fracture toughness in the form:

$$K_{\mathrm{Ic}} = \sigma_{\mathrm{fs}}\frac{h^2}{l^2}\sqrt{l}\sqrt{\frac{\pi}{2}}\frac{1}{\sqrt{\alpha}\left(1+2\alpha\frac{l}{h}\right)} \tag{37}$$

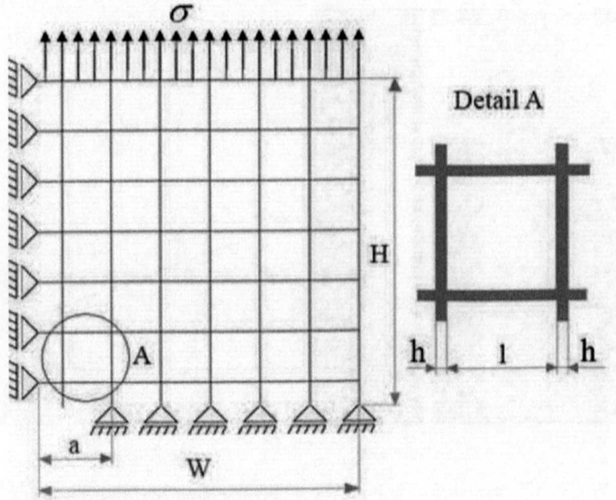

Figure 47. The 2D representative volume

An analytical model for mode II fracture toughness was also proposed based on the shear stress distribution ahead of the crack tip. The first unbroken strut is loaded in pure bending due to shear load of the representative element and the mode II fracture toughness of foam results:

$$K_{\mathrm{IIc}} = \sigma_{\mathrm{fs}} \frac{h^3}{3l^2} \sqrt{\frac{\pi}{2\alpha\, l}} = \sigma_{\mathrm{fs}} \frac{h^3}{3l^{3/2}} \sqrt{\frac{\pi}{2\alpha}} \tag{38}$$

Choi and Sankar (2005) employed Finite Element Analysis modeling on an open cell rectangular model using beam finite elements. The boundary conditions were imposed according to the displacement field of a homogeneous orthotropic material, as proposed by Sih and Liebowitz (1968). Assuming that the first strut in front of the crack fails at the critical tensile stress of the solid material, and performing the analysis considering constant cell length, respectively constant strut thickness, they obtained a proportionality between mode I and mode II fracture toughness and the relative density of the foams.

Another model was proposed by Choi and Lakes (1996), they considered that due to crack blunting the stress field around a crack in the foam is non-singular. The stress field around a crack with length *2a* and a crack tip

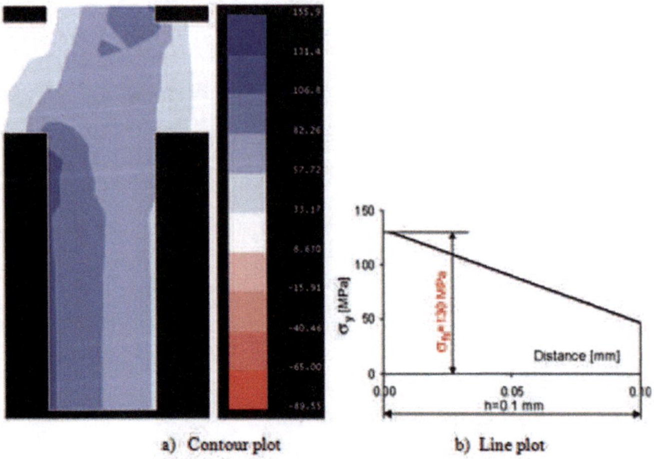

Figure 48. Stress distribution in the first unbroken strut

radius r_{tip} is, at distance $r > r_{\text{tip}}/2$:

$$\sigma = \frac{K_{\text{I}}}{\sqrt{2\pi\, r}} + \frac{K_{\text{I}}}{\sqrt{2\pi\, r}} \left(\frac{r_{tip}}{2r}\right). \tag{39}$$

For a regular tetrakaidecahedron (14-sided polyhedron with 6 square and 8 hexagonal faces) with cell size l the crack tip radius is $r_{\text{tip}} = \sqrt{2}\, l$, and considering that the maximum stress in bending reaches the tensile strength of the solid, it results:

$$\frac{K_{\text{I}}}{\sigma_{\text{fs}} \sqrt{\pi\, l}} = 0.19 \left(\frac{\rho}{\rho_{\text{s}}}\right). \tag{40}$$

All the above micromechanical models relate the foam fracture toughness to the tensile strength of the solid material and microstructure parameters: cell length and relative density. The methodology assumes that the load is transmitted through the foam as a set of discrete forces and moments acting on cell struts. Different integration limits were used in order to determine the forces and moments. Gibson and Ashby (1997), respectively Choi and Sankar (2005) considered a singular stress field, while a non-singular stress field was used by Choi and Lakes (1996) in the vicinity of the crack. The fracture toughness was obtained by considering that the crack extents when the stress in the first strut in front of the crack reaches the tensile strength of the solid. When micromechanical models are used the size effect regarding

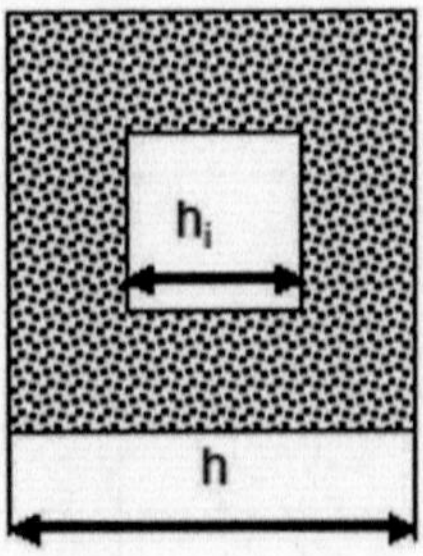

Figure 49. Strut with a box-like central crack

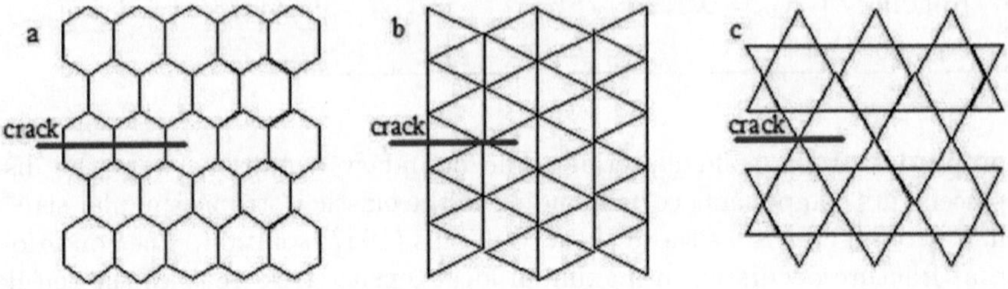

Figure 50. Types of cracked cellular structures: a) hexagonal, b) triangular, c) Kagome

the variation of the tensile strength of the solid material σ_{fs} with strut size should be considered (Green, 1985). Huang and Gibson (1991) proposed a statistical analysis based on Weibull distribution in order to show the effect of cell size on fracture toughness.

Another cracked foam model considered a central void of rectangular shape with size h_i, in the struts thickness h (Fig. 49), the fracture toughness could be expressed (Gibson and Ashby, 1997):

$$K_{\text{Ic}} = C\sigma_{\text{fs}}\sqrt{\pi l}\left(\frac{\rho}{\rho_{\text{s}}}\right)^{3/2}\frac{1+\left(\frac{h_i}{h}\right)^2}{\sqrt{1-\left(\frac{h_i}{h}\right)^2}} \tag{41}$$

The fracture toughness of elastic brittle 2D lattices was determined using the finite element method for three isotropic periodic structures: hexagonal honeycomb, Kagome lattice and triangular honeycomb, Fleck and Qiu (2007), Fig. 50. They investigated the influence of cellular structure topology on the mode I, mode II and mixed mode fracture toughness. Beam elements with cubic interpolation functions were considered, having elastic

Table 11. Micromechanical models to predict Mode I and Mode II fracture toughness, after Fleck and Qiu (2007)

Cellular structure topology	Mode I fracture toughness, K_{Ic}	Mode II fracture toughness, K_{IIc}
Hexagonal (Fig. 50a)	$K_{\mathrm{Ic}} = 0.8\sigma_{\mathrm{fs}}\sqrt{l}\left(\frac{\rho}{\rho_{\mathrm{s}}}\right)^2$	$K_{\mathrm{IIc}} = 0.37\sigma_{\mathrm{fs}}\sqrt{l}\left(\frac{\rho}{\rho_{\mathrm{s}}}\right)^2$
Triangular (Fig. 50b)	$K_{\mathrm{Ic}} = 0.5\sigma_{\mathrm{fs}}\sqrt{l}\left(\frac{\rho}{\rho_{\mathrm{s}}}\right)$	$K_{\mathrm{IIc}} = 0.38\sigma_{\mathrm{fs}}\sqrt{l}\left(\frac{\rho}{\rho_{\mathrm{s}}}\right)$
Kagome (Fig. 50c)	$K_{\mathrm{Ic}} = 0.212\sigma_{\mathrm{fs}}\sqrt{l}\left(\frac{\rho}{\rho_{\mathrm{s}}}\right)^{0.5}$	$K_{\mathrm{IIc}} = 0.133\sigma_{\mathrm{fs}}\sqrt{l}\left(\frac{\rho}{\rho_{\mathrm{s}}}\right)^{0.5}$

properties of the solid material. The boundary conditions were the displacements components corresponding to the elastic isotropic singular stress field around cracks, as those of the Williams (1957) solution. The condition that fracture occurs when maximum local tensile stress reaches the tensile strength of the solid material σ_{fs} was assumed.

The expressions of mode I and mode II fracture toughness of cellular structures are summarized in Table 11. The mode II fracture toughness values are lower than the mode I fracture toughness values for the same relative density and cellular structure. Also, it was highlighted that the Kagome structure has the lowest sensitivity of fracture toughness to relative density.

Lipperman et al. (2007) considered a lattice model consisting in rigidly connected Euler beams which can fail when the stress reaches a critical value. The conventional Mode I and Mode II fracture toughness were calculated and the influence of the relative density was shown. Long finite length cracks, modeled by several broken beams, were considered in the infinite lattice model. Four different layouts were considered: Kagome, triangular, square and hexagonal honeycombs, Fig. 51.

The solution was obtained analytically using discrete Fourier transform, reducing the initial problem for unbounded domain to the analysis of a finite repetitive module in the transform space. The variation of fracture toughness with relative density is investigated for different types of periodic cells. The directional fracture toughness estimates are reported in polar diagrams, which have a circular shape highlighting quasi isotropic fracture behavior. The Mode II fracture toughness is smaller than the Mode I for almost all investigated cases.

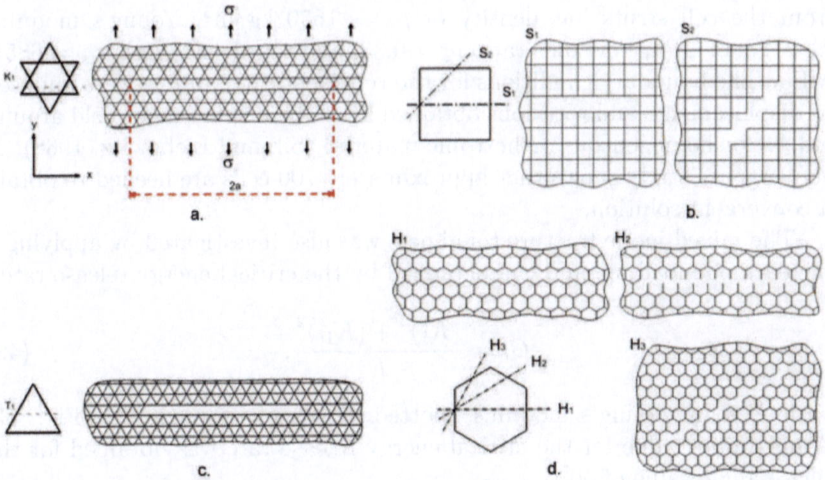

Figure 51. Lattice cracked models: a) Kagome, b) square, c) triangular, d) hexagonal, after Lipperman et al. (2007)

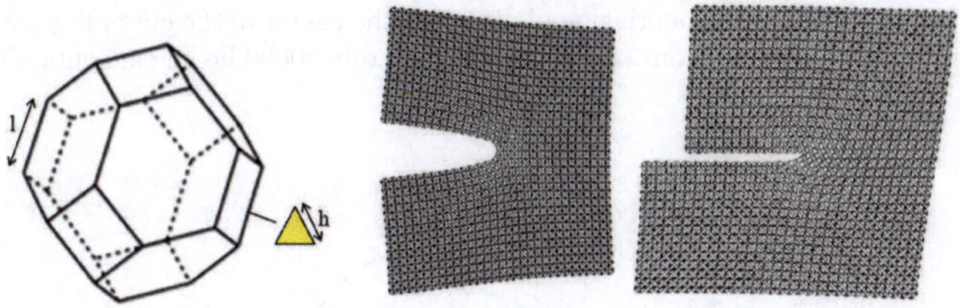

Figure 52. Unit cell and deformed meshes for pure mode I and pure mode II for 30x31 cells model: a) Tetrakaidecahedral unit cell, b) Mode I deformed mesh, c) Mode II deformed mesh

Tetrakaidecahedral unit cell was employed to determine the fracture toughness of open cell foams (Thiyagasundaram et al., 2011). A finite element based micromechanical model was proposed by repeating unit cell with strut length l=1 mm and cross section dimension h=0.06 mm, Fig. 52, resulting a relative density of $\rho/\rho_s = 0.00165$. Cracks were introduced by removing the cells on the crack length, Fig. 52. The solid material from the cell struts has density of $\rho_s = 1650$ kg/m^3, Young's modulus $E_s = 23.42$ GPa, Poisson's ratio $\nu_s = 0.33$, and tensile strength $\sigma_{fs} = 685.5$ MPa. The boundary conditions for the representative volume were imposed as displacements and rotations obtained from the displacement field around cracks for homogeneous orthotropic material (Sih and Liebowitz, 1968). A convergence study shows that approximately 700 cells are needed to obtain a convergent solution.

The mixed mode fracture toughness was also investigated by applying a mixed mode loading, and characterized by the critical energy release rate

$$G_c = \frac{(K_I)^2 + (K_{II})^2}{E}, \tag{42}$$

with E foam Young's modulus, plotted versus phase angle *Me*, Fig. 53. A maximum value for the critical energy release rate was obtained for the phase angle value of 50^0.

Mode I fracture toughness of carbon foams was investigated by Choi and Sankar (2003). The experimental results obtained using SENB specimens subjected to three and four point bending were compared with two finite element micromechanical models:

- a 3D solid model based on a rectangular cube as unit cell from which is extracted a spherical void placed in the center of the cube, Fig. 54a)
- a beam model consisting on approximately 20000 beam elements, Fig. 54b)

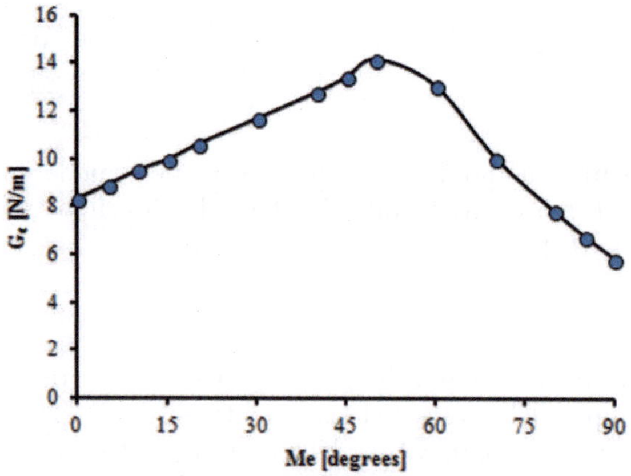

Figure 53. Critical energy release rate for mixed mode loading Me

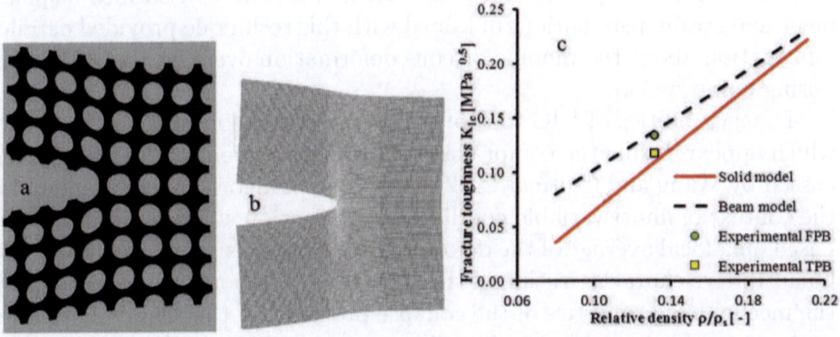

Figure 54. Finite element micromechanical models and fracture toughness results for carbon foam, after Choi and Sankar (2003): a) 3D solid model, b) Beam model, c) Comparison between micromechanical predictions and experimental results

The crack was introduced parallel to one of the principal axis of material, by cutting the ligaments of unit cell. The applied boundary conditions were imposed according with displacement field around crack for an orthotropic solid (Sih and Liebowitz, 1968). The two finite element micromechanical models results are in good agreement with the experimental values obtained for fracture toughness of carbon foams with relative density 0.1312, as shown in Fig. 54c.

4 Damage Identification in Cellular Materials Using Digital Image Correlation (DIC)

When testing a closed-cell aluminum alloy foam (Bastawros et al., 2000), three stages in the deformation response have been identified:

- localized plastic straining at cell nodes;
- bands of concentrated strain containing cell membranes that experience plastic buckling, elastically constrained by surrounding cells - new bands appear in the neighboring regions;
- one of the bands exhibits complete plastic collapse.

Jin et al. (2007) analyzed the tested foam material was a rigid closed-cell polyurethane foam with a nominal density of 320 kg/m^3. The deformation of foam specimens was obtained using the 3-dimensional digital image correlation (3D-DIC) technique. These experiments confirmed that the 3D-DIC technique is able to obtain accurate and full-field large deformation of foam specimens, including strain concentrations. The full-field surface displacement and strain distributions obtained with this technique provided detailed information about the inhomogeneous deformation over the area of interest during compression.

The capabilities of DIC to capture the heterogeneous deformation fields which appear during the compression of ultra-light open-cell foams were discussed by Wang and Cuitino (2002). The present algorithm is formulated in the context of multi-variable non-linear optimization where a merit function based on a local average of the deformation mapping is minimized implicitly. Quantitative characterization of these fields is of importance to understand the mechanical properties of the collapse process and the energy dissipation patterns in this type of materials. The main conclusion is that the collapse of light open-cell foams occurs as a phase transition phenomenon.

4.1 Testing Procedure

Closed-cell polyurethane foam specimens with densities of 100 and 160 kg/m^3 have the dimensions of 50x50x50 mm, while the ones made from the foam of 300 kg/m^3 are of 25x25x25 mm. Tests were done on a Zwick machine having a load cell of 10 kN (Apostol et al., 2014).

DIC method is used, and the Aramis 3D system is calibrated by using a 35x28 mm caliber together with a facet of 27x15 pixels, thus obtaining a 44% facet overlap. At least three tests were done for each of the three densities at speeds of testing as 1 and 5 mm/min. The Young's moduli of elasticity established in compression for the three foams were obtained as: 100 kg/m^3 - E = 25 MPa; 160 kg/m^3 - E = 49 MPa; 300 kg/m^3 - E = 238 MPa. The average values of Poisson's ratio determined with Aramis are: 0.29 for 100 kg/m^3, 0.28 for 160 kg/m^3, and 0.31 for 300 kg/m^3.

For each test a report is generated from Aramis, and presented in the following figures, as having the structure: in the engineering characteristic curve is depicted the moment in which the stage is considered (lower left figure), the vertical displacements over the whole specimen are presented together with the corresponding scale (upper right figure) as well as the corresponding von Mises strains (lower right figure), and some of the characteristic stages of the von Mises strains variation are shown in the vertical middle section of the specimen (upper left figure). About 8 presented stages cover for each test the deformation of each specimen from the linear elastic region till the densification of the foam. In the final stages of compression some of the facets are missing, as calculations were not possible any longer, and the curves of variation are disrupted.

4.2 Evaluation of Results

For the foam of 100 kg/m^3 and a speed of 1 mm/min in Fig. 55 are presented the reports obtained from two different tests in the linear elastic region. Global and local strains are about the same at the same moment of the test, as after about 100 seconds. The vertical displacements are also very similar. However, the report presented on the right shows that the von Mises strains start to concentrate in small deformation bands.

As loading is increased for the same speed of testing, in the plateau region, the 100 kg/m^3 foam may exhibit a different response in the formation of the deformation bands as following the same tests presented in Fig. 55. In Fig. 56 - left report - at about 18% global engineering strain, the von Mises strains form an almost horizontal deformation band with maximum strains of 81%.

For the other test, the same localized bands continue to grow, extending to both vertical edges of the specimen (Fig. 56, right report), and at about 10% global strain we obtain local von Mises strains of 84%. The degradation of the foam continues with the loading and calculations in the facets are not possible any more, obtaining in the last stages discontinuous curves in the variation of von Mises strains (Fig. 56). For the 160 kg/m^3 density foam the influence of the speed of testing (for 1 and 5 mm/min) is not significant (Figs. 57 and 58). At about the same global (technical) strain of 14% for the test in the left report and 12% in the test shown in the right report, the local maximum von Mises strains are 40%, respectively 28%. Bands of deformation form in different locations not due to the influence of different testing conditions, but more as a result of local phenomena.

The density of 300 kg/m^3 behaves as being more influenced by the friction between the material and the plates through which the compressive

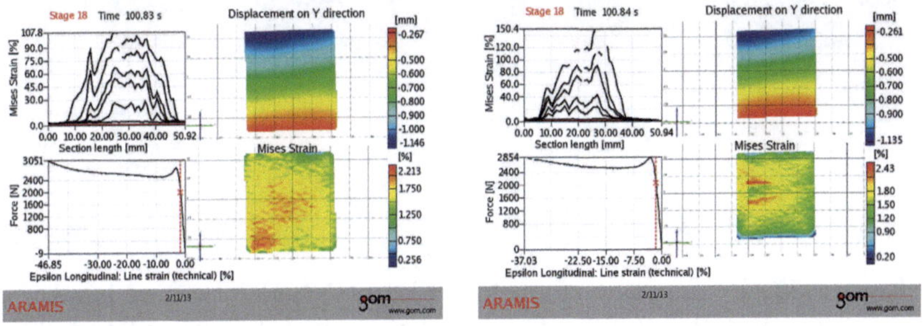

Figure 55. Two stages for different tests in the linear elastic domain for the foam with 100 kg/m^3

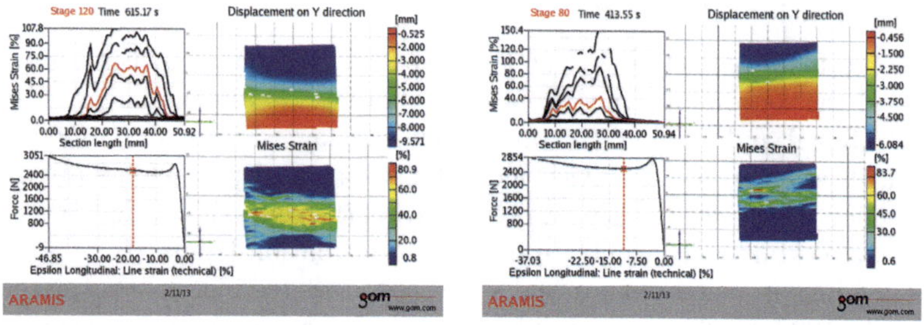

Figure 56. Two stages for different tests in the plateau region for the foam of 100 kg/m^3

force is applied, especially for the speed of 1 mm/min as presented in Figs. 59 and 60. In two tests a V- or X-type of local deformation bands may form as the foam is tested in the plateau region. Most of the crush of the foam is produced in the central part of the specimen, but approximately 45∘ inclined deformation bands tend to get unified with other local stress raisers close to the contact surface in between the foam and the steel plate.

When speed of testing is increased to 5 mm/min, in Figs. 61 and 62 one may observe less evident tendencies to form inclined deformation bands. For the test with results presented in the report from Fig. 61, the crushing of the cells is mostly in the middle of the specimen, with some inclined bands towards the upper face of the specimen. In the other test a clearly defined

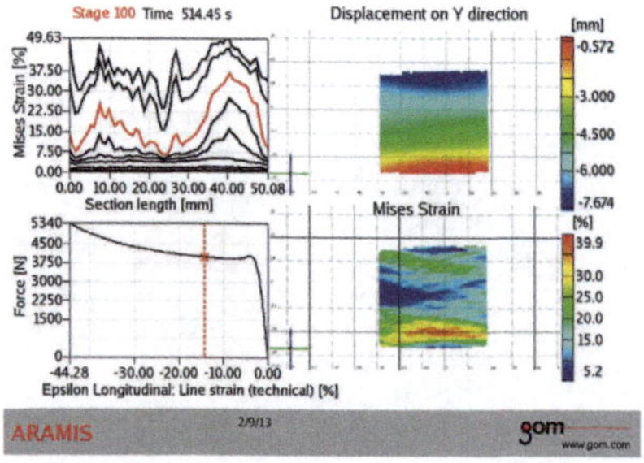

Figure 57. Deformation bands for the foam of 160 kg/m^3 tested at 1 mm/min

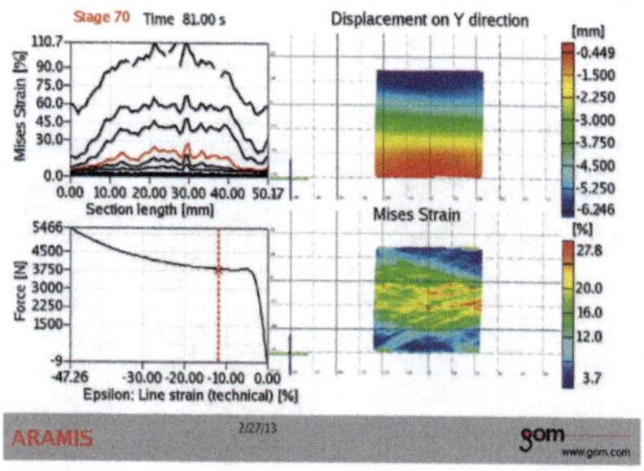

Figure 58. Deformation bands for the foam of 160 kg/m^3 tested at 5 mm/min

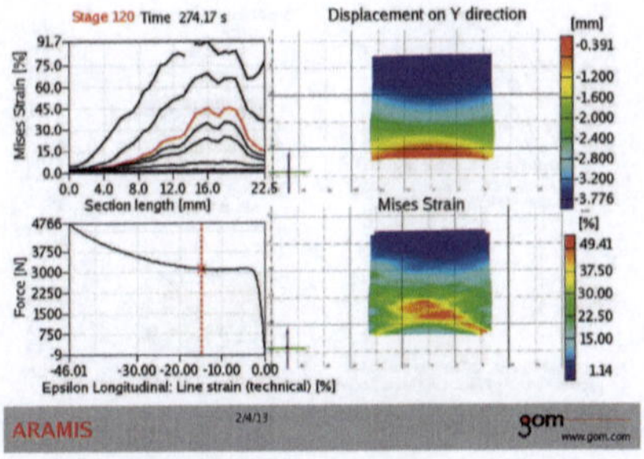

Figure 59. Local deformation for the foam with 300 kg/m^3 tested at 1 mm/min

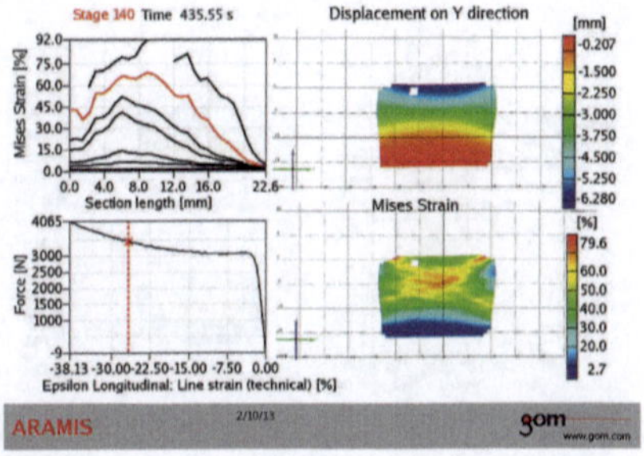

Figure 60. Influence of the boundaries for the foam with 300 kg/m^3 tested at 1 mm/min

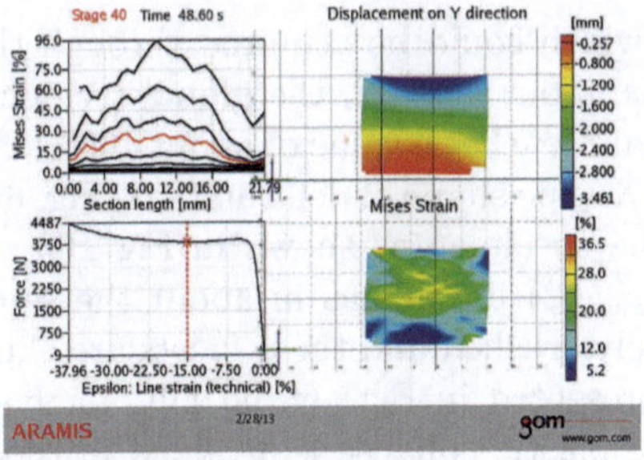

Figure 61. Local deformation for the foam with 300 kg/m^3 tested at 5 mm/min

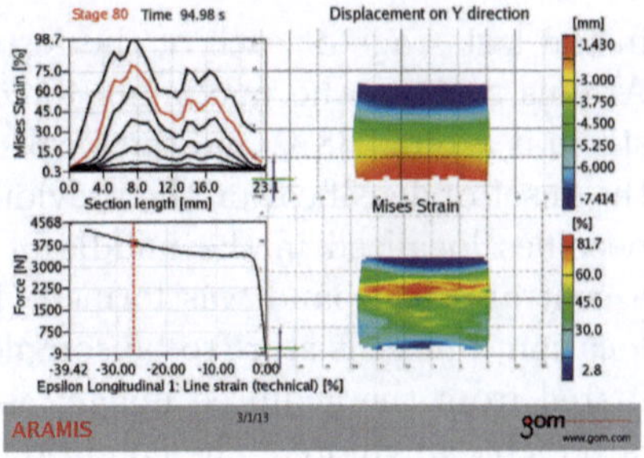

Figure 62. Formation of a central crush band for the foam with 300 kg/m^3 tested at 5 mm/min

horizontal band forms in the middle part of the specimen, from one edge to the other, shown in Fig. 62. The local von Mises strains of about 82% at a global strain of about 27% are consistent with the ones obtained for the test with the results presented in Fig. 60. The difference is given by the form of the deformation bands. These experimental observations on the formation of deformation bands indicate DIC as a powerful full-field tool to monitor the local crushing behaviour, being capable to account for the influence of the foam density and speed of testing.

For the same two tests done at 5 mm/min on the foam of 300 kg/m^3,

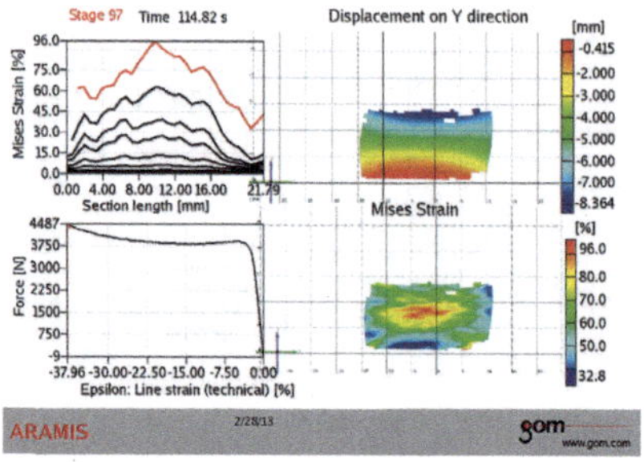

Figure 63. Final stage of acquired deformation localized towards the centre of the specimen

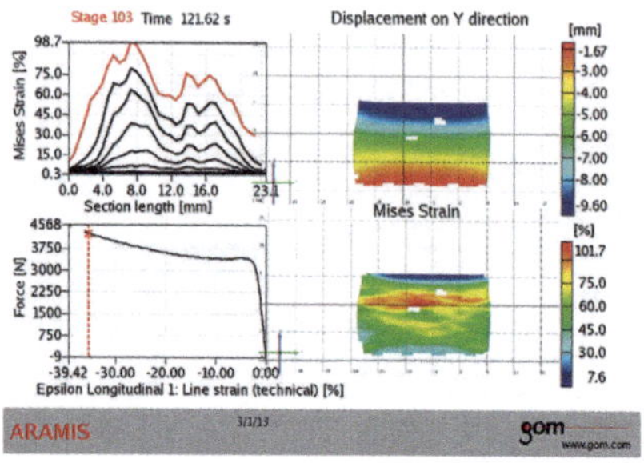

Figure 64. Formation of a central crush band for the foam with 300 kg/m^3 tested at 5 mm/min

in the last stage of each of the tests which could be acquired through the Aramis system, the maximum force is about 4500 N, the global technical strain is around 38%, and the local strain reaches a value of about 100%. At the onset of densification the previous X-type shape of deformation (Fig. 61) becomes localized in the middle of the specimen as shown in Fig. 63. If a central crush band was formed (Fig. 62) it remains in about the same location, but cells start to be completely crushed and those facets are eliminated from the acquired image; as presented in Fig. 64 no DIC analysis is possible anymore. The condition of plastic collapse is attained in those locations.

5 Conclusions

Traction and compression tests on polyurethane foams of three densities were done and results were presented.

In traction, for the 93 kg/m^3 and 200 kg/m^3 densities, the low temperature of -60 °C gave a very fragile foam behavior, opposite to the ductile behavior at 80 °C. For the 35 kg/m^3 foam the differences for the extreme temperatures are reduced in the characteristic curves up to 500 mm/min (Figs. 2-4). For each of the densities the moduli of elasticity decrease about 2 times or even more (especially for the 200 kg/m^3 foam) when the temperature is increased from -60 °C to 80 °C. For each temperature of testing and density of the foam, when the speed of loading is increased from 2 to 500 mm/min, the elongation at failure is somehow reduced, but the modulus of elasticity and maximum stress remain mainly constant.

A considerable amount of tests made available data for the testing in compression of polyurethane foams of the three densities at different temperatures and at speeds of testing starting from 2 mm/min up to 6 m/s. Attention was given to the direction on which testing is done, that is rise direction (out of plane) or direction 3, and in-plane direction or direction 1.

Foam with density 35 kg/m^3 is behaving in overall differently compared to the foams with densities 93 kg/m^3 and 200 kg/m^3. This foam is very sensitive to the increase of speed and temperature of testing when modulus of elasticity is compared. On direction 3 moduli are always greater than on direction 1. When speeds increase from 2 mm/min to 6 m/s moduli are increased in average about 3-4 times when temperature covers the domain from -60 °C to 80 °C. Maximum stress decreases slightly with the increase of temperature for each testing speed, but increases at each temperature level with the increase of speed of testing. Foam recovery varies in an opposite way: increases with the increase of temperature for each testing speed, but somehow decreases at each temperature level with the increase of speed. Maximum stress is greater and foam recovery is smaller for all testing speeds and all test temperatures (values compared for the same parameters) on direction 3 than on direction 1. At all temperatures and testing speeds foam recovery is 4-9% on direction 1 (excepting the speed of 2 mm/min) and 3-6% on direction 3, being much smaller than for the foams of 93 and 200 kg/m. This means that the energy absorption is greater for this grade of foam and damage is produced to a higher extend.

For the foam of 93 kg/m^3 the modulus of elasticity is about the same on both directions being in average close to 10 times greater than for the foam of 35 kg/m^3. At 23 °C for all temperatures the moduli are generally greater (but not too different) on direction 1 than on direction 3, mainly when speed of testing increases. Maximum stress is close to one order of magnitude greater than for the density of 35 kg/m^3. Foam recovery is decreasing with the increase of speed of loading from about 11-13% to 7-9% and is greater than for the foam of 35 kg/m^3, but about the same on both directions 1 and 3. Foam recovery increases with the increase of temperature being 8-9% at -60 °C and 25-30% at 80 °C.

The 200 kg/m^3 foam has at 23 °C moduli of elasticity greater on direction 3 than on direction 1 with about 20-30 %, the difference increasing with the increase of testing speed up to 60% at 6 m/s. On both directions the modulus of elasticity decreases with the increase on temperature at each speed of testing, but increases always with the increase of the speed. Foam recovery decreases slightly with the increase of speed of loading on both directions, being a little bit greater on direction 1 as 13-11%, compared to 11-10% on direction 3 at 23 °C. It is increasing with the increase of temperature at each speed and is greater on direction 1 than on direction 3. At temperatures of -60 °C and 23 °C the recovery is about the same on both directions as 8-12%, but is greater at 80 °C, being 16-19%.

The most important parameter influencing the fracture toughness K_{Ic} and G_{Ic} is the foam density. The type of solid from which the foam is made, also influences the fracture toughness. Cell orientation and dimensions influence also the fracture toughness. As a consequence, the foam has an anisotropic behavior. Loading direction and loading speed influences also the fracture toughness.

Mixed mode fracture of polymeric foams was assessed with different specimen types. Four theoretical fracture criteria were assessed to characterize the failure of rigid PUR foams. The experimental results proof that the ESIF criterion of Richard is most suitable for this type of plastic foams. This is the only criterion which takes into account the ratio between mode I and mode II fracture toughness, $a = K_{Ic}/K_{IIc}$. This parameter decreases with increasing the relative density of the foam. The crack propagation angles measured experimentally using ASCB specimens show good agreement for predominantly mode I loading $M_e < 45^0$.

The results on the size effect for PVC foams (Bažant et al., 2003) and for PUR foams (Marsavina et al., 2014) show that the design of such structures based on strength or plasticity criteria is generally valid only for small structural parts. In the case of large components, the size effect must be taken into account, and for design purposes LEFM concepts should be used.

Dynamic fracture toughness values are higher than the static ones for all investigated foam densities. This could be explained by the increase in pressure of the air entrapped in the close cell structures with increasing loading speed.

The static and dynamic fracture of polyurethane foams is brittle, and no plastic deformations remains after the test and no cushioning occurs during tests.

Micromechanical modeling represents an useful tool for predicting the mechanical properties of cellular materials based on cell topology. However, the obtained results should be validated against experimental data for different types of solid materials, cellular structure. Marsavina et al. (2012) compared the experimental normalized fracture toughness of PUR foam with three of the micromechanical models described above and showed that for low density foams a good agreement between experimental results and the micromechanical model (Choi and Lakes, 1996) $\rho/\rho_s < 0.1$ can be noticed. For higher relative densities ($\rho/\rho_s > 0.1$) the Ashby-Gibson model (Gibson and Ashby, 1997) appears to fit better the experimental results. For the investigated PUR foams the micromechanical model (Green, 1985)

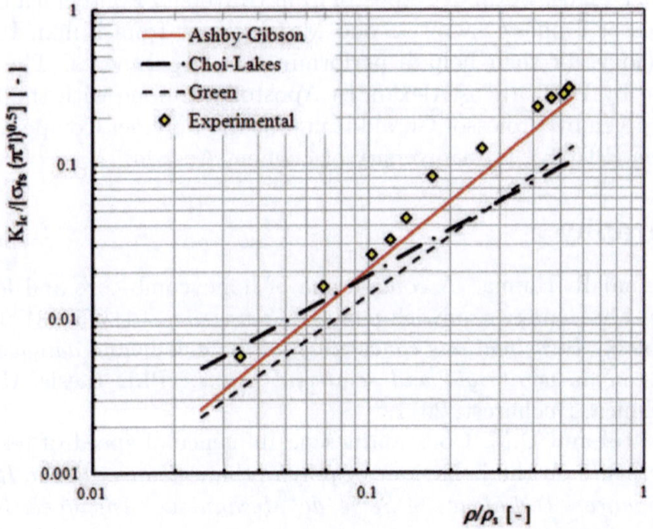

Figure 65. Normalized fracture toughness versus relative density

looks to predict much lower results than the experimental ones, as in Fig. 65. This could be explained by the use of the hollow sphere model, which could not be applied to these foams that have thick cell walls.

At the end we may again emphasize that digital image correlation is a versatile method useful to observe the local deformation bands of polyurethane foams and to assess the collapse of the cells, the onset of densification and the total damage of the foam.

Acknowledgments

Part of the experimental results were obtained in the framework of the project PN-II-ID-PCE-2011-3-0456, contract number 172/2011 financed by the Romanian National Authority for Scientific Research, CNCS – UEFISCDI. Also, the access to the experimental facilities from the *Center of Excellence for Modern Composites Applied in Aerospace and Surface Transport Infrastructure* (European Union Seventh Framework Programme (FP7/2007 – 2013), FP7 - REGPOT – 2009 – 1, under grant agreement No: 245479 at Lublin University of Technology is also acknowledged.

The authors are also grateful to colleagues, Dr. Radu Negru, Dr. Linul Emanoil, Tudor Voiconi, Dr. Dan Şerban from University Politehnica Timişoara, Dr. Dragoş Alexandru Apostol from University Politehnica Bucureşti and Professor Tomasz Sadowski and Marcin Knec from Lublin University of Technology, for their help in performing the experiments. The research carried on by Dr. Dragoş Alexandru Apostol was done with the help and guidance given by Professor Gerald Pinter at the Polymer Competence Centre Leoben (PCCL), the University of Leoben, Austria.

Bibliography

M. Akay and R. Hanna. A comparison of honeycomb-core and foam-core carbon-fibre/epoxy sandwich panels. *Composites*, 21:325–331, 1990.

D.A. Apostol. *Investigations concerning the behaviour and damage of composites made with rigid and semirigid foams.* PhD thesis, University Politehnica, Bucharest, 2011.

D.A. Apostol and D.M. Constantinescu. Influence of speed of testing and temperature on the behaviour of polyurethane foams. *Revue Roumaine des Sciences Techniques – Série de Mechanique Appliquée*, 57:27–61, 2012.

D.A. Apostol, M.C. Miron, and D.M. Constantinescu. Experimental evaluation of the mechanical properties of foams used in sandwich composites. In *Proceedings 24th DANUBIA-ADRIA Symposium on Developments in Experimental Mechanics*, pages 35–36, Cluj-Napoca, 2007.

D.A. Apostol, D.M. Constantinescu, L. Marsavina, and E. Linul. Analysis of deformation bands in polyurethane foams. *Key Engineering Materials*, 601:250–253, 2014.

M.F. Ashby. Cellular solids - scaling of proprieties. In M. Scheffler and P. Colombo, editors, *Cellular Ceramics, Structure, Manufacturing, Properties and Applications*, pages 3–17. Wiley, Weinheim, 2005.

M.R. Ayatollahi, M.R.M. Aliha, and H. Aghaf. An improved semi-circular bend specimen for investigating mixed mode brittle fracture. *Engineering Fracture Mechanics*, 78:110–123, 2011.

A.-F. Bastawros, H. Bart-Smith, and A.G. Evans. Experimental analysis of deformation mechanisms in a closed-cell aluminum alloy foam. *Journal of Mechanics and Physics of Solids*, 48:301–322, 2000.

Z.P. Bažant. *Scaling of Structural Strength.* Hermes-Penton, London, 2002.

Z.P. Bažant, Z. Yong, G. Zi, and I.M. Daniel. Size effect and asymptotic matching analysis of fracture of closed-cell polymeric foam. *International Journal of Solids and Structures*, 40:7197–7217, 2003.

R. Brezny and D.J. Green. Fracture behavior of open-cell ceramics. *Journal of American Ceramic Society*, 72:1145–1152, 1989.

R. Brown. *Handbook of Polymer Testing: Physical Methods.* Marcel Dekker Inc., New York, 1999.

R. Brown. *Handbook of Polymer Testing.* Rapra Technology, Shawbury, 2002.

M. Burman. Fatigue crack initiation and propagation in sandwich structures. Technical Report 98-29, Department of Aeronautical and Vehicle Engineering, Kungliga Tekniska, Högskolan (KTH), Stockholm, 1998.

J.Y. Chen, Y. Huang, and M. Ortiz. Fracture analysis of cellular materials: A strain gradient model. *Journal of Mechanics and Physics of Solids*, 26:789–828, 1998.

W. Chen, F. Lu, and N.A. Winfree. Dynamic compressive response of polyurethane foams of various densities. *Experimental Mechanics*, 42: 65–73, 2002.

J.B. Choi and R.S. Lakes. Fracture toughness of re-entrant foam materials with a negative Poisson's ratio: experiment and analysis. *International Journal of Fracture*, 80:73–83, 1996.

S. Choi and B.V. Sankar. A micromechanics method to predict the fracture toughness of carbon foam. In *Proceeding of the Fourteenth International Conference Materials*, number 1355, Dearborn, Michigan, 2003. Society of Manufacturing Engineering.

S. Choi and B.V. Sankar. A micromechanical method to predict the fracture toughness of cellular materials. *International Journal of Solids and Structures*, 42:1797–1817, 2005.

R.M. Christensen and K.H. Lo. Solutions for effective shear properties in three phase sphere and cylinder models. *Journal of Mechanics and Physics of Solids*, 27:315–330, 1979.

R.M. Christensen and K.H. Lo. Erratum. *Journal of Mechanics and Physics of Solids*, 34:639, 1986.

M. Danielsson. Toughened rigid foam core material for use in sandwich construction. *Cellular Polymers*, 15:417–435, 1996.

F. Erdogan and G. C. Sih. On the crack extension in plates under plane loading and transverse shear. *Journal of Basic Engineering*, 85:519–525, 1963.

N.A. Fleck and X.M. Qiu. The damage tolerance of elastic-brittle, two-dimensional isotropic lattices. *Journal of Mechanics and Physics of Solids*, 55:562–588, 2007.

C.W. Fowlkes. Fracture toughness tests of a rigid polyurethane foam. *International Journal of Fracture*, 10:99–108, 1974.

L.J. Gibson. Modeling the mechanical behaviour of cellular materials. *Mat. Sci. Eng. A*, 110:1–36, 1985.

L.J. Gibson and M.F. Ashby. *Cellular Solids, Structure and Properties.* Cambridge University Press, Cambridge, 2nd edition, 1997.

L. Gong, S. Kyriakides, and W.Y. Jang. Compressive response of open-cell foams. Part II: Initiation and evolution of crushing. *International Journal of Solids and Structures*, 42:1381–1399, 2005a.

L. Gong, S. Kyriakides, and W.Y. Jang. Compressive response of open-cell foams. Part I: Morphology and elastic properties. *International Journal of Solids and Structures*, 42:1355–1379, 2005b.

D.J. Green. Fabrication and mechanical properties of lightweight ceramics produced by sintering of hollow spheres. *Journal of American Ceramic Society*, 68:403–409, 1985.

S. Hallström and J.L. Grenestedt. Mixed mode fracture of cracks and wedge shaped notches in expanded PVC foam. *International Journal of Fracture*, 88:343–358, 1997.

J.S. Huang and L.J. Gibson. Fracture toughness of brittle foams. *Acta Metallurgica et Materialia*, 39:1627–1636, 1991.

M.A. Hussain, S.L. Pu, and J. Underwood. Strain energy release rate for a crack under combined mode I and mode II. In Paris P.C. and Irwin G.R., editors, *Fracture Analysis*, volume STP560, pages 2–28. ASTM, 1974.

H. Jin, W.-Y. Lu, S. Scheffel, T.D. Hinnerichs, and M.K. Neilsen. Full-field characterization of mechanical behavior of polyurethane foams. *International Journal of Solids and Structures*, 44:6930–6944, 2007.

M.E. Kabir, M.C. Saha, and S. Jeelani. Tensile and fracture behavior of polymer foams. *Materials Science and Engineering A*, 429:225–235, 2006.

A.H. Landrock. *Handbook of Plastic Foams. Types, Properties, Manufacture and Applications.* Noyes Publications, Park Ridge, 1995.

Q.M. Li and R.A.W. Mines. Strain measures for rigid crushable foam in uniaxial compression. *Strain*, 38:132–140, 2002.

Q.M. Li, R.A.W. Mines, and R.S. Birch. The crush behaviour of Rohacell-51WF structural foam. *International Journal of Solids and Structures*, 37:6321–6341, 2000.

E. Linul and L. Marsavina. Prediction of fracture toughness for open cell polyurethane foams by finite-element micromechanical analysis. *Iranian Polymer Journal*, 20:735–746, 2011.

F. Lipperman, M. Ryvkin, and M. Fuchs. Fracture toughness of two dimensional cellular material with periodic microstructure. *Int. J. Fracture*, 146:279–290, 2007.

S.K. Maiti, M.F. Ashby, and L.G. Gibson. Fracture toughness of some brittle cellular solids. *Scripta Metal.*, 18:213–217, 1984.

L. Marsavina and E. Linul. Fracture toughness of polyurethane foams. experiments versus micromechanical models. In *Proceedings of the 18th European Conference on Fracture*, Dresden, Germany, 2010.

L. Marsavina and T. Sadowski. Dynamic fracture toughness of polyurethane foam. *Polymer Testing*, 27:941–944, 2008.

L. Marsavina, E. Linul, T. Voiconi, and T. Sadowski. A comparison between dynamic and static fracture toughness of polyurethane foams. *Polymer Testing*, 32:673–680, 2008a.

L. Marsavina, T. Sadowski, D.M. Constantinescu, and R. Negru. Failure of polyurethane foams under different loading conditions. *Key Engineering Materials*, 385-387:205–208, 2008b.

L. Marsavina, T. Sadowski, D.M. Constantinescu, and R. Negru. Polyurethane foam behavior. Experiments versus modeling. *Key Engineering Materials*, 399:123–130, 2009.

L. Marsavina, E. Linul, T. Sadowski, D.M. Constantinescu, M. Knec, and D. Apostol. On fracture toughness of polyurethane foams. In *Proceedings of the 19th European Conference on Fracture*, Kazan, Russia, 2012.

L. Marsavina, D.M. Constantinescu, E. Linul, D.A. Apostol, T. Voiconi, and T. Sadowski. Refinements on fracture toughness of PUR foams. *Engineering Fracture Mechanics*, online: http://dx.doi.org/10.1016/j.engfracmech.2013.12.006, 2014.

A. McIntyre and G.E. Anderson. Fracture properties of a rigid PU foam over a range of densities. *Polymer*, 20:247–253, 1979.

N.J. Mills. *Polymer Foams Handbook.* Butterworth-Heinemann, Oxford, 2007.

N.J. Mills and P. Kang. The effect of water immersion on the fracture toughness of polystirene bead foams. *Journal of Cellular Plastics*, 30: 196–222, 1994.

R.A.W. Mines. Strain rate effects in crushable structural foams. *Applied Mechanics and Materials*, 7-8:231–236, 2007.

R.A.W. Mines and N. Jones. Approximate elastic-plastic analysis of the static and impact behaviour of polymer composite sandwich beams. *Composites*, 26:803–814, 1995.

R.A.W. Mines, C.M. Worral, and A.G. Gibson. The static and impact behaviour of polymer composite sandwich beams. *Composites*, 25:95–110, 1994.

P.M. Noury, R.A. Shenoi, and I. Sinclair. On mixed-mode fracture of PVC foam. *International Journal of Fracture*, 92:131–151, 1998.

P. Poapongsakorn and L. A. Carlsson. Fracture toughness of closed-cell PVC foam: Effect of loading configuration and cell size. *Composite Structures*, 102:1–8, 2013.

X.J. Ren and V.V. Silberschmidt. Numerical modelling of low-density cellular materials. *Computational Materials Science*, 43:65–74, 2008.

H.A. Richard. *Bruchvorhersagen bei überlagerter Normal- und Schubbeanspruchung von Rissen.* VDI-Verlag, 1985.

H.A. Richard, M. Fulland, and M. Sander. Theoretical crack path prediction. *Fatigue and Fracture of Engineering Materials and Structures*, 28: 3–12, 2005.

E.E. Saenz, L.A. Carlsson, and A. Karlsson. Characterization of fracture toughness (Gc) of PVC and PES foams. *Journal of Materials Science*, 46:3207–3215, 2011.

F. Saint-Michel, L. Chazeau, J.-Y. Cavaille, and E. Chabert. Mechanical properties of high density polyurethane foams: I. Effect of the density. *Composite Science and Technology*, 66:2700–2708, 2006.

G.C. Sih. Strain-energy-density factor applied to mixed mode crack problems. *International Journal of Fracture*, 10:305–321, 1974.

G.C. Sih and H. Liebowitz. Mathemetical theories of brittle fracture. In H. Liebowitz, editor, *Fracture - An advanced Treatise*, volume II. Academic Press, New York, Lindon, 1968.

A. Siriruk, D. Penumadu, and K. G. Thomas. Mixed mode fracture behavior of cellular foam cores used in sandwich structures. In *Proceedings of the 18th International Conference on Composite Materials*, Jeju Island, Korea, 2011.

B. Song, W. Chen, T. Yanagita, and D.J. Frew. Temperature effects on the dynamic compressive and failure behaviors of an epoxy syntactic foam. *Composite Structures*, 67:289–298, 2005.

P. Thiyagasundaram, J. Wang, B.V. Sankar, and N.K. Arakere. Fracture toughness of foams with tetrakaidecahedral unit cells using finite element based micromechanics. *Engineering Fracture Mechanics*, 78:1277–1288, 2011.

G.M. Viana and L.A. Carlsson. Mechanical properties and fracture characterisation of cross-linked PVC foams. *Journal of Sandwich Structures and Materials*, 4:91–113, 2002.

P. Viot, F. Beani, and J.-L. Lataillade. Polymeric foam behavior under dynamic compressive loading. *Journal of Material Science*, 40:5829–5837, 2005.

Y. Wang and A.M. Cuitino. Full-field measurements of heterogeneous deformation patterns on polymeric foams using digital image correlation. *International Journal of Solids and Structures*, 39:3777–3796, 2002.

I.M. Ward and J. Sweeney. *An Introduction to the Mechanical Properties of Solid Polymers.* Wiley, Chichester, 2nd edition, 2004.

W.E. Warren and A.M. Kraynik. Linear elastic behavior of a low-density Kelvin foam with open cells. *Trans. ASME. Journal of Applied Mechanics*, 64:787–795, 1997.

M.L. Williams. On the stress distribution at the base of a stationary crack. *Trans. ASME. Journal of Applied Mechanics*, 24:109–114, 1957.

D. Zenkert and J. Backlund. PVC sandwich core materials: mode I fracture toughness. *Composite Science Technology*, 34:225–242, 1989.

Analytical Methods of Predicting Performance of Composite Materials

L. N. McCartney
Materials Division, National Physical Laboratory
Teddington, Middlesex, TW11 0LW, UK

Abstract This paper is a collection of various analytical methods for predicting some of the properties of laminated fibre reinforced composites that can be used when designing composite laminates, or when validating numerical methods of estimating these properties. To begin, convenient methods are given to estimate the properties of undamaged single plies and undamaged symmetric laminates. Methods of predicting fracture in homogenized anisotropic materials are then described, which exploit some very useful properties of orthogonal polynomials. Example solutions are given which are compared with known accurate solutions. The problem is then considered of quantifying, using analytical methods, the dependence of the effective thermoelastic properties of a damaged laminate on the density of ply cracks in the 90^0 ply of a cross-ply laminate. Many very useful inter-relationships are given showing how most of the effective properties of damaged laminates depend on a single damage function. Some example predictions are given for a typical carbon fibre reinforced laminate. Finally, a model is described for predicting the progressive degradation of a unidirectional fibre reinforced composite that is degraded by an aggressive environment causing defect growth in the fibres and eventually the catastrophic failure of the composite. It is also shown how the time dependence of residual strength may be estimated. An example is given of a normalised failure/time curve, and some associated residual strength curves that can be the basis of design methods to avoid the failure of composites that will be exposed to aggressive environments.

1 Introduction

Engineers responsible for the design and maintenance of composite structures will usually be involved with some form of finite element analysis (FEA) so that the stress distributions within and the deformation of the

H. Altenbach, T. Sadowski (Eds.), *Failure and Damage Analysis of Advanced Materials*,
CISM International Centre for Mechanical Sciences
DOI 10.1007/978-3-7091-1835-1_4 © CISM Udine 2015

structure can be estimated. The objective is usually to identify the hot-spots where local stresses are raised to levels that are in danger of initiating damage and structural failure, or to ensure that the deformation during service does not exceed the design limits. A key requirement for finite element analysis is a knowledge of all the relevant materials property data such as the full range of elastic constants, and the thermal expansion coefficients of laminated composites. For an orthotropic material there are twelve independent thermoelastic constants whose values need to be known. Many of the properties cannot easily be measured in the laboratory, and their values are either guessed, or are estimated using a variety of predictive models assuming that fibre and matrix properties, or single ply properties are known.

A challenging objective for engineers is being able to design structural components so that damage and failure can be avoided during service. In structures the presence of stress hot-spots will generate localised damage that can grow progressively as a result of stress increase, or because of fatigue loading. The damage growth locally degrades the material properties leading to load transfer in structures, and to the threat of catastrophic failure. Dealing with the effects of localised damage is exceedingly difficult. One pragmatic approach is to try to design composite structures so that damage formation is avoided. This approach is particularly useful for structures that undergo cyclic loading as it will extend the fatigue damage initiation phase of the component life.

The topics to be considered in this paper describe various recommended methods for assessing the properties of composites, and the resistance of laminated composites to damage formation, by considering a range of analytical modelling methods that are built up from a knowledge of reinforcement and matrix geometry and properties, or from knowledge of the geometry and properties of individual plies. The topics focus on compact analytical formulae that can be used to provide quickly and reliably information on properties needed for FEA, and on damage resistance. In addition, the focus is also on theoretical developments which have not been published in the literature, other than through NPL Reports or Conference Proceedings.

2 Properties of an Undamaged Lamina and Laminates

2.1 Notation for Properties of a Single Lamina

First of all, it is assumed that the fibres of the lamina are aligned exactly in the axial direction forming what is known as a unidirectional fibre reinforced composite. Three different notations will now be introduced, which describe the properties of unidirectional fibre reinforced composites. The

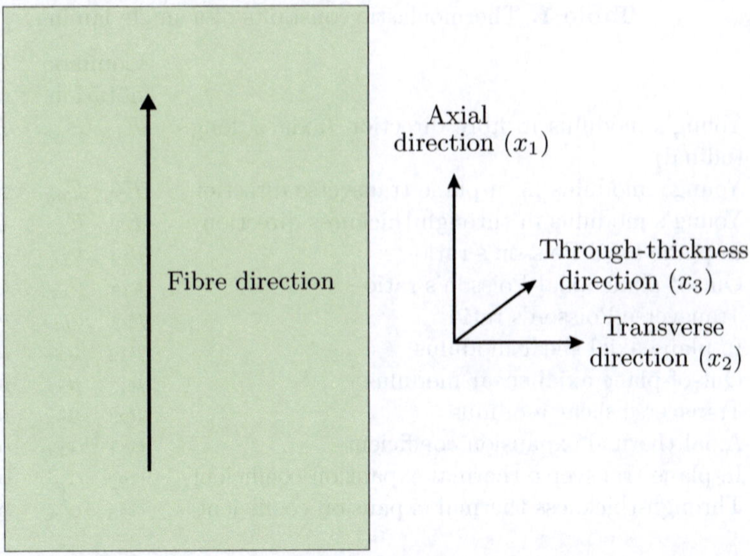

Figure 1. Diagram showing method of defining principal directions and coordinate system for a lamina where fibres are aligned in the axial direction

first two are common notations based on right-handed Cartesian coordinates (x_1, x_2, x_3) or (x, y, z) while the third, to be used in this paper, is a more compact notation that will enable an immediate physical interpretation of each property. Consider a single lamina of a composite material, as shown in Fig. 1, where the fibres are aligned in the x_1-direction (or x-direction). For this case the fibre direction also corresponds with the axial direction of the lamina, while the x_2-direction (or y-direction) corresponds with the in-plane transverse direction, and the x_3-direction (or z-direction) corresponds with the through-thickness direction.

The thermoelastic constants are best defined with respect to stress-strain relations (see Eqs. (1)-(3) below), and each material constant is described as shown in Table 1. The thermal expansion coefficients are associated with a temperature difference $\Delta T = T - T_0$, where T is the uniform temperature of the lamina and T_0 is the uniform reference temperature of the lamina at which all stresses and strains are zero, when the lamina is in an unloaded state.

Table 1. Thermoelastic constants of a single lamina

	Common notation		Compact notation
Young's modulus in fibre direction (axial / longitudinal)	E_{11}	E_{xx}	E_A
Young's modulus in in-plane transverse direction	E_{22}	E_{yy}	E_T
Young's modulus in through-thickness direction	E_{33}	E_{zz}	E_t
In-plane axial Poisson's ratio	ν_{12}	ν_{xy}	ν_A
Out-of-plane axial Poisson's ratio	ν_{13}	ν_{xz}	ν_a
Transverse Poisson's ratio	ν_{23}	ν_{yz}	ν_t
In-plane axial shear modulus	μ_{12}	μ_{xy}	μ_A
Out-of-plane axial shear modulus	μ_{13}	μ_{xz}	μ_a
Transverse shear modulus	μ_{23}	μ_{yz}	μ_t
Axial thermal expansion coefficient	α_{11}	α_{xx}	α_A
In-plane transverse thermal expansion coefficient	α_{22}	α_{yy}	α_T
Through-thickness thermal expansion coefficient	α_{33}	α_{zz}	α_t

2.2 Lamina Stress-Strain Relations

It is assumed that the loading of the lamina is such that the stress and strain distributions are uniform everywhere within the lamina. Such a stress and deformation state occurs when the external surfaces are subject to uniform applied tractions or linear displacements. The stress-strain relations referred to the coordinates (x_1, x_2, x_3) have the following orthotropic form:

$$
\begin{aligned}
\varepsilon_{11} &= \frac{1}{E_{11}}\sigma_{11} - \frac{\nu_{21}}{E_{22}}\sigma_{22} - \frac{\nu_{31}}{E_{33}}\sigma_{33} + \alpha_{11}\Delta T, & \varepsilon_{12} &= \frac{\sigma_{12}}{2\mu_{12}},\\
\varepsilon_{22} &= -\frac{\nu_{12}}{E_{11}}\sigma_{11} + \frac{1}{E_{22}}\sigma_{22} - \frac{\nu_{32}}{E_{33}}\sigma_{33} + \alpha_{22}\Delta T, & \varepsilon_{13} &= \frac{\sigma_{13}}{2\mu_{13}},\\
\varepsilon_{33} &= -\frac{\nu_{13}}{E_{11}}\sigma_{11} - \frac{\nu_{23}}{E_{22}}\sigma_{22} + \frac{1}{E_{33}}\sigma_{33} + \alpha_{33}\Delta T, & \varepsilon_{23} &= \frac{\sigma_{23}}{2\mu_{23}}
\end{aligned}
\tag{1}
$$

When referred to the coordinates (x, y, z), the stress-strain relations are written in the form

$$
\begin{aligned}
\varepsilon_{xx} &= \frac{1}{E_{xx}}\sigma_{xx} - \frac{\nu_{yx}}{E_{yy}}\sigma_{yy} - \frac{\nu_{zx}}{E_{zz}}\sigma_{zz} + \alpha_{xx}\Delta T, & \varepsilon_{xy} &= \frac{\sigma_{xy}}{2\mu_{xy}},\\
\varepsilon_{yy} &= -\frac{\nu_{xy}}{E_{xx}}\sigma_{xx} + \frac{1}{E_{yy}}\sigma_{yy} - \frac{\nu_{zy}}{E_{zz}}\sigma_{zz} + \alpha_{yy}\Delta T, & \varepsilon_{xz} &= \frac{\sigma_{xz}}{2\mu_{xz}},\\
\varepsilon_{zz} &= -\frac{\nu_{xz}}{E_{xx}}\sigma_{xx} - \frac{\nu_{yz}}{E_{yy}}\sigma_{yy} + \frac{1}{E_{zz}}\sigma_{zz} + \alpha_{zz}\Delta T, & \varepsilon_{yz} &= \frac{\sigma_{yz}}{2\mu_{yz}}
\end{aligned}
\tag{2}
$$

Using the compact notation to be used in this paper, involving only independent thermoelastic constants, the stress-strain relations are written:

$$\begin{aligned}
\varepsilon_A &= \frac{1}{E_A}\sigma_A - \frac{\nu_A}{E_A}\sigma_T - \frac{\nu_a}{E_A}\sigma_t + \alpha_A \Delta T, \quad \gamma_A &= \frac{\tau_A}{\mu_A}, \\
\varepsilon_T &= -\frac{\nu_A}{E_A}\sigma_A + \frac{1}{E_T}\sigma_T - \frac{\nu_t}{E_T}\sigma_t + \alpha_T \Delta T, \quad \gamma_a &= \frac{\tau_a}{\mu_a} \\
\varepsilon_t &= -\frac{\nu_a}{E_A}\sigma_A - \frac{\nu_t}{E_T}\sigma_T + \frac{1}{E_t}\sigma_t + \alpha_t \Delta T, \quad \gamma_t &= \frac{\tau_t}{\mu_t}
\end{aligned} \tag{3}$$

The subscripts A, T and t attached to stresses, strains and properties indicate parameters associated respectively with the axial, in-plane transverse and through-thickness directions of the lamina. It should be noted that the upper case subscripts A and T are associated only with in-plane directions and parameters, while the lower case subscripts are associated with the through-thickness direction and parameters. The above three sets of stress-strain relations are equivalent only if:

$$\begin{aligned}
&E_A = E_{11} = E_{xx}, \quad E_T = E_{22} = E_{yy}, \quad E_t = E_{33} = E_{zz}, \\
&\mu_A = \mu_{12} = \mu_{xy}, \quad \mu_a = \mu_{13} = \mu_{xz}, \quad \mu_t = \mu_{23} = \mu_{yz}, \\
&\alpha_A = \alpha_{11} = \alpha_{xx}, \quad \alpha_T = \alpha_{22} = \alpha_{yy}, \quad \alpha_t = \alpha_{33} = \alpha_{zz}
\end{aligned} \tag{4}$$

and

$$\begin{aligned}
&\frac{\nu_a}{E_A} = \frac{\nu_{13}}{E_{11}} = \frac{\nu_{31}}{E_{33}} = \frac{\nu_{xz}}{E_{xx}} = \frac{\nu_{zx}}{E_{yy}} \\
&\text{so that} \quad \nu_{13} = \nu_{xz} = \nu_a, \nu_{31} = \nu_{zx} = \nu_a \frac{E_t}{E_A}, \\
&\frac{\nu_t}{E_T} = \frac{\nu_{23}}{E_{22}} = \frac{\nu_{32}}{E_{33}} = \frac{\nu_{yz}}{E_{yy}} = \frac{\nu_{zy}}{E_{zz}} \\
&\text{so that} \quad \nu_{23} = \nu_{yz} = \nu_t, \nu_{32} = \nu_{zy} = \nu_t \frac{E_t}{E_T}, \\
&\frac{\nu_A}{E_A} = \frac{\nu_{21}}{E_{22}} = \frac{\nu_{12}}{E_{11}} = \frac{\nu_{yx}}{E_{yy}} = \frac{\nu_{xy}}{E_{xx}} \\
&\text{so that} \quad \nu_{12} = \nu_{xy} = \nu_A, \nu_{21} = \nu_{yx} = \nu_A \frac{E_T}{E_A}
\end{aligned} \tag{5}$$

It should be noted that

$$\begin{aligned}
&\gamma_A = 2\varepsilon_{12} = 2\varepsilon_{xy}, \quad \gamma_a = 2\varepsilon_{13} = 2\varepsilon_{xz}, \quad \gamma_t = 2\varepsilon_{23} = 2\varepsilon_{yz}, \\
&\tau_A = \sigma_{12} = \sigma_{xy}, \quad \tau_a = \sigma_{13} = \sigma_{xz}, \quad \tau_t = \sigma_{23} = \sigma_{yz}
\end{aligned} \tag{6}$$

The parameters $\gamma_A, \gamma_a, \gamma_t$ are known as engineering shear strains which are twice the corresponding shear strains introduced when using tensor notation. It should be noted that just 9 lamina elastic properties appear in the compact version (3) of the stress-strain relations whereas 12 elastic properties appear when using the conventional approach of the relations (1) or (2). The relations (5) indicate that the 9 lamina elastic properties of the compact notation are independent, adding a further reason for their use in this paper. The thermoelastic constants of individual plies in a laminate are usually assumed to be transverse isotropic so that

$$E_t = E_T, \nu_a = \nu_A, \mu_a = \mu_A, \alpha_t = \alpha_T \quad \text{and} \quad E_T = 2\mu_t\,(1+\nu_t)$$

The number of independent thermoelastic constants then reduces from 12 to 7.

2.3 Inverted Form of Lamina Stress-Strain Relations

The inverted form of the stress-strain relations (3) may be written

$$\begin{aligned} \sigma_A &= \bar{E}_A\varepsilon_A + \bar{\nu}_A\bar{E}_T\varepsilon_T + \bar{\nu}_a\bar{E}_t\varepsilon_t - \bar{E}_A\bar{\alpha}_A\Delta T, & \tau_A &= \mu_A\gamma_A, \\ \sigma_T &= \bar{\nu}_A\bar{E}_T\varepsilon_A + \bar{E}_T\varepsilon_T + \bar{\nu}_t\bar{E}_t\varepsilon_t - \bar{E}_T\bar{\alpha}_T\Delta T, & \tau_a &= \mu_a\gamma_a, \\ \sigma_t &= \bar{\nu}_a\bar{E}_t\varepsilon_A + \bar{\nu}_t\bar{E}_t\varepsilon_T + \bar{E}_t\varepsilon_t - \bar{E}_t\bar{\alpha}_t\Delta T, & \tau_t &= \mu_t\gamma_t, \end{aligned} \tag{7}$$

where

$$\begin{aligned} \bar{E}_A &= \frac{E_A}{\Lambda}\left(1-\nu_t^2\frac{E_t}{E_T}\right), \\ \bar{E}_T &= \frac{E_T}{\Lambda}\left(1-\nu_a^2\frac{E_t}{E_A}\right), \\ \bar{E}_t &= \frac{E_t}{\Lambda}\left(1-\nu_A^2\frac{E_T}{E_A}\right), \\ \bar{\nu}_A\bar{E}_T &= \frac{E_T}{\Lambda}\left(\nu_A+\nu_a\nu_t\frac{E_t}{E_T}\right), \\ \bar{\nu}_a\bar{E}_t &= \frac{E_t}{\Lambda}\left(\nu_a+\nu_t\nu_A\right), \\ \bar{\nu}_t\bar{E}_t &= \frac{E_t}{\Lambda}\left(\nu_t+\nu_a\nu_A\frac{E_T}{E_A}\right), \\ \Lambda &= 1-\nu_a^2\frac{E_t}{E_A}-\nu_t^2\frac{E_t}{E_T}-\nu_A^2\frac{E_T}{E_A}-2\nu_a\nu_t\nu_A\frac{E_t}{E_A} \end{aligned} \tag{8}$$

$$\begin{aligned} \bar{E}_A\bar{\alpha}_A &= \bar{E}_A\alpha_A + \bar{\nu}_A\bar{E}_T\alpha_T + \bar{\nu}_a\bar{E}_t\alpha_t, \\ \bar{E}_T\bar{\alpha}_T &= \bar{\nu}_A\bar{E}_T\alpha_A + \bar{E}_T\alpha_T + \bar{\nu}_t\bar{E}_t\alpha_t, \\ \bar{E}_t\bar{\alpha}_t &= \bar{\nu}_a\bar{E}_t\alpha_A + \bar{\nu}_t\bar{E}_t\alpha_T + \bar{E}_t\alpha_t \end{aligned} \tag{9}$$

It should be noted that the elastic coefficients of the three stress-strain relations (7) are symmetric as required.

2.4 Using the Contracted Notation for Tensors

The components of the stress and strain tensors are now assembled in column vectors so that

$$
\begin{bmatrix} \sigma_A \\ \sigma_T \\ \sigma_t \\ \tau_t \\ \tau_a \\ \tau_A \end{bmatrix} \equiv \begin{bmatrix} \sigma_1 \\ \sigma_2 \\ \sigma_3 \\ \sigma_4 \\ \sigma_5 \\ \sigma_6 \end{bmatrix} \equiv \begin{bmatrix} \sigma_{11} \\ \sigma_{22} \\ \sigma_{33} \\ \sigma_{23} \\ \sigma_{13} \\ \sigma_{12} \end{bmatrix} \equiv \begin{bmatrix} \sigma_{xx} \\ \sigma_{yy} \\ \sigma_{zz} \\ \sigma_{yz} \\ \sigma_{xz} \\ \sigma_{xy} \end{bmatrix},
$$

$$
\begin{bmatrix} \varepsilon_A \\ \varepsilon_T \\ \varepsilon_t \\ \gamma_t \\ \gamma_a \\ \gamma_A \end{bmatrix} \equiv \begin{bmatrix} \varepsilon_1 \\ \varepsilon_2 \\ \varepsilon_3 \\ \varepsilon_4 \\ \varepsilon_5 \\ \varepsilon_6 \end{bmatrix} \equiv \begin{bmatrix} \varepsilon_{11} \\ \varepsilon_{22} \\ \varepsilon_{33} \\ 2\varepsilon_{23} \\ 2\varepsilon_{13} \\ 2\varepsilon_{12} \end{bmatrix} \equiv \begin{bmatrix} \varepsilon_{xx} \\ \varepsilon_{yy} \\ \varepsilon_{zz} \\ 2\varepsilon_{yz} \\ 2\varepsilon_{xz} \\ 2\varepsilon_{xy} \end{bmatrix} \tag{10}
$$

General linear elastic stress-strain relations, including thermal expansion terms, have the contracted matrix form

$$
\begin{bmatrix} \sigma_A \\ \sigma_T \\ \sigma_t \\ \tau_t \\ \tau_a \\ \tau_A \end{bmatrix} = \begin{bmatrix} C_{11} & C_{12} & C_{13} & C_{14} & C_{15} & C_{16} \\ C_{21} & C_{22} & C_{23} & C_{24} & C_{25} & C_{26} \\ C_{31} & C_{32} & C_{33} & C_{34} & C_{35} & C_{36} \\ C_{41} & C_{42} & C_{43} & C_{44} & C_{45} & C_{46} \\ C_{51} & C_{52} & C_{53} & C_{54} & C_{55} & C_{56} \\ C_{61} & C_{62} & C_{63} & C_{64} & C_{65} & C_{66} \end{bmatrix} \begin{bmatrix} \varepsilon_A \\ \varepsilon_T \\ \varepsilon_t \\ \gamma_t \\ \gamma_a \\ \gamma_A \end{bmatrix} - \begin{bmatrix} U_1 \\ U_2 \\ U_3 \\ U_4 \\ U_5 \\ U_6 \end{bmatrix} \Delta T, \tag{11}
$$

where C_{IJ} are elastic constants which are components of the second order matrix $\boldsymbol{C}$ and where U_I are thermal expansion constants which are components of the vector $\boldsymbol{U}$, the indexes I and J ranging from 1 to 6. For orthotropic materials the stress-strain relations have the simpler matrix form

$$
\begin{bmatrix} \sigma_A \\ \sigma_T \\ \sigma_t \\ \tau_t \\ \tau_a \\ \tau_A \end{bmatrix} = \begin{bmatrix} C_{11} & C_{12} & C_{13} & 0 & 0 & 0 \\ C_{21} & C_{22} & C_{23} & 0 & 0 & 0 \\ C_{31} & C_{32} & C_{33} & 0 & 0 & 0 \\ 0 & 0 & 0 & C_{44} & 0 & 0 \\ 0 & 0 & 0 & 0 & C_{55} & 0 \\ 0 & 0 & 0 & 0 & 0 & C_{66} \end{bmatrix} \begin{bmatrix} \varepsilon_A \\ \varepsilon_T \\ \varepsilon_t \\ \gamma_t \\ \gamma_a \\ \gamma_A \end{bmatrix} - \begin{bmatrix} U_1 \\ U_2 \\ U_3 \\ 0 \\ 0 \\ 0 \end{bmatrix} \Delta T \tag{12}
$$

By comparing (12) with (7) using (8) and (9), it follows that the non-zero components of the $\boldsymbol{C}$ matrix are related to the elastic constants defined by the stress-strain relations (3) as follows

$$
\begin{aligned}
&C_{11} = \frac{E_A}{\Lambda}\left(1 - \nu_t^2 \frac{E_t}{E_T}\right) = \bar{E}_A, \quad C_{12} = \frac{E_T}{\Lambda}\left(\nu_A + \nu_a \nu_t \frac{E_t}{E_T}\right) = \bar{\nu}_A \bar{E}_T, \\
&C_{13} = \frac{E_t}{\Lambda}\left(\nu_a + \nu_t \nu_A\right) = \bar{\nu}_a \bar{E}_t, \quad C_{21} = \frac{E_T}{\Lambda}\left(\nu_A + \nu_a \nu_t \frac{E_t}{E_T}\right) = \bar{\nu}_A \bar{E}_T, \\
&C_{22} = \frac{E_T}{\Lambda}\left(1 - \nu_a^2 \frac{E_t}{E_A}\right) = \bar{E}_T, \quad C_{23} = \frac{E_t}{\Lambda}\left(\nu_t + \nu_a \nu_A \frac{E_T}{E_A}\right) = \bar{\nu}_t \bar{E}_t, \\
&C_{31} = \frac{E_t}{\Lambda}\left(\nu_a + \nu_t \nu_A\right) = \bar{\nu}_a \bar{E}_t, \quad C_{32} = \frac{E_t}{\Lambda}\left(\nu_t + \nu_a \nu_A \frac{E_T}{E_A}\right) = \bar{\nu}_t \bar{E}_t, \\
&C_{33} = \frac{E_t}{\Lambda}\left(1 - \nu_A^2 \frac{E_T}{E_A}\right) = \bar{E}_t, \quad C_{44} = \mu_t, C_{55} = \mu_a, C_{66} = \mu_A
\end{aligned} \tag{13}
$$

with

$$
\Lambda = 1 - \nu_a^2 \frac{E_t}{E_A} - \nu_t^2 \frac{E_t}{E_T} - \nu_A^2 \frac{E_T}{E_A} - 2\nu_a \nu_t \nu_A \frac{E_t}{E_A}
$$

The non-zero components of the $\boldsymbol{U}$ vector are related to the thermoelastic constants defined by the stress-strain relations (3) as follows

$$
U_1 = \bar{E}_A \bar{\alpha}_A, \quad U_2 = \bar{E}_T \bar{\alpha}_T, \quad U_3 = \bar{E}_t \bar{\alpha}_t \tag{14}
$$

The inverse matrix form of (12) is of the form

$$
\begin{bmatrix} \varepsilon_A \\ \varepsilon_T \\ \varepsilon_t \\ \gamma_t \\ \gamma_a \\ \gamma_A \end{bmatrix} = \begin{bmatrix} S_{11} & S_{12} & S_{13} & 0 & 0 & 0 \\ S_{21} & S_{22} & S_{23} & 0 & 0 & 0 \\ S_{31} & S_{32} & S_{33} & 0 & 0 & \\ 0 & 0 & 0 & S_{44} & 0 & 0 \\ 0 & 0 & 0 & 0 & S_{55} & 0 \\ 0 & 0 & 0 & 0 & 0 & S_{66} \end{bmatrix} \begin{bmatrix} \sigma_A \\ \sigma_T \\ \sigma_t \\ \tau_t \\ \tau_a \\ \tau_A \end{bmatrix} + \begin{bmatrix} \alpha_A \\ \alpha_T \\ \alpha_t \\ 0 \\ 0 \\ 0 \end{bmatrix} \Delta T, \tag{15}
$$

and a comparison with (3) shows that

$$
\begin{aligned}
S_{11} &= \frac{1}{E_A}, & S_{12} &= -\frac{\nu_A}{E_A}, & S_{13} &= -\frac{\nu_a}{E_A}, \\
S_{21} &= -\frac{\nu_A}{E_A}, & S_{22} &= \frac{1}{E_T}, & S_{23} &= -\frac{\nu_t}{E_T}, \\
S_{31} &= -\frac{\nu_a}{E_A}, & S_{32} &= -\frac{\nu_t}{E_T}, & S_{33} &= \frac{1}{E_t}, \\
S_{44} &= \frac{1}{\mu_t}, & S_{55} &= \frac{1}{\mu_a}, & S_{66} &= \frac{1}{\mu_A}
\end{aligned} \tag{16}
$$

It is often useful to be able to estimate lamina properties from those of the reinforcements and matrix that have been used to make the composite. For multi-phase composites, where the matrix is reinforced with N different types of fibre such that the fibre volume fraction of the ith fibre is denoted by V_{f}^i and that of the matrix by V_{m}, effective properties denoted with the superscript eff may be estimated using the following formulae for transverse isotropic materials derived using Maxwell's methodology (Maxwell, 1873; McCartney and Kelly, 2008; McCartney, 2010) (see also reference Hashin (1983))

$$\frac{1}{\kappa_T^{\text{eff}} + \kappa_T^{\text{m}}} = \sum_{i=1}^{N} \frac{V_{\text{f}}^i}{\kappa_T^{\text{f}(i)} + \kappa_T^{\text{m}}} + \frac{V_{\text{m}}}{\kappa_T^{\text{m}} + \kappa_T^{\text{m}}}, \tag{17}$$

$$\frac{1}{k_T^{\text{eff}} + \mu_T^{\text{m}}} = \sum_{i=1}^{N} \frac{V_{\text{f}}^i}{k_T^i + \mu_T^{\text{m}}} + \frac{V_{\text{m}}}{k_T^{\text{m}} + \mu_T^{\text{m}}}, \tag{18}$$

$$\nu_A^{\text{eff}} = \frac{\displaystyle\sum_{i=1}^{N} \frac{V_{\text{f}}^i k_T^i \nu_A^i}{k_T^i + \mu_T^{\text{m}}} + \frac{V_{\text{m}} k_T^{\text{m}} \nu_A^{\text{m}}}{k_T^{\text{m}} + \mu_T^{\text{m}}}}{\displaystyle\sum_{i=1}^{N} \frac{V_f^i k_T^i}{k_T^i + \mu_T^m} + \frac{V_m k_T^m}{k_T^m + \mu_T^m}}, \tag{19}$$

$$\begin{aligned} E_A^{\text{eff}} + \frac{4k_T^{\text{eff}} \left(\nu_A^{\text{eff}}\right)^2 \mu_{\text{m}}}{k_T^{\text{eff}} + \mu_{\text{m}}} &= \sum_{i=1}^{N} V_{\text{f}}^i \left(E_A^{\text{f}(i)} + \frac{4k_T^{\text{f}(i)} \left(\nu_A^{\text{f}(i)}\right)^2 \mu_{\text{m}}}{k_T^{\text{f}(i)} + \mu_{\text{m}}} \right) \\ &+ V_{\text{m}} \left(E_{\text{m}} + \frac{4k_T^{\text{m}} \nu_{\text{m}}^2 \mu_{\text{m}}}{k_T^{\text{m}} + \mu_{\text{m}}} \right), \end{aligned} \tag{20}$$

$$\frac{1}{\mu_A^{\text{eff}} + \mu_A^{\text{m}}} = \sum_{i=1}^{N} \frac{V_{\text{f}}^i}{\mu_A^i + \mu_A^{\text{m}}} + \frac{V_{\text{m}}}{\mu_A^{\text{m}} + \mu_A^{\text{m}}}, \tag{21}$$

$$\frac{1}{\mu_T^{\text{eff}} + \mu_m^*} = \sum_{i=1}^{N} \frac{V_{\text{f}}^i}{\mu_T^i + \mu_m^*} + \frac{V_{\text{m}}}{\mu_T^m + \mu_m^*}, \quad \text{where} \quad \mu_{\text{m}}^* = \frac{k_T^{\text{m}} \mu_T^{\text{m}}}{k_T^{\text{m}} + 2\mu_T^{\text{m}}}, \tag{22}$$

$$\begin{aligned} E_A^{\text{eff}} \alpha_A^{\text{eff}} + \frac{4\nu_A^{\text{eff}} (\alpha_T^{\text{eff}} + \nu_A^{\text{eff}} \alpha_A^{\text{eff}})}{\dfrac{1}{k_T^{\text{eff}}} + \dfrac{1}{\mu_{\text{m}}}} &= V_{\text{f}} E_A^{\text{f}} \alpha_A^{\text{f}} + V_{\text{m}} E_{\text{m}} \alpha_{\text{m}} \\ &+ V_{\text{f}} \frac{4\nu_A^{\text{f}} (\alpha_T^{\text{f}} + \nu_A^{\text{f}} \alpha_A^{\text{f}})}{\dfrac{1}{k_T^{\text{f}}} + \dfrac{1}{\mu_{\text{m}}}} + V_{\text{m}} \frac{4\nu_{\text{m}} (\alpha_{\text{m}} + \nu_{\text{m}} \alpha_{\text{m}})}{\dfrac{1}{k_T^{\text{m}}} + \dfrac{1}{\mu_{\text{m}}}}, \end{aligned} \tag{23}$$

$$\alpha_T^{\mathrm{eff}} + \nu_A^{\mathrm{eff}}\alpha_A^{\mathrm{eff}} = \frac{\displaystyle\sum_{i=1}^{N} \frac{V_{\mathrm{f}}^i k_T^i \left(\alpha_T^i + \nu_A^i \alpha_A^i\right)}{k_T^i + \mu_T^m} + \frac{V_{\mathrm{m}} k_T^{\mathrm{m}} \left(\alpha_T^{\mathrm{m}} + \nu_A^{\mathrm{m}} \alpha_A^{\mathrm{m}}\right)}{k_T^{\mathrm{m}} + \mu_T^{\mathrm{m}}}}{\displaystyle\sum_{i=1}^{N} \frac{V_{\mathrm{f}}^i k_T^i}{k_T^i + \mu_T^{\mathrm{m}}} + \frac{V_{\mathrm{m}} k_T^{\mathrm{m}}}{k_T^{\mathrm{m}} + \mu_T^{\mathrm{m}}}}, \qquad (24)$$

where $k_T^{\mathrm{m}} = k_{\mathrm{m}} + \frac{1}{3}\mu_{\mathrm{m}}$. Since

$$\frac{1}{E_T^{\mathrm{eff}}} = \frac{1}{4\mu_T^{\mathrm{eff}}} + \frac{1}{4k_T^{\mathrm{eff}}} + \frac{(\nu_A^{\mathrm{eff}})^2}{E_A^{\mathrm{eff}}}, \quad \frac{\nu_T^{\mathrm{eff}}}{E_T^{\mathrm{eff}}} = \frac{1}{4\mu_T^{\mathrm{eff}}} - \frac{1}{4k_T^{\mathrm{eff}}} - \frac{4(\nu_A^{\mathrm{eff}})^2}{E_A^{\mathrm{eff}}} \qquad (25)$$

It is possible to use the relations (17)-(24) to estimate values of E_T^{eff} and ν_T^{eff}.

2.5 Thermoelastic Constants for Angled Laminae

It is required now to consider how these properties are used to derive those of angled plies. One very important parameter that must be clearly defined is the angle ϕ defining the fibre directions in an angled lamina. Figure 2 illustrates the definition that is used, and it should be noted that this angle differs in sign from that used in some previous publications by the author. The change of sign affects only the signs of shear coupling parameters.

Figure 2 also defines the three orthogonal principal directions of the laminate, and associates these directions with a right handed system (x_1, x_2, x_3) of Cartesian coordinates defining a set of global axes. The angle ϕ defining the fibre direction of the lamina is measured from the global x_1-axis in a clockwise direction when viewing from a point situated on the negative part of the global x_3-axis.

Following rotation of the lamina the stress-strain equations, including thermal expansion terms, relating global strain components to global stress

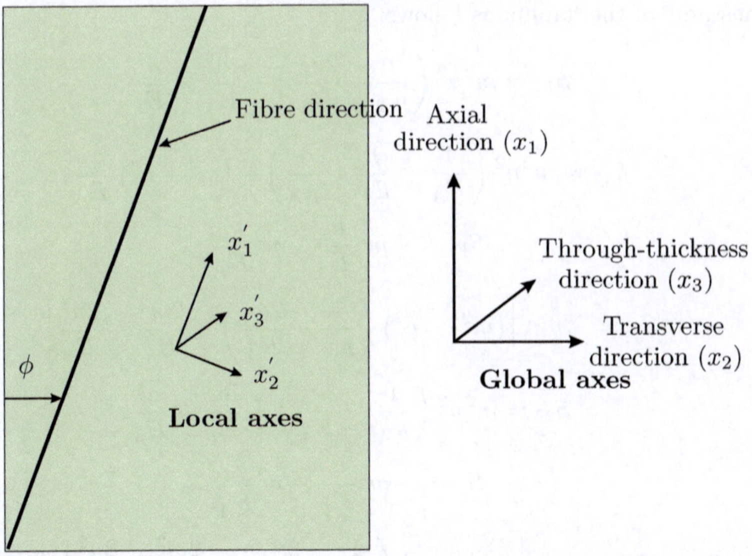

Figure 2. Diagram showing method of defining principal directions and coordinate system for an angled lamina, and the angle ϕ specifying the fibre direction

components are given, using matrix notation, by

$$\begin{bmatrix} \varepsilon_A \\ \varepsilon_T \\ \varepsilon_t \\ \gamma_t \\ \gamma_a \\ \gamma_A \end{bmatrix} \equiv \begin{bmatrix} \varepsilon_{11} \\ \varepsilon_{22} \\ \varepsilon_{33} \\ 2\varepsilon_{23} \\ 2\varepsilon_{13} \\ 2\varepsilon_{12} \end{bmatrix}$$

$$= \begin{bmatrix} S_{11} & S_{12} & S_{13} & 0 & 0 & S_{16} \\ S_{12} & S_{22} & S_{23} & 0 & 0 & S_{26} \\ S_{13} & S_{23} & S_{33} & 0 & 0 & S_{36} \\ 0 & 0 & 0 & S_{44} & S_{45} & 0 \\ 0 & 0 & 0 & S_{45} & S_{55} & 0 \\ S_{16} & S_{26} & S_{36} & 0 & 0 & S_{66} \end{bmatrix} \begin{bmatrix} \sigma_A \\ \sigma_T \\ \sigma_t \\ \tau_t \\ \tau_a \\ \tau_A \end{bmatrix} + \begin{bmatrix} V_1 \\ V_2 \\ V_3 \\ 0 \\ 0 \\ V_6 \end{bmatrix} \Delta T, \tag{26}$$

where on setting $m = \cos\phi$ and $n = \sin\phi$ it can be shown that the coefficients in these stress-strain relations are related to the thermoelastic constants of the lamina as follows:

$$S_{11} = m^2n^2\left(\frac{1}{\mu_A} - \frac{2\nu_A}{E_A}\right) + \frac{m^4}{E_A} + \frac{n^4}{E_T}, \tag{27}$$

$$S_{12} = m^2n^2\left(\frac{1}{E_A} + \frac{1}{E_T} - \frac{1}{\mu_A}\right) - \left(m^4 + n^4\right)\frac{\nu_A}{E_A}, \tag{28}$$

$$S_{13} = -\,m^2\frac{\nu_a}{E_A} - n^2\frac{\nu_t}{E_T}, \tag{29}$$

$$S_{16} = -\,mn\left[\left(m^2 - n^2\right)\left(\frac{1}{\mu_A} - \frac{2\nu_A}{E_A}\right) - \frac{2m^2}{E_A} + \frac{2n^2}{E_T}\right], \tag{30}$$

$$S_{22} = m^2n^2\left(\frac{1}{\mu_A} - \frac{2\nu_A}{E_A}\right) + \frac{m^4}{E_T} + \frac{n^4}{E_A}, \tag{31}$$

$$S_{23} = -m^2\frac{\nu_t}{E_T} - n^2\frac{\nu_a}{E_A}, \tag{32}$$

$$S_{26} = mn\left[\left(m^2 - n^2\right)\left(\frac{1}{\mu_A} - \frac{2\nu_A}{E_A}\right) - \frac{2m^2}{E_T} + \frac{2n^2}{E_A}\right], \tag{33}$$

$$S_{33} = \frac{1}{E_t}, \tag{34}$$

$$S_{36} = 2mn\left(\frac{\nu_t}{E_T} - \frac{\nu_a}{E_A}\right), \tag{35}$$

$$S_{44} = \frac{m^2}{\mu_t} + \frac{n^2}{\mu_a}, \tag{36}$$

$$S_{45} = mn\left(\frac{1}{\mu_a} - \frac{1}{\mu_t}\right), \tag{37}$$

$$S_{55} = \frac{m^2}{\mu_a} + \frac{n^2}{\mu_t}, \tag{38}$$

$$S_{66} = 4m^2n^2\left(\frac{1}{E_A} + \frac{1}{E_T} + \frac{2\nu_A}{E_A}\right) + \frac{\left(m^2 - n^2\right)^2}{\mu_A}, \tag{39}$$

$$\begin{aligned} V_1 &= m^2\alpha_A + n^2\alpha_T, \\ V_2 &= m^2\alpha_T + n^2\alpha_A, \\ V_3 &= \alpha_t, V_4 = 0, V_5 = 0 \\ V_6 &= 2mn\left(\alpha_A - \alpha_T\right) \end{aligned} \tag{40}$$

It should be noted that the expressions (26)-(40) reduce to (15) and (16) when $\phi = 0$ so that $m = 1$ and $n = 0$.

2.6 Inverse Approach

The inverted form for the stress-strain relations of an angled ply is given by

$$
\begin{bmatrix} \sigma_A \\ \sigma_T \\ \sigma_t \\ \tau_t \\ \tau_a \\ \tau_A \end{bmatrix} \equiv \begin{bmatrix} \sigma_{11} \\ \sigma_{22} \\ \sigma_{33} \\ \sigma_{23} \\ \sigma_{13} \\ \sigma_{12} \end{bmatrix}
$$

$$
= \begin{bmatrix} C_{11} & C_{12} & C_{13} & 0 & 0 & C_{16} \\ C_{12} & C_{22} & C_{23} & 0 & 0 & C_{26} \\ C_{13} & C_{23} & C_{33} & 0 & 0 & C_{36} \\ 0 & 0 & 0 & C_{44} & C_{45} & 0 \\ 0 & 0 & 0 & C_{45} & C_{55} & 0 \\ C_{16} & C_{26} & C_{36} & 0 & 0 & C_{66} \end{bmatrix} \begin{bmatrix} \varepsilon_A \\ \varepsilon_T \\ \varepsilon_t \\ \gamma_t \\ \gamma_a \\ \gamma_A \end{bmatrix} - \begin{bmatrix} U_1 \\ U_2 \\ U_3 \\ 0 \\ 0 \\ U_6 \end{bmatrix} \Delta T, \tag{41}
$$

where

$$
\begin{aligned}
C_{11} &= m^4 \bar{E}_A + n^4 \bar{E}_T + 2m^2 n^2 \left(\bar{\nu}_A \bar{E}_T + 2\mu_A\right), \\
C_{12} &= \left(m^4 + n^4\right) \bar{\nu}_A \bar{E}_T + m^2 n^2 \left(\bar{E}_A + \bar{E}_T - 4\mu_A\right), \\
C_{13} &= \left(m^2 \bar{\nu}_a + n^2 \bar{\nu}_t\right) \bar{E}_t, \\
C_{16} &= mn \left[m^2 \bar{E}_A - n^2 \bar{E}_T - \left(m^2 - n^2\right)\left(\bar{\nu}_A \bar{E}_T + 2\mu_A\right)\right], \\
C_{22} &= m^4 \bar{E}_T + n^4 \bar{E}_A + 2m^2 n^2 \left(\bar{\nu}_A \bar{E}_T + 2\mu_A\right), \\
C_{23} &= \left(m^2 \bar{\nu}_t + n^2 \bar{\nu}_a\right) \bar{E}_t, \\
C_{26} &= mn \left[n^2 \bar{E}_A - m^2 \bar{E}_T + \left(m^2 - n^2\right)\left(\bar{\nu}_A \bar{E}_T + 2\mu_A\right)\right], \\
C_{33} &= \bar{E}_t \\
C_{36} &= mn \left(\bar{\nu}_a - \bar{\nu}_t\right) \bar{E}_t, \\
C_{44} &= m^2 \mu_t + n^2 \mu_a, \\
C_{45} &= mn \left(\mu_a - \mu_t\right), \\
C_{55} &= m^2 \mu_a + n^2 \mu_t, \\
C_{66} &= m^2 n^2 \left(\bar{E}_A + \bar{E}_T - 2\bar{\nu}_A \bar{E}_T\right) + \left(m^2 - n^2\right)^2 \mu_A,
\end{aligned} \tag{42}
$$

and

$$
\begin{aligned}
U_1 &= m^2 \bar{E}_A \bar{\alpha}_A + n^2 \bar{E}_T \bar{\alpha}_T, \\
U_2 &= n^2 \bar{E}_A \bar{\alpha}_A + m^2 \bar{E}_T \bar{\alpha}_T, \\
U_3 &= \bar{E}_t \bar{\alpha}_t, \\
U_6 &= mn \left(\bar{E}_A \bar{\alpha}_A - \bar{E}_T \bar{\alpha}_T\right)
\end{aligned} \tag{43}
$$

It should be noted that (42) and (43) reduce to (13)-(14) when $\phi = 0$ so that $m = 1$ and $n = 0$.

2.7 Shear Coupling Parameters and Reduced Stress-Strain Relations

The final task of Sect. 2 is to introduce the concept of shear coupling and to define the thermoelastic constants that characterise the phenomenon. It can be shown that

$$
\begin{aligned}
\varepsilon_A &= \frac{1}{E_A^{(P)}}\sigma_A - \frac{\nu_A^{(P)}}{E_A^{(P)}}\sigma_T - \frac{\nu_a^{(P)}}{E_A^{(P)}}\sigma_t - \frac{\lambda_A^{(P)}}{\mu_A^{(P)}}\tau_A + \alpha_A^{(P)}\Delta T, \\
\varepsilon_T &= -\frac{\nu_A^{(P)}}{E_A^{(P)}}\sigma_A + \frac{1}{E_T^{(P)}}\sigma_T - \frac{\nu_t^{(P)}}{E_T^{(P)}}\sigma_t - \frac{\lambda_T^{(P)}}{\mu_A^{(P)}}\tau_A + \alpha_T^{(P)}\Delta T, \\
\varepsilon_t &= -\frac{\nu_a^{(P)}}{E_A^{(P)}}\sigma_A - \frac{\nu_t^{(P)}}{E_T^{(P)}}\sigma_T + \frac{1}{E_t^{(P)}}\sigma_t - \frac{\lambda_t^{(P)}}{\mu_A^{(P)}}\tau_A + \alpha_t^{(P)}\Delta T, \\
\gamma_A &= -\frac{\lambda_A^{(P)}}{\mu_A^{(P)}}\sigma_A - \frac{\lambda_T^{(P)}}{\mu_A^{(P)}}\sigma_T - \frac{\lambda_t^{(P)}}{\mu_A^{(P)}}\sigma_t + \frac{1}{\mu_A^{(P)}}\tau_A + \alpha_S^{(P)}\Delta T,
\end{aligned}
\tag{44}
$$

where the superscript P denotes that the thermoelastic constant refers to an angled ply. By comparing (26) and (44) it is clear that

$$
\begin{aligned}
&S_{11} = \frac{1}{E_A^{(P)}}, \quad S_{12} = -\frac{\nu_A^{(P)}}{E_A^{(P)}}, \quad S_{13} = -\frac{\nu_a^{(P)}}{E_A^{(P)}}, \quad S_{16} = -\frac{\lambda_A^{(P)}}{\mu_A^{(P)}}, \\
&S_{22} = \frac{1}{E_T^{(P)}}, \quad S_{23} = -\frac{\nu_t^{(P)}}{E_T^{(P)}}, \quad S_{26} = -\frac{\lambda_T^{(P)}}{\mu_A^{(P)}}, \quad S_{33} = \frac{1}{E_t^{(P)}}, \\
&S_{36} = -\frac{\lambda_t^{(P)}}{\mu_A^{(P)}}, \quad S_{66} = \frac{1}{\mu_A^{(P)}}, \\
&V_1 = \alpha_A^{(P)}, \quad V_2 = \alpha_T^{(P)}, \quad V_3 = \alpha_t^{(P)}, \quad V_6 = \alpha_S^{(P)}
\end{aligned}
\tag{45}
$$

The parameters $\lambda_A, \lambda_T, \lambda_t$ are dimensionless shear coupling properties as they characterise the coupling of the shear stress τ_A to the non-shear strains $\varepsilon_A, \varepsilon_T$ and ε_t. The parameter α_S characterises a shear deformation response to temperature changes. When the fibres are aligned in the axial and in-plane transverse directions all four parameters have zero values. It should be noted that the signs of the shear coupling parameters $\lambda_A, \lambda_T, \lambda_t$ and the expansion coefficient α_S depend upon the sign of the orientation angle ϕ, indicating why it is essential to define exactly how this angle is defined.

From $(44)_4$

$$
\tau_A = \lambda_A^{(P)}\sigma_A + \lambda_T^{(P)}\sigma_T + \lambda_t^{(P)}\sigma_t + \mu_A^{(P)}\gamma_A - \mu_A^{(P)}\alpha_S^{(P)}\Delta T, \tag{46}
$$

and on substituting in the remaining relations of (44)

$$\begin{aligned}
\tilde{\varepsilon}_A \equiv \varepsilon_A + \lambda_A^{(P)} \gamma_A &= \frac{1}{\tilde{E}_A^{(P)}}\sigma_A - \frac{\tilde{\nu}_A^{(P)}}{\tilde{E}_A^{(P)}}\sigma_T - \frac{\tilde{\nu}_a^{(P)}}{\tilde{E}_A^{(P)}}\sigma_t + \tilde{\alpha}_A^{(P)}\Delta T,\\
\tilde{\varepsilon}_T \equiv \varepsilon_T + \lambda_T^{(P)} \gamma_A &= -\frac{\tilde{\nu}_A^{(P)}}{\tilde{E}_A^{(P)}}\sigma_A + \frac{1}{\tilde{E}_T^{(P)}}\sigma_T - \frac{\tilde{\nu}_t^{(P)}}{\tilde{E}_T^{(P)}}\sigma_t + \tilde{\alpha}_T^{(P)}\Delta T, \qquad (47)\\
\tilde{\varepsilon}_t \equiv \varepsilon_t + \lambda_t^{(P)} \gamma_A &= -\frac{\tilde{\nu}_a^{(P)}}{\tilde{E}_A^{(P)}}\sigma_A - \frac{\tilde{\nu}_t^{(P)}}{\tilde{E}_T^{(P)}}\sigma_T + \frac{1}{\tilde{E}_t^{(P)}}\sigma_t + \tilde{\alpha}_t^{(P)}\Delta T,
\end{aligned}$$

where

$$\begin{aligned}
\frac{1}{\tilde{E}_A^{(P)}} &= \frac{1}{E_A^{(P)}} - \left(\lambda_A^{(P)}\right)^2 \frac{1}{\mu_A^{(P)}}, & \frac{1}{\tilde{E}_T^{(P)}} &= \frac{1}{E_T^{(P)}} - \left(\lambda_T^{(P)}\right)^2 \frac{1}{\mu_A^{(P)}},\\
\frac{1}{\tilde{E}_t^{(P)}} &= \frac{1}{E_t^{(P)}} - \left(\lambda_t^{(P)}\right)^2 \frac{1}{\mu_A^{(P)}}, & \frac{\tilde{\nu}_t^{(P)}}{\tilde{E}_T^{(P)}} &= \frac{\nu_t^{(P)}}{E_T^{(P)}} + \frac{\lambda_t^{(P)}\lambda_T^{(P)}}{\mu_A^{(P)}},\\
\frac{\tilde{\nu}_A^{(P)}}{\tilde{E}_A^{(P)}} &= \frac{\nu_A^{(P)}}{E_A^{(P)}} + \frac{\lambda_A^{(P)}\lambda_T^{(P)}}{\mu_A^{(P)}}, & \frac{\tilde{\nu}_a^{(P)}}{\tilde{E}_A^{(P)}} &= \frac{\nu_a^{(P)}}{E_A^{(P)}} + \frac{\lambda_t^{(P)}\lambda_A^{(P)}}{\mu_A^{(P)}}, \qquad (48)\\
\tilde{\alpha}_A^{(P)} &= \alpha_A^{(P)} + \lambda_A^{(P)}\alpha_S^{(P)}, & \tilde{\alpha}_T^{(P)} &= \alpha_T^{(P)} + \lambda_T^{(P)}\alpha_S^{(P)},\\
\tilde{\alpha}_t^{(P)} &= \alpha_t^{(P)} + \lambda_t^{(P)}\alpha_S^{(P)}
\end{aligned}$$

The relations (47) are known as the reduced stress-strain relations for the angled lamina as they have exactly the same form as three of the stress-strain relations (3) which apply when $\phi = 0$.

2.8 Mixed Form of Stress-Strain Relations

The final steps of this Sect. 2 are to manipulate the stress-strain equations (46) and (47) so that they are in a form that will be useful when considering the effective properties of laminates. The objective is to express stresses and strains in terms of the parameters $\varepsilon_A, \varepsilon_T, \gamma_A, \sigma_T$ and ΔT which will have the same values in all plies of a laminate. It can be shown that

$$\begin{aligned}
\sigma_A &= \Omega_{11}^{(P)}\varepsilon_A + \Omega_{12}^{(P)}\varepsilon_T + \Omega_{13}^{(P)}\sigma_t + \Omega_{16}^{(P)}\gamma_A - \omega_1^{(P)}\Delta T,\\
\sigma_T &= \Omega_{12}^{(P)}\varepsilon_A + \Omega_{22}^{(P)}\varepsilon_T + \Omega_{23}^{(P)}\sigma_t + \Omega_{26}^{(P)}\gamma_A - \omega_2^{(P)}\Delta T, \qquad (49)\\
\varepsilon_t &= -\Omega_{13}^{(P)}\varepsilon_A - \Omega_{23}^{(P)}\varepsilon_T + \Omega_{33}^{(P)}\sigma_t - \Omega_{36}^{(P)}\gamma_A + \omega_3^{(P)}\Delta T,\\
\tau_A &= \Omega_{16}^{(P)}\varepsilon_A + \Omega_{26}^{(P)}\varepsilon_T + \Omega_{36}^{(P)}\sigma_t + \Omega_{66}^{(P)}\gamma_A - \omega_6^{(P)}\Delta T,
\end{aligned}$$

where

$$
\begin{aligned}
&\Omega_{11}^{(P)} = \frac{\tilde{E}_A^{(P)}}{\Psi^{(P)}}, \quad \Omega_{12}^{(P)} = \frac{\tilde{\nu}_A^{(P)}\tilde{E}_T^{(P)}}{\Psi^{(P)}}, \quad \Omega_{13}^{(P)} = \frac{\hat{\nu}_a^{(P)}}{\Psi^{(P)}},\\
&\Omega_{16}^{(P)} = \frac{\tilde{E}_A^{(P)}}{\Psi^{(P)}}\lambda_A^{(P)} + \frac{\tilde{\nu}_A^{(P)}\tilde{E}_T^{(P)}}{\Psi^{(P)}}\lambda_T^{(P)}, \quad \Omega_{22}^{(P)} = \frac{\tilde{E}_T^{(P)}}{\Psi^{(P)}},\\
&\Omega_{23}^{(P)} = \frac{\hat{\nu}_t^{(P)}}{\Psi^{(P)}}, \quad \Omega_{26}^{(P)} = \frac{\tilde{\nu}_A^{(P)}\tilde{E}_T^{(P)}}{\Psi^{(P)}}\lambda_A^{(P)} + \frac{\tilde{E}_T^{(P)}}{\Psi^{(P)}}\lambda_T^{(P)},\\
&\Omega_{33}^{(P)} = \frac{\Lambda^{(P)}}{\Psi^{(P)}\tilde{E}_t^{(P)}}, \quad \Omega_{36}^{(P)} = \frac{\hat{\nu}_a^{(P)}}{\Psi^{(P)}}\lambda_A^{(P)} + \frac{\hat{\nu}_t^{(P)}}{\Psi^{(P)}}\lambda_T^{(P)} + \lambda_t^{(P)},\\
&\Omega_{66}^{(P)} = \frac{\tilde{E}_A^{(P)}}{\Psi^{(P)}}\lambda_A^{(P)}\hat{\lambda}_A^{(P)} + \frac{\tilde{E}_T^{(P)}}{\Psi^{(P)}}\lambda_T^{(P)}\hat{\lambda}_T^{(P)} + \mu_A^{(P)},
\end{aligned}
\tag{50}
$$

$$
\begin{aligned}
\omega_1^{(P)} &= \frac{\tilde{E}_A^{(P)}}{\Psi^{(P)}}\tilde{\alpha}_A^{(P)} + \frac{\tilde{\nu}_A^{(P)}\tilde{E}_T^{(P)}}{\Psi^{(P)}}\tilde{\alpha}_T^{(P)},\\
\omega_2^{(P)} &= \frac{\tilde{E}_T^{(P)}}{\Psi^{(P)}}\left(\tilde{\alpha}_T^{(P)} + \tilde{\nu}_A^{(P)}\tilde{\alpha}_A^{(P)}\right),\\
\omega_3^{(P)} &= \frac{\hat{\nu}_a^{(P)}}{\Psi^{(P)}}\tilde{\alpha}_A^{(P)} + \frac{\hat{\nu}_t^{(P)}}{\Psi^{(P)}}\tilde{\alpha}_T^{(P)} + \tilde{\alpha}_t^{(P)},\\
\omega_6^{(P)} &= \frac{\tilde{E}_A^{(P)}}{\Psi^{(P)}}\hat{\lambda}_A^{(P)}\tilde{\alpha}_A^{(P)} + \frac{\tilde{E}_T^{(P)}}{\Psi^{(P)}}\hat{\lambda}_T^{(P)}\tilde{\alpha}_T^{(P)} + \mu_A^{(P)}\alpha_S^{(P)},
\end{aligned}
\tag{51}
$$

and where

$$
\begin{aligned}
\Psi^{(P)} &= 1 - \left(\tilde{\nu}_A^{(P)}\right)^2\frac{\tilde{E}_T^{(P)}}{\tilde{E}_A^{(P)}},\\
\hat{\nu}_a^{(P)} &= \tilde{\nu}_a^{(P)} + \tilde{\nu}_t^{(P)}\tilde{\nu}_A^{(P)},\\
\hat{\nu}_t^{(P)} &= \tilde{\nu}_t^{(P)} + \tilde{\nu}_a^{(P)}\tilde{\nu}_A^{(P)}\frac{\tilde{E}_T^{(P)}}{\tilde{E}_A^{(P)}}
\end{aligned}
\tag{52}
$$

$$
\begin{aligned}
\Lambda^{(P)} &= \Psi^{(P)} - \tilde{\nu}_a^{(P)}\frac{\tilde{E}_t^{(P)}}{\tilde{E}_A^{(P)}}\hat{\nu}_a^{(P)} - \tilde{\nu}_t^{(P)}\frac{\tilde{E}_t^{(P)}}{\tilde{E}_T^{(P)}}\hat{\nu}_t^{(P)}\\
&= 1 - \left(\tilde{\nu}_a^{(P)}\right)^2\frac{\tilde{E}_t^{(P)}}{\tilde{E}_A^{(P)}} - \left(\tilde{\nu}_t^{(P)}\right)^2\frac{\tilde{E}_t^{(P)}}{\tilde{E}_T^{(P)}} - \left(\tilde{\nu}_A^{(P)}\right)^2\frac{\tilde{E}_T^{(P)}}{\tilde{E}_A^{(P)}}\\
&\quad -2\tilde{\nu}_a^{(P)}\tilde{\nu}_t^{(P)}\tilde{\nu}_A^{(P)}\frac{\tilde{E}_t^{(P)}}{\tilde{E}_A^{(P)}},
\end{aligned}
\tag{53}
$$

$$
\hat{\lambda}_A^{(P)} = \lambda_A^{(P)} + \lambda_T^{(P)}\tilde{\nu}_A^{(P)}\frac{\tilde{E}_T^{(P)}}{\tilde{E}_A^{(P)}}, \quad \hat{\lambda}_T^{(P)} = \lambda_T^{(P)} + \lambda_A^{(P)}\tilde{\nu}_A^{(P)}
\tag{54}
$$

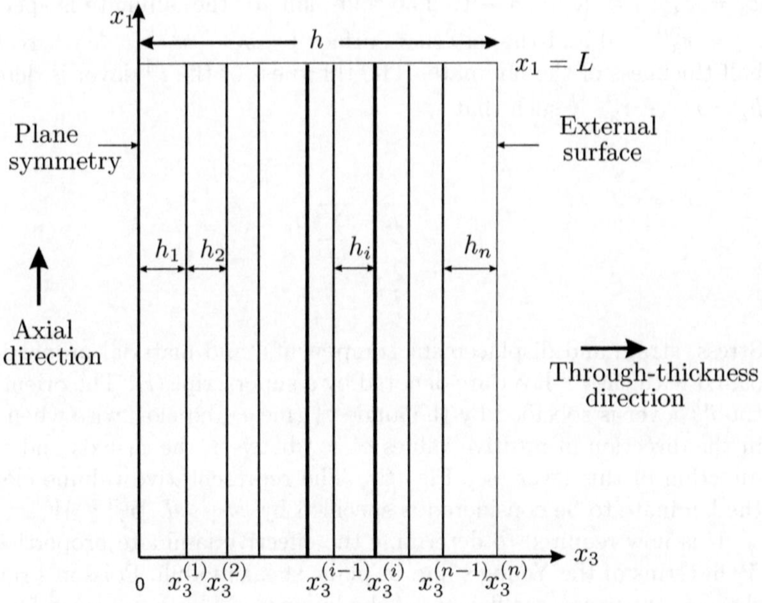

Figure 3. Schematic diagram of geometry for one half of a general symmetric laminate

2.9 Effective Thermoelastic Properties of Undamaged Symmetric Laminates

Consider now a laminate that is made by perfectly bonding together various laminae (i.e. plies) having different orientations so that there are no defects (i.e. the laminate is undamaged). The situation under consideration concerns the deformation of a symmetric multi-layered laminate of total thickness $2h$ constructed of $2n$ perfectly bonded plies that can have any combination of orientations, such that symmetry about the mid-plane of the laminate is preserved. The plies in each half of the laminate can be made of different materials and each can have a different thickness as illustrated in Fig. 3. As laminate symmetry is assumed, it is necessary to consider only the right hand set of n layers (see Fig. 3). A global right handed set of Cartesian coordinates is chosen having the origin at the centre of the mid-plane of the laminate. The x_1-direction defines the longitudinal or axial direction, the x_2-direction defines the in-plane transverse direction and the x_3-direction defines the through-thickness direction. The locations of the $n-1$ interfaces in one half of the laminate ($x_3 > 0$) are specified by

$x_3 = x_3^{(i)}$, $i = 1, \ldots, n-1$. The mid-plane of the laminate is specified by $x_3 = x_3^{(0)} = 0$ and the external surface by $x_3 = x_3^{(n)} = h$ where h is the half-thickness of the laminate. The thickness of the ith layer is denoted by $h_i = x_3^{(i)} - x_3^{(i-1)}$ such that

$$h = \sum_{i=1}^{n} h_i. \tag{55}$$

Stress, strain and displacement components, and material properties associated with the ith layer are denoted by a superscript (i). The orientation of the ith layer is specified by the angle ϕ_i (measured clockwise when looking in the direction of positive values of x_3) between the x_1-axis and the fibre direction of this layer (see Fig. 2). The representative volume element of the laminate to be considered is specified by $|x_1| \leq L, |x_2| \leq W, |x_3| \leq h$.

It is now required to determine the effective laminate properties $\boldsymbol{S}$ and $\boldsymbol{V}$ in terms of the Young's and in-plane shear moduli, Poisson's ratios and thermal expansion coefficients of the laminate. This is achieved by modifying the relations (26) derived for a single angled lamina so that they apply to a laminate rather than to an individual angled ply. When the laminate is considered as a homogenised plate, the effective stress-strain relations must be of the following form analogous to the corresponding single ply relations

$$\begin{bmatrix} \varepsilon_A \\ \varepsilon_T \\ \varepsilon_t \\ \gamma_t \\ \gamma_a \\ \gamma_A \end{bmatrix} \equiv \begin{bmatrix} \varepsilon_{11} \\ \varepsilon_{22} \\ \varepsilon_{33} \\ 2\varepsilon_{23} \\ 2\varepsilon_{13} \\ 2\varepsilon_{12} \end{bmatrix}$$

$$= \begin{bmatrix} S_{11}^{(L)} & S_{12}^{(L)} & S_{13}^{(L)} & 0 & 0 & S_{16}^{(L)} \\ S_{12}^{(L)} & S_{22}^{(L)} & S_{23}^{(L)} & 0 & 0 & S_{26}^{(L)} \\ S_{13}^{(L)} & S_{23}^{(L)} & S_{33}^{(L)} & 0 & 0 & S_{36}^{(L)} \\ 0 & 0 & 0 & S_{44}^{(L)} & S_{45}^{(L)} & 0 \\ 0 & 0 & 0 & S_{45}^{(L)} & S_{55}^{(L)} & 0 \\ S_{16}^{(L)} & S_{26}^{(L)} & S_{36}^{(L)} & 0 & 0 & S_{66}^{(L)} \end{bmatrix} \begin{bmatrix} \sigma_A \\ \sigma_T \\ \sigma_t \\ \tau_t \\ \tau_a \\ \tau_A \end{bmatrix} + \begin{bmatrix} V_1^{(L)} \\ V_2^{(L)} \\ V_3^{(L)} \\ 0 \\ 0 \\ V_6^{(L)} \end{bmatrix} \Delta T, \tag{56}$$

where the superscript (L) is used to denote effective thermoelastic constants, and where

$$
\begin{aligned}
&S_{11}^{(L)} = \frac{1}{E_A^{(L)}}, S_{12}^{(L)} = -\frac{\nu_A^{(L)}}{E_A^{(L)}}, S_{13}^{(L)} = -\frac{\nu_a^{(L)}}{E_A^{(L)}}, S_{16}^{(L)} = -\frac{\lambda_A^{(L)}}{\mu_A^{(L)}}, \\
&S_{22}^{(L)} = \frac{1}{E_T^{(L)}}, S_{23}^{(L)} = -\frac{\nu_t^{(L)}}{E_T^{(L)}}, S_{26}^{(L)} = -\frac{\lambda_T^{(L)}}{\mu_A^{(L)}}, S_{33}^{(L)} = \frac{1}{E_t^{(L)}}, \\
&S_{36}^{(L)} = -\frac{\lambda_t^{(L)}}{\mu_A^{(L)}}, S_{44}^{(L)} = \frac{1}{\mu_t^{(L)}}, S_{45}^{(L)} = \Phi^{(L)}, S_{55}^{(L)} = \frac{1}{\mu_a^{(L)}}, S_{66}^{(L)} = \frac{1}{\mu_A^{(L)}}, \\
&V_1^{(L)} = \alpha_A^{(L)}, V_2^{(L)} = \alpha_T^{(L)}, V_3^{(L)} = \alpha_t^{(L)}, V_6^{(L)} = \alpha_S^{(L)}
\end{aligned} \tag{57}
$$

It can be shown that the laminate properties appearing in (57) are calculated using the following relations

$$
\begin{aligned}
&\frac{1}{E_A^{(L)}} = \frac{1}{\tilde{E}_A^{(L)}} + \frac{\left(\lambda_A^{(L)}\right)^2}{\mu_A^{(L)}}, && \frac{1}{E_T^{(L)}} = \frac{1}{\tilde{E}_T^{(L)}} + \frac{\left(\lambda_T^{(L)}\right)^2}{\mu_A^{(L)}}, \\
&\frac{1}{E_t^{(L)}} = \frac{1}{\tilde{E}_t^{(L)}} + \frac{\left(\lambda_t^{(L)}\right)^2}{\mu_A^{(L)}}, && \nu_t^{(L)} = E_T^{(L)} \left(\frac{\tilde{\nu}_t^{(L)}}{\tilde{E}_T^{(L)}} - \frac{\lambda_t^{(L)}\lambda_T^{(L)}}{\mu_A^{(L)}} \right), \\
&\nu_A^{(L)} = E_A^{(L)} \left(\frac{\tilde{\nu}_A^{(L)}}{\tilde{E}_A^{(L)}} - \frac{\lambda_A^{(L)}\lambda_T^{(L)}}{\mu_A^{(L)}} \right), && \nu_a^{(L)} = E_A^{(L)} \left(\frac{\tilde{\nu}_a^{(L)}}{\tilde{E}_A^{(L)}} - \frac{\lambda_t^{(L)}\lambda_A^{(L)}}{\mu_A^{(L)}} \right), \\
&\alpha_A^{(L)} = \tilde{\alpha}_A^{(L)} - \lambda_A^{(L)}\alpha_S^{(L)}, && \alpha_T^{(L)} = \tilde{\alpha}_T^{(L)} - \lambda_T^{(L)}\alpha_S^{(L)}, \\
&\alpha_t^{(L)} = \tilde{\alpha}_t^{(L)} - \lambda_t^{(L)}\alpha_S^{(L)},
\end{aligned} \tag{58}
$$

where

$$
\begin{aligned}
&\tilde{E}_A^{(L)} = \Omega_{11}^{(L)} - \frac{\left(\Omega_{12}^{(L)}\right)^2}{\Omega_{22}^{(L)}}, \tilde{\nu}_A^{(L)} = \frac{\Omega_{12}^{(L)}}{\Omega_{22}^{(L)}}, \tilde{E}_T^{(L)} = \Omega_{22}^{(L)} - \frac{\left(\Omega_{12}^{(L)}\right)^2}{\Omega_{11}^{(L)}}, \\
&\tilde{\nu}_a^{(L)} = \Omega_{13}^{(L)} - \tilde{\nu}_A^{(L)}\Omega_{23}^{(L)}, \tilde{\nu}_t^{(L)} = \Omega_{23}^{(L)} - \tilde{\nu}_A^{(L)}\frac{\tilde{E}_T^{(L)}}{\tilde{E}_A^{(L)}}\Omega_{13}^{(L)}, \\
&\frac{1}{\tilde{E}_t^{(L)}} = \left(\frac{\tilde{\nu}_a^{(L)}}{\tilde{E}_A^{(L)}}\right)^2 \Omega_{11}^{(L)} + 2\frac{\tilde{\nu}_a^{(L)}}{\tilde{E}_A^{(L)}}\frac{\tilde{\nu}_t^{(L)}}{\tilde{E}_T^{(L)}}\Omega_{12}^{(L)} + \left(\frac{\tilde{\nu}_t^{(L)}}{\tilde{E}_T^{(L)}}\right)^2 \Omega_{22}^{(L)} + \Omega_{33}^{(L)},
\end{aligned} \tag{59}
$$

$$
\lambda_A^{(L)} = \frac{1}{\tilde{E}_A^{(L)}}\Omega_{16}^{(L)} - \frac{\tilde{\nu}_A^{(L)}}{\tilde{E}_A^{(L)}}\Omega_{26}^{(L)}, \lambda_T^{(L)} = \frac{1}{\tilde{E}_T^{(L)}}\Omega_{26}^{(L)} - \frac{\tilde{\nu}_A^{(L)}}{\tilde{E}_A^{(L)}}\Omega_{16}^{(L)}, \tag{60}
$$

$$\lambda_t^{(L)} = \Omega_{36}^{(L)} - \Omega_{13}^{(L)}\lambda_A^{(L)} - \Omega_{23}^{(L)}\lambda_T^{(L)}, \tag{61}$$

$$\mu_A^{(L)} = \Omega_{66}^{(L)} - \Omega_{11}^{(L)}\left(\lambda_A^{(L)}\right)^2 - 2\Omega_{12}^{(L)}\lambda_A^{(L)}\lambda_T^{(L)} - \Omega_{22}^{(L)}\left(\lambda_T^{(L)}\right)^2 \tag{62}$$

$$\tilde{\alpha}_A^{(L)} = \frac{1}{\tilde{E}_A^{(L)}}\omega_1^{(L)} - \frac{\tilde{\nu}_A^{(L)}}{\tilde{E}_A^{(L)}}\omega_2^{(L)}, \tilde{\alpha}_T^{(L)} = \frac{1}{\tilde{E}_T^{(L)}}\omega_2^{(L)} - \frac{\tilde{\nu}_A^{(L)}}{\tilde{E}_A^{(L)}}\omega_1^{(L)}, \tag{63}$$

$$\begin{aligned} \tilde{\alpha}_t^{(L)} &= \omega_3^{(L)} - \Omega_{13}^{(L)}\tilde{\alpha}_A^{(L)} - \Omega_{23}^{(L)}\tilde{\alpha}_T^{(L)}, \\ \alpha_S^{(L)} &= \frac{1}{\mu_A^{(L)}}\left(\omega_6^{(L)} - \Omega_{16}^{(L)}\tilde{\alpha}_A^{(L)} - \Omega_{26}^{(L)}\tilde{\alpha}_T^{(L)}\right) \end{aligned} \tag{64}$$

The quantities $\Omega_{IJ}^{(L)}$ and $\omega_I^{(L)}$ are calculated using the following summations

$$\begin{aligned} \Omega_{11}^{(L)} &= \frac{1}{h}\sum_{i=1}^{n} h_i\Omega_{11}^{(i)}, \Omega_{12}^{(L)} = \frac{1}{h}\sum_{i=1}^{n} h_i\Omega_{12}^{(i)}, \\ \Omega_{22}^{(L)} &= \frac{1}{h}\sum_{i=1}^{n} h_i\Omega_{22}^{(i)}, \Omega_{33}^{(L)} = \frac{1}{h}\sum_{i=1}^{n} h_i\Omega_{33}^{(i)}, \\ \Omega_{16}^{(L)} &= \frac{1}{h}\sum_{i=1}^{n} h_i\Omega_{16}^{(i)}, \Omega_{26}^{(L)} = \frac{1}{h}\sum_{i=1}^{n} h_i\,\Omega_{26}^{(i)}, \\ \Omega_{66}^{(L)} &= \frac{1}{h}\sum_{i=1}^{n} h_i\Omega_{66}^{(i)}, \end{aligned} \tag{65}$$

$$\Omega_{13}^{(L)} = \frac{1}{h}\sum_{i=1}^{n} h_i\Omega_{13}^{(i)}, \Omega_{23}^{(L)} = \frac{1}{h}\sum_{i=1}^{n} h_i\Omega_{23}^{(i)}, \Omega_{36}^{(L)} = \frac{1}{h}\sum_{i=1}^{n} h_i\Omega_{36}^{(i)}, \tag{66}$$

$$\begin{aligned} \omega_1^{(L)} &= \frac{1}{h}\sum_{i=1}^{n} h_i\omega_1^{(i)}, \omega_2^{(L)} = \frac{1}{h}\sum_{i=1}^{n} h_i\omega_2^{(i)}, \\ \omega_3^{(L)} &= \frac{1}{h}\sum_{i=1}^{n} h_i\omega_3^{(i)}, \omega_6^{(L)} = \frac{1}{h}\sum_{i=1}^{n} h_i\omega_6^{(i)} \end{aligned} \tag{67}$$

These relations are derived by applying the following expressions for effective stresses and strains to the relations (50) that apply to individual plies

$$\sigma_A = \frac{1}{h}\sum_{i=1}^{n} h_i\sigma_A^{(i)}, \sigma_T = \frac{1}{h}\sum_{i=1}^{n} h_i\sigma_T^{(i)}, \tau_A = \frac{1}{h}\sum_{i=1}^{n} h_i\tau_A^{(i)}, \varepsilon_t = \frac{1}{h}\sum_{i=1}^{n} h_i\varepsilon_t^{(i)} \tag{68}$$

The effective properties characterising the out-of-plane shear of the laminate are given by

$$\begin{aligned} S_{44}^{(L)} &= \frac{1}{h}\sum_{i=1}^{n} h_i\left(\frac{m_i^2}{\mu_t^{(i)}}+\frac{n_i^2}{\mu_a^{(i)}}\right) = \frac{1}{\mu_t^{(L)}}, \\ S_{45}^{(L)} &= \frac{1}{h}\sum_{i=1}^{n} h_i m_i n_i\left(\frac{1}{\mu_a^{(i)}}-\frac{1}{\mu_t^{(i)}}\right) = \Phi^{(L)}, \\ S_{55}^{(L)} &= \frac{1}{h}\sum_{i=1}^{n} h_i\left(\frac{m_i^2}{\mu_a^{(i)}}+\frac{n_i^2}{\mu_t^{(i)}}\right) = \frac{1}{\mu_a^{(L)}}, \end{aligned} \tag{69}$$

where

$$S_{44}^{(L)} = \frac{1}{h}\sum_{i=1}^{n} h_i S_{44}^{(i)}, \quad S_{45}^{(L)} = \frac{1}{h}\sum_{i=1}^{n} h_i S_{45}^{(i)}, \quad S_{55}^{(L)} = \frac{1}{h}\sum_{i=1}^{n} h_i S_{55}^{(i)} \tag{70}$$

3 Fracture in Homogenised Anisotropic Materials

This Section will introduce a very convenient and elegant method of solving crack problems in plates based on orthogonal polynomials. The methodology will then be applied to bridged cracks, especially those arising in laminates where ply cracks are confined to a limited number of plies (usually the 90° plies). The use of orthogonal functions enables very simple expressions for stress intensity factors and energy release rates. The first objective is to consider a very useful method of analysing isolated crack problems for composite materials that have been homogenised as an anisotropic continuum.

3.1 Stress-Strain Relations

It is assumed that there is no through-thickness loading and that the axial loading direction is parallel to the y-axis. The plate is assumed to be orthotropic having the following stress-strain relations

$$\begin{aligned} \varepsilon_{xx} &= a_{11}\sigma_{xx} + a_{12}\sigma_{yy} + a_{16}\sigma_{xy}, \\ \varepsilon_{yy} &= a_{12}\sigma_{xx} + a_{22}\sigma_{yy} + a_{26}\sigma_{xy}, \\ 2\varepsilon_{xy} &= a_{16}\sigma_{xx} + a_{26}\sigma_{yy} + a_{66}\sigma_{xy}, \end{aligned} \tag{71}$$

where

$$a_{11} = \frac{1}{E_T}, a_{22} = \frac{1}{E_A}, a_{12} = -\frac{\nu_A}{E_A}, a_{66} = \frac{1}{\mu_A}, a_{16} = a_{26} = 0 \tag{72}$$

For conditions of plane strain such that $\varepsilon_{zz} \equiv 0$,

$$a_{11} = \frac{1-\nu_T^2}{E_T}, a_{22} = \frac{1}{E_A}\left(1-\nu_A^2\frac{E_T}{E_A}\right), a_{12} = -\frac{\nu_A(1+\nu_T)}{E_A}, \\ a_{66} = \frac{1}{\mu_A}, a_{16} = a_{26} = 0 \tag{73}$$

For a 0°-90°-0° cross-ply laminate the values of E_A and E_T (the axial and transverse Young's moduli), ν_A (the axial Poisson's ratio) and μ_A (the axial shear modulus) are calculated from single ply properties. It should be noted that thermal expansion effects have been neglected in (71).

3.2 A Representation for Stress and Displacement Fields

Consider a straight through-crack of length $2c$ embedded within an infinite transverse isotropic plate so that the crack plane is normal to the surfaces of the plate. For conditions of generalised plane stress and plane strain, crack problems are two dimensional. Introduce a set of rectangular Cartesian coordinates (x, y) such that the crack is parallel to the x-axis and occupies the region $|x-a| \leq c$, $y = b$. The complex variable z is introduced such that

$$z = x + iy \tag{74}$$

The locations of the two crack tips are specified by

$$z = t_1 = a - c + ib, \tag{75}$$

$$z = t_2 = a + c + ib \tag{76}$$

The stress and displacement representation developed by Sih and Liebowitz (1968) is used, which is valid for plane strain deformations in rectilinearly anisotropic bodies containing a crack. In the absence of body forces, as is assumed in the present model, this representation may also be applied to the conditions of generalised plane stress that are being considered.

The representation involves two analytic functions $\phi(z_1)$ and $\psi(z_2)$ of the complex variables

$$z_1 = x - a + s_1(y-b), \quad z_2 = x - a + s_2(y-b), \tag{77}$$

where s_1 and s_2 are the two roots having positive imaginary parts (see below) of the following quadratic equation (since $a_{16} = a_{26} = 0$)

$$a_{11}s^4 + (2a_{12} + a_{66})s^2 + a_{22} = 0 \tag{78}$$

The representation for the stress components is given by

$$\sigma_{xx} = \Re\left[s_1^2\phi'(z_1) + s_2^2\psi'(z_2)\right], \tag{79}$$

$$\sigma_{yy} = \Re\left[\phi'(z_1) + \psi'(z_2)\right], \tag{80}$$

$$\sigma_{xy} = \Re\left[s_1\phi'(z_1) + s_2\psi'(z_2)\right], \tag{81}$$

and that for the displacement components is

$$u = \Re\left[p_1\,\phi(z_1) + p_2\,\psi(z_2)\right], \tag{82}$$

$$v = \Re\left[q_1\,\phi(z_1) + q_2\,\psi(z_2)\right], \tag{83}$$

where (since $a_{16} = a_{26} = 0$)

$$\left.\begin{array}{lcl} p_k & = & a_{11}s_k^2 + a_{12} \\ q_k & = & a_{12}s_k + a_{22}/s_k \end{array}\right\} \text{ for } \quad k = 1, 2 \tag{84}$$

Putting

$$\gamma = \frac{a_{22}}{a_{11}}, \quad \delta = \frac{2a_{12} + a_{66}}{2a_{11}}, \tag{85}$$

it follows from (78) that

$$s^2 = -\delta \pm \sqrt{\delta^2 - \gamma} \tag{86}$$

It can be shown that for both GRP and CFRP, the values of s^2 are real and negative. The following distinct pure imaginary roots are, therefore, obtained

$$s_1 = i\sqrt{\delta - \sqrt{\delta^2 - \gamma}}, s_2 = i\sqrt{\delta + \sqrt{\delta^2 - \gamma}} \tag{87}$$

The other two roots of (78) are given by $\bar{s}_1 = -s_1, \bar{s}_2 = -s_2$. The stress and displacement representation automatically satisfies the equilibrium equations and the stress-strain relations (71) for any analytic functions $\phi(z)$ and $\psi(z)$ of the complex variable z. They are now assumed to take the following form:

$$\phi(z) \equiv \frac{1}{2\pi i}\int_{t_1}^{t_2} w(t)\hat{\rho}(t)\ln\frac{1}{z-t}\mathrm{d}t, \tag{88}$$

$$\psi(z) \equiv \frac{1}{2\pi i}\int_{t_1}^{t_2} w(t)\hat{\sigma}(t)\ln\frac{1}{z-t}\mathrm{d}t \tag{89}$$

The density functions $\hat{\rho}(t)$ and $\hat{\sigma}(t)$ are assumed to be polynomials and

$$w(t) \equiv \frac{t_2 - t_1}{\left[\,(t-t_1)(t-t_2)\,\right]^{1/2}} \tag{90}$$

In order that the displacement components u and v are single-valued functions at all points in the complex plane lying outside the crack, the following crack tip closure conditions must be satisfied:

$$\int_{t_1}^{t_2} w(t)\hat{\rho}(t)\mathrm{d}t = 0, \qquad \int_{t_1}^{t_2} w(t)\hat{\sigma}(t)\mathrm{d}t = 0 \tag{91}$$

Differentiating (88) and (89) leads to

$$\phi'(z) \equiv \frac{1}{2\pi i}\int_{t_1}^{t_2} \frac{w(t)\hat{\rho}(t)}{t-z}\mathrm{d}t, \tag{92}$$

$$\psi'(z) \equiv \frac{1}{2\pi i}\int_{t_1}^{t_2} \frac{w(t)\hat{\sigma}(t)}{t-z}\mathrm{d}t \tag{93}$$

Provided the conditions (91) are satisfied

$$\phi'(z) = O(z^{-2}), \quad \psi'(z) = O(z^{-2}) \quad \text{as} \quad |z| \to \infty, \tag{94}$$

indicating that the stress field arising from the representation (79)-(81) has zero net force applied at infinity.

The algebra may be simplified by changing variables to

$$\zeta = \xi + i\eta, \tag{95}$$

where

$$\zeta = \zeta(z) = \frac{2z - (t_1 + t_2)}{t_2 - t_1}, \tag{96}$$

$$\xi = \frac{x-a}{c}, \quad \eta = \frac{y-b}{c} \tag{97}$$

The crack is then described by $-1 < \xi < 1, \eta = 0$ and it follows from (92) and (93) that

$$\phi'(z) \equiv \frac{1}{\pi}\int_{-1}^{1} \frac{\rho(s)}{\sqrt{1-s^2}}\frac{\mathrm{d}s}{\zeta(z)-s}, \tag{98}$$

$$\psi'(z) \equiv \frac{1}{\pi}\int_{-1}^{1} \frac{\sigma(s)}{\sqrt{1-s^2}}\frac{\mathrm{d}s}{\zeta(z)-s}, \tag{99}$$

where $\rho(\xi) \equiv \hat{\rho}(t), \sigma(\xi) \equiv \hat{\sigma}(t)$. From (91) the crack closure conditions are written

$$\int_{-1}^{1} \frac{\rho(s)\mathrm{d}s}{\sqrt{1-s^2}} = 0, \quad \int_{-1}^{1} \frac{\sigma(s)\mathrm{d}s}{\sqrt{1-s^2}} = 0 \tag{100}$$

3.3 Chebyshev Polynomial Expansion

Chebyshev polynomials of the first kind are defined over the interval $[-1, 1]$ by

$$T_n(\cos\alpha) \equiv \cos n\alpha, 0 \leq \alpha \leq \pi, n \geq 0 \tag{101}$$

Chebyshev polynomials of the second kind are defined by

$$U_n(\cos\alpha) = \frac{\sin(n+1)\alpha}{\sin\alpha}, 0 \leq \alpha \leq \pi, n \geq 0 \tag{102}$$

The functions $T_n(z)$ and $U_n(z)$ can both be analytically continued to the entire complex plane, simply by considering the usual analytic continuation of the functions $\sin(z)$ and $\cos(z)$. Use will be made of the following two identities:

$$T_n\left((\lambda + 1/\lambda)/2\right) \equiv (\lambda^n + 1/\lambda^n)/2, n \geq 0, \tag{103}$$

$$U_{n-1}\left((\lambda + 1/\lambda)/2\right) \equiv (\lambda^n - 1/\lambda^n)/(\lambda - 1/\lambda), n \geq 1 \tag{104}$$

The density functions $\rho(\xi)$ and $\sigma(\xi)$ are now assumed to be of the form

$$\rho(\xi) \equiv \sum_{n=1}^{N} A_n T_n(\xi), \sigma(\xi) \equiv \sum_{n=1}^{N} B_n T_n(\xi), \tag{105}$$

where A_n and B_n are complex coefficients. The crack closure conditions (100) are automatically satisfied and substitution of (105) in (98) and (99) leads to

$$\phi'(z) \equiv \sum_{n=1}^{N} A_n H_n(\zeta), \quad \psi'(z) \equiv \sum_{n=1}^{N} B_n H_n(\zeta), \tag{106}$$

where, using a result established by Gladwell and England (1977),

$$H_n(\zeta) \equiv \frac{1}{\pi} \int_{-1}^{1} \frac{T_n(s)}{\sqrt{1-s^2}} \frac{\mathrm{d}s}{\zeta - s} \equiv \frac{\left[\zeta - (\zeta^2 - 1)^{1/2}\right]^n}{(\zeta^2 - 1)^{1/2}}, n \geq 0 \tag{107}$$

By selecting the branch of $(\zeta^2 - 1)^{1/2}$ which is asymptotic to ζ as $|\zeta| \to \infty$, it can be shown that for $n \geq 1$,

$$H_n(\zeta) \equiv \frac{\left[\zeta - (\zeta^2 - 1)^{1/2}\right]^n}{(\zeta^2 - 1)^{1/2}} \equiv \frac{T_n(\zeta)}{(\zeta^2 - 1)^{1/2}} - U_{n-1}(\zeta) \tag{108}$$

3.4 Traction Distribution on the Crack

Let $S_n^+, S_t^+, S_n^-, S_t^-$ be the normal and transverse tractions on the upper and lower surfaces of the crack. Since

$$
\left(\zeta^2-1\right)^{1/2}=\begin{cases}\sqrt{\xi^2-1}, \xi>1, \eta=0,\\ i\sqrt{1-\xi^2}, |\xi|<1, \eta=0+,\\ -i\sqrt{1-\xi^2}, |\xi|<1, \eta=0-,\\ -\sqrt{\xi^2-1}, \xi<-1, \eta=0,\end{cases} \tag{109}
$$

it follows from (108) that

$$
H_n^{\pm}(\xi)=\pm i\frac{T_n(\xi)}{\sqrt{1-\xi^2}}-U_{n-1}(\xi), |\xi|<1, n\geq 1, \tag{110}
$$

where H_n^+ and H_n^- are the limiting values of $H_n(\zeta)$ on the positive and negative sides of the crack respectively. Substituting (106) into (80) and (81) using (110) leads to the following relations valid only for $|\xi|<1$:

$$
S_n^{\pm}(\xi)=\sum_{n=1}^{N}\left(\pm\Im\left[A_n+B_n\right]\frac{T_n(\xi)}{\sqrt{1-\xi^2}}-\Re\left[A_n+B_n\right]U_{n-1}(\xi)\right), \tag{111}
$$

$$
S_t^{\pm}(\xi)=\sum_{n=1}^{N}\left(\pm\Im\left[s_1A_n+s_2B_n\right]\frac{T_n(\xi)}{\sqrt{1-\xi^2}}+\Re\left[s_1A_n+s_2B_n\right]\, U_{n-1}(\xi)\right) \tag{112}
$$

The tractions $S_n^+, S_t^+, S_n^-, S_t^-$ must be bounded as $\xi\to\pm 1$, for values of ξ in the range $|\xi|<1$ leading to the following conditions for the complex coefficients A_n, B_n:

$$
\Im\left[A_n+B_n\right]=\Im\left[s_1A_n+s_2B_n\right]=0 \tag{113}
$$

These conditions may easily be satisfied by choosing real α_n, β_n such that

$$
\alpha_n=A_n+B_n, \beta_n=-(s_1A_n+s_2B_n), \tag{114}
$$

and

$$
A_n=-\frac{\alpha_n s_2+\beta_n}{s_1-s_2}, \quad B_n=\frac{\alpha_n s_1+\beta_n}{s_1-s_2}, \tag{115}
$$

which, furthermore, yields the following simple expression for the tractions on the crack surfaces:

$$
\left(S_n^{\pm}+iS_t^{\pm}\right)(\xi)=-\sum_{n=1}^{N}\left(\alpha_n+i\,\beta_n\right)U_{n-1}(\xi), |\xi|<1 \tag{116}
$$

3.5 Stress and Displacement Fields Around the Crack

Define

$$G_n(\zeta) \equiv \left[\zeta - (\zeta^2 - 1)^{1/2}\right]^n, n \geq 1, \tag{117}$$

noting that numerical calculation is convenient by making use of the fact that

$$G_n(\zeta) \equiv \exp\left[-n(\alpha + i\beta)\right], \quad \text{when} \quad \zeta = \cosh(\alpha + i\beta) \tag{118}$$

It is easily shown that

$$\frac{\mathrm{d}}{\mathrm{d}\zeta} G_n(\zeta) = -nH_n(\zeta), n \geq 1 \tag{119}$$

and it then follows from (106) that

$$\phi(z) = -c \sum_{n=1}^{N} \frac{A_n}{n} G_n(\zeta), \tag{120}$$

$$\psi(z) = -c \sum_{n=1}^{N} \frac{B_n}{n} G_n(\zeta) \tag{121}$$

Let

$$\zeta_1 = \xi + s_1\eta \quad \text{and} \quad \zeta_2 = \xi + s_2\eta \tag{122}$$

On substituting (106), (115), (120) and (121) into the representation (79)-(83), the stresses and displacement components may be expressed

$$\sigma_{xx} = -\sum_{n=1}^{N} \Re\left[\left(\alpha_n s_1 s_2 + \beta_n(s_1 + s_2)\right) H_n(\zeta_1) + s_2^2(\alpha_n s_1 + \beta_n)\Delta H_n(\zeta_1, \zeta_2)\right], \tag{123}$$

$$\sigma_{yy} = \sum_{n=1}^{N} \Re\left[\alpha_n H_n(\zeta_1) - (\alpha_n s_1 + \beta_n)\Delta H_n(\zeta_1, \zeta_2)\right], \tag{124}$$

$$\sigma_{xy} = \sum_{n=1}^{N} \Re\left[\beta_n H_n(\zeta_1) + s_2(\alpha_n s_1 + \beta_n)\Delta H_n(\zeta_1, \zeta_2)\right], \tag{125}$$

$$\begin{aligned} u &= c\sum_{n=1}^{N} \frac{1}{n}\Re\left[\left(\alpha_n(a_{11}s_1 s_2 - a_{12}) + \beta_n\, a_{11}(s_1 + s_2)\right) G_n(\zeta_1)\right. \\ &+ \left.(a_{11}s_2^2 + a_{12})(\alpha_n s_1 + \beta_n)\Delta G_n(\zeta_1, \zeta_2)\right], \end{aligned} \tag{126}$$

$$v = c\sum_{n=1}^{N}\frac{1}{n}\Re\left[\left(-\alpha_n a_{22}\left(\frac{s_1+s_2}{s_1 s_2}\right)+\beta_n\left(a_{12}-\frac{a_{22}}{s_1 s_2}\right)\right)G_n(\zeta_1)\right.$$
$$\left. + \left(a_{12}s_2+\tfrac{a_{22}}{s_2}\right)(\alpha_n s_1+\beta_n)\,\Delta G_n(\zeta_1,\zeta_2)\right], \tag{127}$$

where

$$\Delta H_n(\zeta_1,\zeta_2) \equiv \begin{cases} \left[\dfrac{H_n(\zeta_1)-H_n(\zeta_2)}{s_1-s_2}\right], & \text{for} \quad s_1 \neq s_2, \\ \lim\limits_{s_1\to s_2}\left[\dfrac{H_n(\zeta_1)-H_n(\zeta_2)}{s_1-s_2}\right] & \\ = -\eta\left(\dfrac{n}{(\zeta_1^2-1)^{1/2}}+\dfrac{\zeta_1}{\zeta_1^2-1}\right)H(\zeta_1), & \text{for } s_1 = s_2 \end{cases} \tag{128}$$

and

$$\Delta G_n(\zeta_1,\zeta_2) \equiv \begin{cases} \left[\dfrac{G_n(\zeta_1)-G_n(\zeta_2)}{s_1-s_2}\right], & \text{for} \quad s_1 \neq s_2, \\ \lim\limits_{s_1\to s_2}\left[\dfrac{G_n(\zeta_1)-G_n(\zeta_2)}{s_1-s_2}\right] = -n\eta H(\zeta_1), & \text{for} \quad s_1 = s_2, \end{cases} \tag{129}$$

One limiting situation, $s_1 = s_2 = i$, occurs when the material is isotropic, and in this case the expressions (123)-(127) coincide with those of McCartney and Gorley (1987) for the case of parallel cracks. It should be noted that

$$\Delta G_n(\zeta_1,\,\zeta_2) \;\to\; 0\,, \quad \Delta H_n(\zeta_1,\,\zeta_2) \;\to\; 0 \;\text{ as } y \to 0$$

3.6 Displacement Discontinuity Across the Crack

From (108) and (117)

$$G_n(\zeta) \equiv H_n(\zeta)\left(\zeta^2-1\right)^{1/2} \equiv T_n(\xi) - \left(\zeta^2-1\right)^{1/2}U_{n-1}(\zeta) \tag{130}$$

It is deduced from (109) that the limiting values of $G_n(\zeta)$ on the crack faces are given by

$$G_n^{\pm}(\xi) = T_n(\xi) \pm i\sqrt{1-\xi^2}\,U_{n-1}(\xi) \tag{131}$$

By considering the limiting distributions for the normal and tangential displacements along the upper and lower surfaces of the crack, which are denoted by $V_n^+, V_n^-, U_t^+, U_t^-$, use can be made of (126) and (127) together with (131) to obtain an expression for the displacement discontinuities across the crack:

$$\Delta v(\xi) \equiv (V_n^+ - V_n^-)(\xi) = 4c\sqrt{a_{22}g}\sqrt{1-\xi^2}\sum_{n=1}^{N}\frac{\alpha_n}{n}U_{n-1}(\xi), \qquad (132)$$

$$\Delta u(\xi) \equiv (U_t^+ - U_t^-)(\xi) = 4c\sqrt{a_{11}g}\sqrt{1-\xi^2}\sum_{n=1}^{N}\frac{\beta_n}{n}U_{n-1}(\xi), \qquad (133)$$

where

$$g = \frac{1}{4}\left(2\sqrt{a_{11}a_{22}} + 2a_{12} + a_{66}\right) \qquad (134)$$

In the isotropic limit, this result agrees with the corresponding result of McCartney and Gorley (1987).

3.7 Stress Intensity Factors

For the crack tip at t_1, the mode I and mode II stress intensity factors $K_{\mathrm{I}}^1, K_{\mathrm{II}}^1$ are defined by

$$K_{\mathrm{I}}^1 + iK_{\mathrm{II}}^1 = \lim_{\xi\to-1}\left[\sqrt{2\pi c(-\xi-1)}\,(S_n(\xi) + iS_t(\xi))\right], \qquad (135)$$

where $S_n(\xi)$ and $S_t(\xi)$ are the normal and tangential tractions acting on $\xi < -1, \eta = 0$. It follows from (80), (81), (106), (108) and (115) that on $\xi < -1, \eta = 0$ the tractions are given by

$$S_n(\xi) + iS_t(\xi) = -\sum_{n=1}^{N}(\alpha_n + i\,\beta_n)\left[\frac{T_n(\xi)}{\sqrt{\xi^2-1}} + U_{n-1}(\xi)\right] \qquad (136)$$

Substituting into (135) leads immediately to the simple result

$$K_{\mathrm{I}}^1 + iK_{\mathrm{II}}^1 = \sqrt{\pi c}\sum_{n=1}^{N}(-1)^{n+1}(\alpha_n + i\,\beta_n), \qquad (137)$$

since $T_n(-1) = (-1)^n$. Similarly for the crack tip at t_2,

$$K_{\mathrm{I}}^2 + iK_{\mathrm{II}}^2 = \lim_{\xi\to1}\left[\sqrt{2\pi c(\xi-1)}\,(S_n(\xi) + iS_t(\xi))\right], \qquad (138)$$

where $S_n(\xi)$ and $S_t(\xi)$ are the normal and tangential tractions acting on $\xi > 1, \eta = 0$. For $\xi > 1, \eta = 0$,

$$S_n(\xi) + iS_t(\xi) = \sum_{n=1}^{N}(\alpha_n + i\,\beta_n)\left[\frac{T_n(\xi)}{\sqrt{\xi^2-1}} - U_{n-1}(\xi)\right] \qquad (139)$$

Since $T_n(1) = 1$, substitution into (138) yields the following corresponding simple result

$$K_{\mathrm{I}}^2 + iK_{\mathrm{II}}^2 = \sqrt{\pi c}\sum_{n=1}^{N}(\alpha_n + i\,\beta_n) \tag{140}$$

3.8 Example Prediction

Consider now a test example of two collinear cracks of equal length embedded in an isotropic material, where the exact solution is known and can be compared with the predictions obtained using orthogonal polynomials. Figure 4 shows the geometry where the two cracks have length $2c$ and the separation of the inner crack tips is $2s$. A uniaxial stress σ is applied in a direction normal to the crack planes. The crack problem is such that the deformation at the crack tips is mode I, i.e. the mode II stress intensity factors are zero.

The exact solution for the mode I stress intensity factors is given by Rooke and Cartwright (1976)

$$\begin{aligned} K_{\mathrm{I}}(x=s) &= \sigma\sqrt{\pi c}\sqrt{\frac{s+2c}{\alpha c}}\frac{1}{k}\left[\frac{E(k)}{K(k)} - \alpha^2\right], \\ K_{\mathrm{I}}(x=s+2c) &= \sigma\sqrt{\pi c}\sqrt{\frac{s+2c}{c}}\frac{1}{k}\left[1 - \frac{E(k)}{K(k)}\right] \end{aligned} \tag{141}$$

with $k = \sqrt{1-\alpha^2}$ and $\alpha = c/(s+2c)$. For a unit applied stress and when $c = 0.45\,c_0, s = 0.1c_0$ and on selecting $N = 100$, for any normalising crack length c_0, the methodology based on orthogonal polynomials leads to the results

$$K_{\mathrm{I}}(x=s) = 1.4923379, \quad K_{\mathrm{I}}(x=s+2c) = 1.2916506$$

The corresponding values obtained from the exact solution (141) is

$$K_{\mathrm{I}}(x=s) = 1.4923379, \quad K_{\mathrm{I}}(x=s+2c) = 1.2916506,$$

which are identical to the numerical estimates, thus confirming the validity of the methodology based on orthogonal polynomials.

It should be noted that the relation (139) can be used to investigate magnitude of the tractions on the crack surfaces. For the example considered the tractions, which should be zero, have the order of 10^{-16} indicating the very high accuracy of the methodology used.

Consider now a test example of three equally spaced vertically stacked cracks of equal length embedded in an infinite isotropic material. Figure

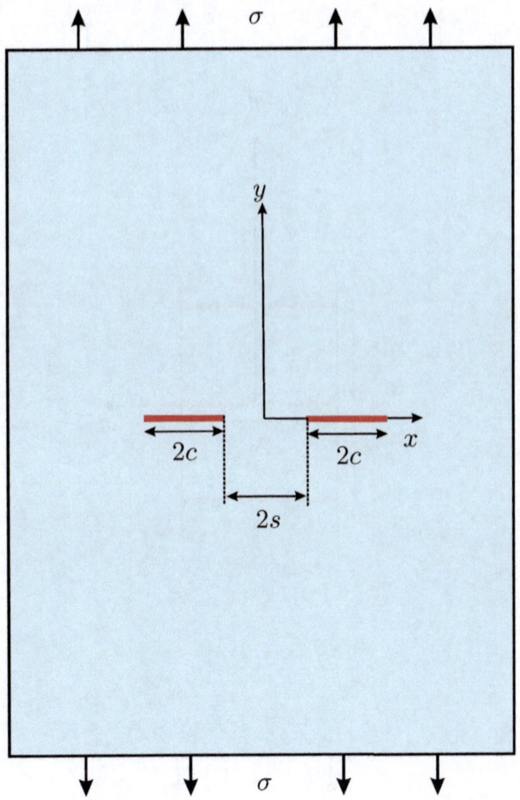

Figure 4. Diagram illustrating geometry and loading of part of an infinite isotropic plate having two collinear cracks

5 illustrates the geometry where the three cracks have length $2c$ and the vertical separation of the cracks is s. A uniaxial stress σ is applied in a direction normal to the crack planes. The crack problem is such that the deformation at the tips of the central crack is mode I, and the deformation at the other tips is mixed mode. The magnitudes of the model I and mode II stress intensity factors for the upper and lower cracks are expected to be the same.

For a unit applied stress and when $c = 0.5\, c_0, s = 0.5c_0$ and on selecting $N = 100$, for any normalising crack length c_0, the methodology based on

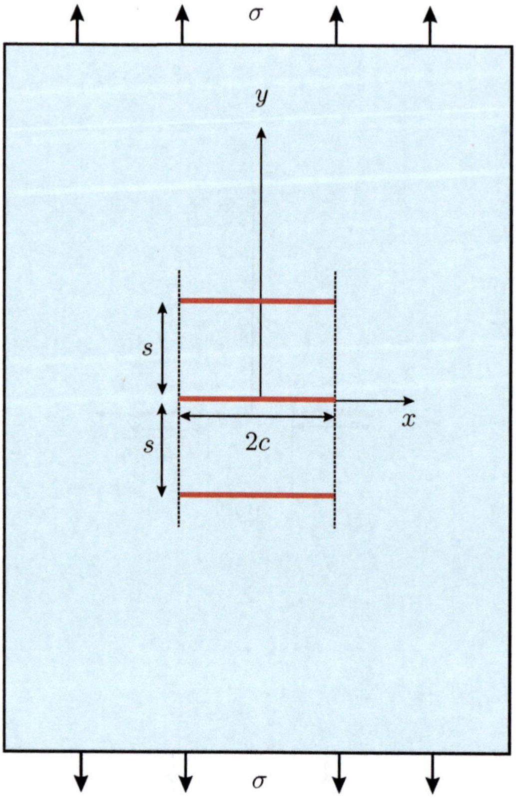

Figure 5. Diagram illustrating geometry and loading of part of an infinite isotropic plate having three equally spaced vertically stacked cracks

orthogonal polynomials leads to the results

$$\begin{array}{lll} \text{Top crack:} & K_{\mathrm{I}} = 0.92924447, & K_{\mathrm{II}} = 0.14265165, \\ \text{Central crack:} & K_{\mathrm{I}} = 0.70031026, & K_{\mathrm{II}} = 0.0, \\ \text{Bottom crack:} & K_{\mathrm{I}} = 0.92924447, & K_{\mathrm{II}} = -0.14265165 \end{array}$$

It is seen that the mode I stress intensity factor for the central crack is less than that of the upper and lower cracks. This illustrates the shielding effect on the central crack because of the presence of the other two cracks. The model II stress intensity of the lowest crack is seen to be negative because the local shear stress is negative. For this example the tractions, which should be zero, have the order of 10^{-16} again indicating the very high accuracy of the methodology used.

4 Generalised Plane Strain Theory for Cross-Ply Laminates

Stress transfer phenomena in cross-ply laminates have been examined extensively in the literature using an approximate method of stress analysis, first developed for UD composites, that is known as 'shear-lag theory'. An improved approach was developed by the author many years ago (McCartney, 1992) by removing various approximations that had to be made when developing shear-lag solutions. The objective of this Section is to show how the methodology described in reference (McCartney, 1992) may be further extended.

Using a simple cross-ply laminate as an example, alternative methods will be described for determining the stress and displacement distributions when there is a regularly spaced array of ply cracks in the 90° ply. It will be shown how the effective elastic and thermal constants can be estimated. The example considered here, given in detail for the first time, is a very good way of presenting the important physical principles that have already been applied to the more complex case of general symmetric laminates. The approach to be described and developed in detail here has been extended in previous work by the author recently summarised in one of the publications (McCartney, 2013a) of an International Exercise (Kaddour et al., 2013a).

Consider the model of a simple $[0/90]_s$ cross-ply laminate illustrated in Fig. 6 where two possible representative volume elements (RVEs) are shown. For the first shown in Fig. 6(a) one ply crack in the inner 90° ply is located on the plane $x_1 = 0$ and neighbouring ply cracks (not shown) are on the planes $x_1 = \pm 2L$. For the second RVE shown in Fig. 6(b), the plane $x_1 = 0$ is mid-way between two neighbouring ply cracks in the inner 90° ply on the planes $x_1 = \pm L$. The inner 90° ply has total thickness $2a$ and the two outer 0° plies each have thickness denoted by b so that the total thickness of the laminate is $2h$ where $h = a + b$. A set of Cartesian coordinates (x_1, x_2, x_3) is introduced such that the origin lies on the mid-plane of the laminate at the mid-point between two neighbouring cracks in the 90° ply. The x_1-axis is directed along the principal loading direction and the x_3-axis is directed in the through-thickness direction.

The following equilibrium equations must be satisfied for both the 0°

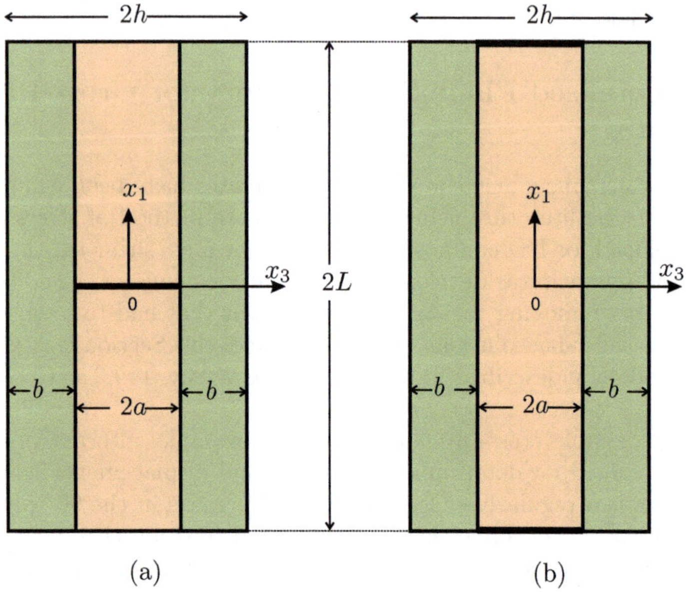

Figure 6. Representative volume elements for a cracked cross-ply laminate

and 90°plies

$$\begin{aligned}\frac{\partial\sigma_{11}}{\partial x_1}+\frac{\partial\sigma_{12}}{\partial x_2}+\frac{\partial\sigma_{13}}{\partial x_3}&=0,\\ \frac{\partial\sigma_{12}}{\partial x_1}+\frac{\partial\sigma_{22}}{\partial x_2}+\frac{\partial\sigma_{23}}{\partial x_3}&=0,\\ \frac{\partial\sigma_{13}}{\partial x_1}+\frac{\partial\sigma_{23}}{\partial x_2}+\frac{\partial\sigma_{33}}{\partial x_3}&=0,\end{aligned}\tag{142}$$

where σ_{ij} are the stress components. The plies are regarded as transverse isotropic solids so that the stress-strain-temperature relations involve the axial and transverse values of the Young's modulus E, Poisson's ratio ν, shear modulus μ and thermal expansion coefficient α. Superscripts '0' or '90' will be used to denote the ply to which a stress, strain and displacement component refers. For the 0° plies

$$\begin{aligned}
\varepsilon_{11}^0 &= \frac{\partial u_1^0}{\partial x_1} = \frac{1}{E_A^0}\sigma_{11}^0 - \frac{\nu_A^0}{E_A^0}\sigma_{22}^0 - \frac{\nu_a^0}{E_A^0}\sigma_{33}^0 + \alpha_A^0 \Delta T, \\
\varepsilon_{22}^0 &= \frac{\partial u_2^0}{\partial x_2} = -\frac{\nu_A^0}{E_A^0}\sigma_{11}^0 + \frac{1}{E_T^0}\sigma_{22}^0 - \frac{\nu_t^0}{E_T^0}\sigma_{33}^0 + \alpha_T^0 \Delta T, \\
\varepsilon_{33}^0 &= \frac{\partial u_3^0}{\partial x_3} = -\frac{\nu_a^0}{E_A^0}\sigma_{11}^0 - \frac{\nu_t^0}{E_T^0}\sigma_{22}^0 + \frac{1}{E_t^0}\sigma_{33}^0 + \alpha_t^0 \Delta T, \\
2\varepsilon_{12}^0 &= \frac{\partial u_1^0}{\partial x_2} + \frac{\partial u_2^0}{\partial x_1} = \frac{\sigma_{12}^0}{\mu_A^0}, \\
2\varepsilon_{13}^0 &= \frac{\partial u_1^0}{\partial x_3} + \frac{\partial u_3^0}{\partial x_1} = \frac{\sigma_{13}^0}{\mu_a^0}, \\
2\varepsilon_{23}^0 &= \frac{\partial u_2}{\partial x_3} + \frac{\partial u_3^0}{\partial x_2} = \frac{\sigma_{23}^0}{\mu_t^0},
\end{aligned} \tag{143}$$

while for the 90° plies

$$\begin{aligned}
\varepsilon_{11}^{90} &= \frac{\partial u_1^{90}}{\partial x_1} = \frac{1}{E_T^{90}}\sigma_{11}^{90} - \frac{\nu_A^{90}}{E_A^{90}}\sigma_{22}^{90} - \frac{\nu_t^{90}}{E_T^{90}}\sigma_{33}^{90} + \alpha_T^{90} \Delta T, \\
\varepsilon_{22}^{90} &= \frac{\partial u_2^{90}}{\partial x_2} = -\frac{\nu_A^{90}}{E_A^{90}}\sigma_{11}^{90} + \frac{1}{E_A^{90}}\sigma_{22}^{90} - \frac{\nu_a^{90}}{E_A^{90}}\sigma_{33}^{90} + \alpha_A^{90} \Delta T, \\
\varepsilon_{33}^{90} &= \frac{\partial u_3^{90}}{\partial x_3} = -\frac{\nu_t^{90}}{E_T^{90}}\sigma_{11}^{90} - \frac{\nu_a^{90}}{E_A^{90}}\sigma_{22}^{90} + \frac{1}{E_t^{90}}\sigma_{33}^{90} + \alpha_t^{90} \Delta T, \\
2\varepsilon_{12}^{90} &= \frac{\partial u_1^{90}}{\partial x_2} + \frac{\partial u_2^{90}}{\partial x_1} = \frac{\sigma_{12}^{90}}{\mu_A^{90}}, \\
2\varepsilon_{13}^{90} &= \frac{\partial u_1^{90}}{\partial x_3} + \frac{\partial u_3^{90}}{\partial x_1} = \frac{\sigma_{13}^{90}}{\mu_t^{90}}, \\
2\varepsilon_{23}^{90} &= \frac{\partial u_2^{90}}{\partial x_3} + \frac{\partial u_3^{90}}{\partial x_2} = \frac{\sigma_{23}^{90}}{\mu_a^{90}},
\end{aligned} \tag{144}$$

where the strain and displacement components are denoted by ε_{ij} and u_i respectively. The subscripts A, T and t are attached to the properties to associate them respectively with the axial, in-plane transverse and through-thickness directions of the lamina. It should be noted that the upper case subscripts A and T are associated only with in-plane directions, while the lower case subscripts are associated with the through-thickness direction. The relations (144) are either obtained by modifying directly the relations (143) for the 0° plies, or by using the relations (27)-(40) to rotate the ply by an angle $\pm 90^o$. The thermoelastic constants of individual plies in a laminate are usually assumed to be transverse isotropic so that $E_t = E_T, \nu_a = \nu_A, \mu_a = \mu_A, \alpha_t = \alpha_T$ and $E_T = 2\mu_t(1+\nu_t)$.

4.1 Free Surface, Interface, Edge and Symmetry Conditions

In order that the field equations can be solved uniquely, it is necessary to impose a sufficient number of boundary and interface conditions. The free surface ($x_3 = \pm h$) and interface ($x_3 = \pm a$) conditions will first be considered. On the free surfaces

$$\sigma_{33}^0 = \sigma_t, \sigma_{13}^0 = \sigma_{23}^0 = 0, \quad \text{on} \quad x_3 = \pm h, \tag{145}$$

and on the interfaces

$$\begin{matrix} \sigma_{33}^0 = \sigma_{33}^{90}, & \sigma_{13}^0 = \sigma_{13}^{90}, & \sigma_{23}^0 = \sigma_{23}^{90}, \\ u_1^0 = u_1^{90}, & u_2^0 = u_2^{90}, & u_3^0 = u_3^{90}, \end{matrix} \quad \text{on} \quad x_3 = \pm a \tag{146}$$

The edges $x_2 = \pm W$ are such that in-plane transverse displacement is uniform having the following values

$$u_2^0 = u_2^{90} = \pm W\varepsilon_T, \quad \text{on} \quad x_2 = \pm W, \tag{147}$$

where ε_T is the in-plane transverse strain that is uniform everywhere in the laminate when generalised plane strain conditions are imposed. The edges $x_2 = \pm W$ are assumed to have zero shear stresses so that

$$\sigma_{12}^0 = \sigma_{12}^{90} = 0, \sigma_{23}^0 = \sigma_{23}^{90} = 0, \quad \text{on} \quad x_2 = \pm W \tag{148}$$

For the above boundary conditions, and because of the symmetric nature of the laminate, there will be symmetry about $x_3 = 0$ of the stress, strain and displacement distributions such that the following conditions are satisfied

$$\sigma_{13}^0 = \sigma_{13}^{90} = 0, \sigma_{23}^0 = \sigma_{23}^{90} = 0, u_3^0 = 0, \quad \text{on} x_3 = 0 \tag{149}$$

When applying laminate edge conditions applied on planes normal to the x_1-axis, two possible approaches can be made. Consider first of all the RVE shown in Fig. 6(a) which can be used for undamaged laminates, and for damaged laminates where a ply crack in the 90° ply is located at $x_1 = 0$. The edges $x_1 = \pm L$ are such that in-plane axial displacement is uniform having the following values

$$u_1^0 = u_1^{90} = \pm L\varepsilon_A, \quad \text{on} \quad x_1 = \pm L, \tag{150}$$

where ε_A is the effective axial applied strain. The edges $x_1 = \pm L$ are assumed to have zero shear stresses so that

$$\sigma_{12}^0 = \sigma_{12}^{90} = 0, \quad \sigma_{13}^0 = \sigma_{13}^{90} = 0, \quad \text{on} \quad x_1 = \pm L \tag{151}$$

These conditions imply that there is symmetry about the plane $x_1 = 0$ so that

$$\sigma_{12}^0 = \sigma_{12}^{90} = 0, \quad \sigma_{13}^0 = \sigma_{13}^{90} = 0, \quad u_1^0 = u_1^{90} = 0, \quad \text{on} \quad x_1 = 0 \tag{152}$$

Consider now the RVE shown in Fig. 6(b) where the 90° ply cracks are located on the planes $x_1 = \pm L$. The boundary conditions applied are given by

$$\sigma_{12}^0 = \sigma_{12}^{90} = 0, \sigma_{13}^0 = \sigma_{13}^{90} = 0, \sigma_{11}^{90} = 0, u_1^0 = \pm L\varepsilon_A, \quad \text{on} \quad x_1 = \pm L \tag{153}$$

It is clear from both the RVEs shown in Fig. 6 and the boundary conditions applied on planes normal to the x_1-axis that for uniformly arrays of ply cracks there is symmetry about the planes $x_1 = 0$ and $x_1 = \pm L$.

4.2 Key Results for Undamaged Laminates

For undamaged laminates and in regions away from laminate edges, the axial and transverse strains in both 0° and 90° plies have the uniform values $\hat{\varepsilon}_A$ and $\hat{\varepsilon}_T$ respectively. The 'hat' symbol is attached to the in-plane strains to distinguish them from the differing values ε_A and ε_T that will arise when the laminate is damaged. The through-thickness stress in both plies has the uniform value σ_t. For undamaged laminates it is useful to define the following laminate constants for the 0° and 90° plies

$$\begin{aligned}
&\frac{1}{\tilde{E}_A^0} = \frac{1}{E_A^0}\left(1 - (\nu_A^0)^2 \frac{E_T^0}{E_A^0}\right), \frac{1}{\tilde{E}_T^0} = \frac{1}{E_T^0}\left(1 - (\nu_A^0)^2 \frac{E_T^0}{E_A^0}\right), \\
&\frac{\tilde{\nu}_a^0}{\tilde{E}_A^0} = \frac{\nu_a^0 + \nu_t^0 \nu_A^0}{E_A^0}, \frac{\tilde{\nu}_t^0}{\tilde{E}_T^0} = \frac{\nu_t^0}{E_T^0} + \frac{\nu_a^0 \nu_A^0}{E_A^0}, \\
&\tilde{\alpha}_A^0 = \alpha_A^0 + \nu_A^0 \frac{E_T^0}{E_A^0} \alpha_T^0, \tilde{\alpha}_T^0 = \alpha_T^0 + \nu_A^0 \alpha_A^0,
\end{aligned} \tag{154}$$

$$\begin{aligned}
&\frac{1}{\tilde{E}_A^{90}} = \frac{1}{E_A^{90}}\left(1 - (\nu_A^{90})^2 \frac{E_T^{90}}{E_A^{90}}\right), && \frac{1}{\tilde{E}_T^{90}} = \frac{1}{E_T^{90}}\left(1 - (\nu_A^{90})^2 \frac{E_T^{90}}{E_A^{90}}\right), \\
&\frac{\tilde{\nu}_a^{90}}{\tilde{E}_A^{90}} = \frac{\nu_a^{90}}{E_A^{90}} + \frac{\nu_t^{90} \nu_A^{90}}{E_A^{90}}, && \frac{\tilde{\nu}_t^{90}}{\tilde{E}_T^{90}} = \frac{\nu_t^{90}}{E_T^{90}} + \frac{\nu_a^{90} \nu_A^{90}}{E_A^{90}}, \\
&\tilde{\alpha}_A^{90} = \alpha_A^{90} + \nu_A^{90} \frac{E_T^{90}}{E_A^{90}} \alpha_T^{90}, && \tilde{\alpha}_T^{90} = \alpha_T^{90} + \nu_A^{90} \alpha_A^{90}
\end{aligned} \tag{155}$$

From a consideration of mechanical equilibrium the uniform ply stresses can be used to define, for an undamaged laminate, the effective axial stress $\hat{\sigma}_A$ and the effective in-plane transverse stress $\hat{\sigma}_T$ as follows

$$h\hat{\sigma}_A = b\hat{\sigma}_{11}^0 + a\hat{\sigma}_{11}^{90}, \tag{156}$$

$$h\,\hat{\sigma}_T = b\hat{\sigma}^0_{22} + a\hat{\sigma}^{90}_{22} \tag{157}$$

The 'hat' symbol is used to distinguish these effective stresses from those that will result when the laminate is damaged. Corresponding to the uniform through-thickness stress σ_t, an effective through-thickness strain $\hat{\varepsilon}_t$ can be defined by the relation

$$h\hat{\varepsilon}_t = b\hat{\varepsilon}^0_{33} + a\hat{\varepsilon}^{90}_{33} \tag{158}$$

It should be noted that the value σ_t for the through-thickness stress of an undamaged laminate corresponds to the effective value when the laminate is damaged. It can be shown that on defining the constants

$$\begin{aligned}
&A = \frac{b}{h}\tilde{E}^0_A + \frac{a}{h}\tilde{E}^{90}_T, B = \frac{b}{h}\nu^0_A\tilde{E}^0_T + \frac{a}{h}\nu^{90}_A\tilde{E}^{90}_T, C = \frac{b}{h}\tilde{\nu}^0_a + \frac{a}{h}\tilde{\nu}^{90}_t,\\
&F = \frac{b}{h}\tilde{E}^0_T + \frac{a}{h}\tilde{E}^{90}_A, G = \frac{b}{h}\tilde{\nu}^0_t + \frac{a}{h}\tilde{\nu}^{90}_a,\\
&P = \frac{b}{h}\tilde{E}^0_A\tilde{\alpha}^0_A + \frac{a}{h}\tilde{E}^{90}_T\tilde{\alpha}^{90}_T, Q = \frac{b}{h}\tilde{E}^0_T\tilde{\alpha}^0_T + \frac{a}{h}\tilde{E}^{90}_A\tilde{\alpha}^{90}_A,
\end{aligned} \tag{159}$$

the thermoelastic constants of an undamaged cross-ply laminate are given by

$$\begin{aligned}
&E^{(L)}_A = A - \frac{B^2}{F}, \quad E^{(L)}_T = F - \frac{B^2}{A}, \quad \nu^{(L)}_A = \frac{B}{F},\\
&\nu^{(L)}_a = C - \frac{BG}{F}, \quad \nu^{(L)}_t = G - \frac{BC}{A},\\
&\alpha^{(L)}_A = \frac{1}{E^{(L)}_A}\left(P - \nu^{(L)}_A Q\right), \quad \alpha^{(L)}_T = \frac{1}{E^{(L)}_T}\left(Q - \nu^{(L)}_A\frac{E^{(L)}_T}{E^{(L)}_A}P\right),
\end{aligned} \tag{160}$$

such that the in-plane non-shear stress-strain relations for an undamaged laminate are

$$\begin{aligned}
\hat{\varepsilon}_A &= \frac{1}{E^{(L)}_A}\hat{\sigma}_A - \frac{\nu^{(L)}_A}{E^{(L)}_A}\hat{\sigma}_T - \frac{\nu^{(L)}_a}{E^{(L)}_A}\sigma_t + \alpha^{(L)}_A\Delta T,\\
\hat{\varepsilon}_T &= -\frac{\nu^{(L)}_A}{E^{(L)}_A}\hat{\sigma}_A + \frac{1}{E^{(L)}_T}\hat{\sigma}_T - \frac{\nu^{(L)}_t}{E^{(L)}_T}\sigma_t + \alpha^{(L)}_T\Delta T
\end{aligned} \tag{161}$$

It should be noted that

$$\begin{aligned}
&\tilde{E}^{(L)}_A = A = \frac{b}{h}\tilde{E}^0_A + \frac{a}{h}\tilde{E}^{90}_T, \qquad \nu^{(L)}_A\frac{E^{(L)}_T}{E^{(L)}_A} = \frac{B}{A} = \frac{b\,\nu^0_A\tilde{E}^0_T + a\nu^{90}_A\tilde{E}^{90}_T}{b\tilde{E}^0_A + a\,\tilde{E}^{90}_T},\\
&\tilde{E}^{(L)}_T = F = \frac{b}{h}\tilde{E}^0_T + \frac{a}{h}\tilde{E}^{90}_A, \qquad \nu^{(L)}_A = \frac{B}{F} = \frac{b\nu^0_A\tilde{E}^0_T + a\nu^{90}_A\tilde{E}^{90}_T}{b\tilde{E}^0_T + a\tilde{E}^{90}_A},
\end{aligned} \tag{162}$$

where

$$\tilde{E}_A^{(L)} = \frac{E_A^{(L)}}{1-(\nu_A^{(L)})^2 E_T^{(L)}/E_A^{(L)}}, \quad \tilde{E}_T^{(L)} = \frac{E_T^{(L)}}{1-(\nu_A^{(L)})^2 E_T^{(L)}/E_A^{(L)}} \tag{163}$$

To calculate the through-thickness properties the following constants are first defined

$$\begin{aligned}
A' &= \frac{b}{h}\frac{\nu_a^0+\nu_t^0\nu_A^0}{E_A^0}\tilde{E}_A^0 + \frac{a}{h}\left(\frac{\nu_t^{90}}{E_T^{90}} + \frac{\nu_a^{90}\nu_A^{90}}{E_A^{90}}\right)\tilde{E}_T^{90} = \frac{b}{h}\tilde{\nu}_a^0 + \frac{a}{h}\tilde{\nu}_t^{90}, \\
B' &= \frac{b}{h}\left(\frac{\nu_t^0}{E_T^0} + \frac{\nu_a^0\nu_A^0}{E_A^0}\right)\tilde{E}_T^0 + \frac{a}{h}\frac{\nu_a^{90}+\nu_t^{90}\nu_A^{90}}{E_A^{90}}\tilde{E}_A^{90} = \frac{b}{h}\tilde{\nu}_t^0 + \frac{a}{h}\tilde{\nu}_a^{90}, \\
C' &= \frac{b}{h}\left(\frac{1}{E_t^0} - \frac{\nu_a^0\tilde{\nu}_a^0}{E_A^0} - \frac{\nu_t^0\tilde{\nu}_t^0}{E_T^0}\right) + \frac{a}{h}\left(\frac{1}{E_t^{90}} - \frac{\nu_t^{90}\tilde{\nu}_t^{90}}{E_T^{90}} - \frac{\nu_a^{90}\tilde{\nu}_a^{90}}{E_A^{90}}\right), \\
P' &= \frac{b}{h}\left(\alpha_t^0 + \frac{\nu_a^0}{E_A^0}\tilde{E}_A^0\tilde{\alpha}_A^0 + \frac{\nu_t^0}{E_T^0}\tilde{E}_T^0\tilde{\alpha}_T^0\right) \\
&+ \frac{a}{h}\left(\alpha_t^{90} + \frac{\nu_a^{90}}{E_A^{90}}\tilde{E}_A^{90}\tilde{\alpha}_A^{90} + \frac{\nu_t^{90}}{E_T^{90}}\tilde{E}_T^{90}\tilde{\alpha}_T^{90}\right)
\end{aligned} \tag{164}$$

It can be shown that the through-thickness properties are given by

$$\begin{aligned}
&\nu_a^{(L)} = A' - B'\nu_A^{(L)}, \quad \nu_t^{(L)} = B' - A'\nu_A^{(L)}\frac{E_T^{(L)}}{E_A^{(L)}}, \\
&\frac{1}{E_t^{(L)}} = C' + A'\frac{\nu_a^{(L)}}{E_A^{(L)}} + B'\frac{\nu_t^{(L)}}{E_T^{(L)}}, \quad \alpha_t^{(L)} = P' - A'\alpha_A^{(L)} - B'\alpha_T^{(L)},
\end{aligned} \tag{165}$$

such that

$$\hat{\varepsilon}_t = -\frac{\nu_a^{(L)}}{E_A^{(L)}}\hat{\sigma}_A - \frac{\nu_t^{(L)}}{E_T^{(L)}}\hat{\sigma}_T + \frac{1}{E_t^{(L)}}\sigma_t + \alpha_t^{(L)}\Delta T, \tag{166}$$

It can be shown that the same values result for the minor Poisson's ratios $\nu_a^{(L)}$ and $\nu_t^{(L)}$ when using the relations (160) or (165).

4.3 Effective Applied Stresses and Strains

Consider a symmetric damaged laminate of length $2L$, width $2W$ and total thickness $2h$, for which the plies are uniformly thick and perfectly bonded together. In-plane loading is applied by imposing uniform axial and transverse displacements, denoted by $\pm U_A$ and $\pm U_T$, on the edges of the laminate thus defining an effective axial strain $\varepsilon_A = U_A/L$ and an effective transverse strain $\varepsilon_T = U_T/W$. The faces of the laminate are subjected to a uniform applied stress denoted by σ_t. Corresponding to the effective in-plane strains ε_A and ε_T, axial and transverse effective applied stresses can be defined for damaged laminates by

$$\sigma_A = \frac{1}{4Wh}\int_{-W}^{W}\int_{-h}^{h}\sigma_{11}\mathrm{d}x_2\mathrm{d}x_3, \quad \sigma_T = \frac{1}{4Lh}\int_{-L}^{L}\int_{-h}^{h}\sigma_{22}\mathrm{d}x_1\mathrm{d}x_3 \tag{167}$$

Corresponding to the applied through-thickness stress σ_t an effective through-thickness strain ε_t can be defined for damaged and undamaged laminates by

$$\varepsilon_t = \frac{1}{8LWh}\int_{-W}^{W}\int_{-L}^{L}\left[u_3(x_1,x_2,h) - u_3(x_1,x_2,-h)\right]\mathrm{d}x_1\mathrm{d}x_2 \tag{168}$$

For an undamaged laminate subject to these displacements, the axial and transverse strains will be uniform throughout all plies of the laminate having the values ε_A and ε_T respectively, and the through-thickness stress will be uniform throughout the laminate having the value σ_t in each ply. It then follows from (167) and (168) that $\sigma_A = \hat{\sigma}_A$, $\sigma_T = \hat{\sigma}_T$ and $\varepsilon_t = \hat{\varepsilon}_t$ where $\hat{\sigma}_A$, $\hat{\sigma}_T$ and $\hat{\varepsilon}_t$ are defined by the relations (156)-(158).

4.4 Generalised Plane Strain Solution

Generalised plane strain conditions are assumed so that the stress and strain distributions do not depend on the x_3-coordinate. This situation occurs when the displacement field is of the form

$$u_1 = u_1(x_1,x_3), \quad u_2 = \varepsilon_T x_2, \quad u_3 = u_3(x_1,x_3), \tag{169}$$

where $\varepsilon_{22} = \varepsilon_T$ is the uniform in-plane transverse strain in both of the 0° and 90° plies. In addition, it is assumed that on the laminate faces $x_3 = \pm h$ that $\sigma_{33}^0 \equiv \sigma_{33}^{90} \equiv \sigma_t$ where σ_t is a uniform applied through-thickness stress and $\sigma_{13} \equiv \sigma_{23} \equiv 0$. If $\varepsilon_T = 0$ then the laminate is highly constrained in the transverse direction leading to well known conditions of plane strain deformation. It should be noted that the solution for undamaged laminates derived in the previous section is automatically one of generalised plane strain as the transverse strain ε_T is uniform everywhere in those regions of the laminate that are sufficiently well away from the edges.

It can be shown that for the 0° plies

$$\sigma_{22}^0 = \frac{\nu_A^0 E_T^0}{E_A^0}\sigma_{11}^0 + \nu_t^0\sigma_{33}^0 - E_T^0\alpha_T^0\Delta T + E_T^0\varepsilon_T, \tag{170}$$

$$\varepsilon_{11}^0 = \frac{1}{\tilde{E}_A^0}\sigma_{11}^0 - \frac{\tilde{\nu}_a^0}{\tilde{E}_A^0}\sigma_{33}^0 + \tilde{\alpha}_A^0 \Delta T - \nu_A^0 \frac{E_T^0}{E_A^0}\varepsilon_T, \tag{171}$$

$$\varepsilon_{33}^0 = -\frac{\tilde{\nu}_a^0}{\tilde{E}_A^0}\sigma_{11}^0 + \frac{1}{\tilde{E}_t^0}\sigma_{33}^0 + \tilde{\alpha}_t^0 \Delta T - \nu_t^0 \varepsilon_T, \tag{172}$$

where

$$\frac{1}{\tilde{E}_t^0} = \frac{1}{E_t^0} - \frac{(\nu_t^0)^2}{E_T^0}, \tilde{\alpha}_t^0 = \alpha_t^0 + \nu_t^0 \alpha_T^0 \tag{173}$$

For the 90° ply

$$\sigma_{22}^{90} = \nu_A^{90}\sigma_{11}^{90} + \nu_a^{90}\sigma_{33}^{90} - E_A^{90}\alpha_A^{90}\Delta T + E_A^{90}\varepsilon_T, \tag{174}$$

$$\varepsilon_{11}^{90} = \frac{1}{\tilde{E}_T^{90}}\sigma_{11}^{90} - \frac{\tilde{\nu}_t^{90}}{\tilde{E}_T^{90}}\sigma_{33}^{90} + \tilde{\alpha}_T^{90}\Delta T - \nu_A^{90}\varepsilon_T, \tag{175}$$

$$\varepsilon_{33}^{90} = -\frac{\tilde{\nu}_t^{90}}{\tilde{E}_T^{90}}\sigma_{11}^{90} + \frac{1}{\tilde{E}_t^{90}}\sigma_{33}^{90} + \tilde{\alpha}_t^{90}\Delta T - \nu_a^{90}\varepsilon_T, \tag{176}$$

where

$$\frac{1}{\tilde{E}_t^{90}} = \frac{1}{E_t^{90}} - \frac{(\nu_a^{90})^2}{E_A^{90}}, \quad \tilde{\alpha}_t^{90} = \alpha_t^{90} + \nu_a^{90}\alpha_A^{90} \tag{177}$$

4.5 Key Results for Damaged Laminates

The easiest approach is to consider the stress and displacement representation involving an unknown stress transfer function $C(x_1)$ and to show that it satisfies required field equations and boundary conditions. For generalised plane strain conditions where it is assumed $\sigma_{12} = \sigma_{23} = 0$ and $u_2 = \varepsilon_T x_2$, the stress distribution is assumed to have the following form

$$\sigma_{11}^0(x_1, x_3) = C(x_1) + \hat{\sigma}_{11}^0, \quad \sigma_{11}^{90}(x_1, x_3) = -\frac{b}{a}C(x_1) + \hat{\sigma}_{11}^{90}, \tag{178}$$

$$\sigma_{13}^0(x_1, x_3) = C'(x_1)\,(h - x_3), \quad \sigma_{13}^{90}(x_1, x_3) = \frac{b}{a}C'(x_1)x_3, \tag{179}$$

$$\sigma_{22}^0(x_1, x_3) = \frac{\nu_A^0 E_T^0}{E_A^0}C(x_1) + \frac{1}{2}\nu_t^0 C''(x_1)\,(h - x_3)^2 + \hat{\sigma}_{22}^0, \tag{180}$$

$$\sigma_{22}^{90}(x_1, x_3) = -\frac{b}{a}\nu_A^{90}C(x_1) + \frac{b}{2a}\nu_a^{90}C''(x_1)\left(ah - x_3^2\right) + \hat{\sigma}_{22}^{90}, \tag{181}$$

$$\begin{aligned} \sigma_{33}^0(x_1, x_3) &= \frac{1}{2}C''(x_1)\,(h - x_3)^2 + \sigma_t, \\ \sigma_{33}^{90}(x_1, x_3) &= \frac{b}{2a}C''(x_1)\left(ah - x_3^2\right) + \sigma_t, \end{aligned} \tag{182}$$

and the corresponding displacement representation is

$$u_1^0(x_1,x_3) = \left(\frac{\tilde{\nu}_a^0}{2\tilde{E}_A^0}(x_3-a)^2 - \frac{\tilde{\nu}_t^{90}}{\tilde{E}_T^{90}}b(x_3-a) + \frac{b^2-(h-x_3)^2}{2\mu_a^0}\right)C'(x_1)$$
$$-\left(\frac{(h-x_3)^4 - b^4 + 4b^3(x_3-a)}{24\tilde{E}_t^0} + \frac{ab\,(2h+b)\,(x_3-a)}{6\tilde{E}_t^{90}}\right)C'''(x_1) + A(x_1), \tag{183}$$

$$u_1^{90}(x_1,x_3) = \frac{b}{2a}\left(\frac{1}{\mu_t^{90}} - \frac{\tilde{\nu}_t^{90}}{\tilde{E}_T^{90}}\right)(x_3^2-a^2)C'(x_1)$$
$$+ \frac{b}{24a}\frac{1}{\tilde{E}_t^{90}}\left(x_3^4 - a^4 - 6ah(x_3^2-a^2)\right)C'''(x_1) + A(x_1) \tag{184}$$

$$u_3^0(x_1,x_3) = \hat{\varepsilon}_{33}^0(x_3-a) + \hat{\varepsilon}_{33}^{90}a + \left(b\frac{\tilde{\nu}_t^{90}}{\tilde{E}_T^{90}} - \frac{\tilde{\nu}_a^0}{\tilde{E}_A^0}(x_3-a)\right)C(x_1)$$
$$+ \frac{1}{6}\left[\frac{1}{\tilde{E}_t^0}\left(b^3-(h-x_3)^3\right) + \frac{ab}{\tilde{E}_t^{90}}(2h+b)\right]C''(x_1), \tag{185}$$

$$u_3^{90}(x_1,x_3) = \frac{b}{a}\frac{\tilde{\nu}_t^{90}}{\tilde{E}_T^{90}}C(x_1)x_3 + \frac{b}{6a}\frac{1}{\tilde{E}_t^{90}}C''(x_1)\left(3ah - x_3^2\right)x_3 + \hat{\varepsilon}_{33}^{90}x_3 \tag{186}$$

It follows that

$$h\sigma_A = b\sigma_{11}^0(x_1,x_3) + a\sigma_{11}^{90}(x_1,x_3) = b\hat{\sigma}_{11}^0 + a\hat{\sigma}_{11}^{90} = h\hat{\sigma}_A, \tag{187}$$

so that the effective applied axial stress for damaged laminate σ_A is equal to the effective axial stress $\hat{\sigma}_A$ for the corresponding undamaged laminate. The function $A(x_1)$ appearing in (183) and (184) is for the moment arbitrary.

The representation automatically satisfies the equilibrium equations (142) and the required interface continuity conditions. In addition, all the stress-strain relations (143) and (144), except for the two relations $(143)_1$ and $(144)_1$ involving the axial strains ε_{11}^0 and ε_{11}^{90} respectively, for any functions $C(x_1)$ and $A(x_1)$. It is possible, however, to satisfy these axial relations after they are averaged through the thickness of the 0° and 90° plies respectively, as will now be described.

Assuming symmetry about the mid-plane $x_3 = 0$, the average of any quantities $f_0(x_1,x_3)$ and $f_{90}(x_1,x_3)$ associated with the 0° and 90° plies are defined respectively by

$$\bar{f}_0(x_1) = \frac{1}{b}\int_a^h f_0(x_1,x_3)\,\mathrm{d}x_3\,, \quad \bar{f}_{90}(x_1) = \frac{1}{a}\int_0^a f_{90}(x_1,x_3)\,\mathrm{d}x_3 \tag{188}$$

On averaging (183) and (184) using (188) it can be shown that

$$\begin{aligned}\bar{u}_1^0(x_1) &= \left(\frac{\tilde{\nu}_a^0}{6\tilde{E}_A^0} - \frac{\tilde{\nu}_t^{90}}{2\tilde{E}_T^{90}} + \frac{1}{3\mu_a^0}\right) b^2 C'(x_1) \\ &- \left(\frac{1}{20\tilde{E}_t^0} + \frac{a\,(2a+3b)}{12b^2\tilde{E}_t^{90}}\right) b^4 C'''(x_1) + A(x_1),\end{aligned} \tag{189}$$

$$\begin{aligned}\bar{u}_1^{90}(x_1) &= \frac{a}{3b}\left(\frac{\tilde{\nu}_t^{90}}{\tilde{E}_T^{90}} - \frac{1}{\mu_t^{90}}\right) b^2\, C'(x_1) \\ &+ \frac{a^2(5b+4a)}{30b^3}\frac{1}{\tilde{E}_t^{90}} b^4 C'''(x_1) + A(x_1)\end{aligned} \tag{190}$$

On substituting into the averaged axial stress-strain relations for the 0° and 90° plies, and on eliminating the terms $\hat{\sigma}_{11}^0$ and $\hat{\sigma}_{11}^{90}$, and the function $A(x_1)$, it can be shown that the stress transfer function must satisfy the following fourth order ordinary differential equation

$$Fb^4C''''(x_1) - Gb^2C''(x_1) + HC(x_1) = 0, \tag{191}$$

where

$$\begin{aligned}F &= \frac{1}{20\tilde{E}_t^0} + \frac{2}{15\tilde{E}_t^{90}}\frac{a}{b}\left(\frac{a^2}{b^2} + \frac{5}{2}\frac{a}{b} + \frac{15}{8}\right) > 0, \\ G &= \frac{1}{3}\left(\frac{1}{\mu_a^0} + \frac{1}{\mu_t^{90}}\frac{a}{b} + \frac{\tilde{\nu}_a^0}{\tilde{E}_A^0} - \frac{\tilde{\nu}_t^{90}}{\tilde{E}_T^{90}}\frac{2a+3b}{b}\right), \\ H &= \frac{1}{\tilde{E}_A^0} + \frac{1}{\tilde{E}_T^{90}}\frac{b}{a} > 0\end{aligned} \tag{192}$$

The function $A(x_1)$ can be calculated using either of the following two equivalent relations

$$\begin{aligned}A(x_1) &= \left(\frac{1}{20\tilde{E}_t^0} + \frac{a(2a+3b)}{12b^2\tilde{E}_t^{90}}\right) b^4 C'''(x_1) \\ &- \left(\frac{\tilde{\nu}_a^0}{3\tilde{E}_A^0} - \frac{\tilde{\nu}_t^{90}}{2\tilde{E}_T^{90}} + \frac{1}{3\mu_a^0}\right) b^2 C'(x_1) + \frac{1}{\tilde{E}_A^0}\bar{C}(x_1) + \hat{\varepsilon}_A x_1,\end{aligned} \tag{193}$$

$$\begin{aligned}A(x_1) &= -\frac{a^2(5b+4a)}{30b^3\tilde{E}_t^{90}} b^4 C'''(x_1) \\ &- \frac{a}{3b}\left(\frac{4a+3b}{2a}\frac{\tilde{\nu}_t^{90}}{\tilde{E}_T^{90}} - \frac{1}{\mu_t^{90}}\right) b^2 C'(x_1) - \frac{b}{a}\frac{1}{\tilde{E}_T^{90}}\bar{C}(x_1) + \hat{\varepsilon}_A x_1,\end{aligned} \tag{194}$$

where

$$\bar{C}(x_1) \equiv \int_0^{x_1} C(x)\mathrm{d}x. \tag{195}$$

When there is symmetry about $x_1 = 0$, it follows that $C'(x_1) = 0$ and $C'''(x_1) = 0$. The integration constant has been selected so that $A(x_1) = 0$.

On substituting (193) in (189)

$$\bar{u}_1^0(x_1) = \frac{1}{\tilde{E}_A^0}\bar{C}(x_1) - \frac{\tilde{\nu}_a^0}{6\tilde{E}_A^0}b^2 C'(x_1) + \hat{\varepsilon}_A x_1, \tag{196}$$

and on substituting (194) in (190)

$$\bar{u}_1^{90}(x_1) = -\frac{b}{a}\frac{1}{\tilde{E}_T^{90}}\bar{C}(x_1) - \frac{2a+3b}{6b}\frac{\tilde{\nu}_t^{90}}{\tilde{E}_T^{90}}b^2 C'(x_1) + \hat{\varepsilon}_A x_1 \tag{197}$$

It should be noted that $\bar{u}_1^0(0) = \bar{u}_1^{90}(0) = 0$ consistent with the conditions $(152)_3$. Because the function $A(x_1)$ is now known in terms of $C(x_1)$, the u_1 displacement distributions (183) and (184) are fully specified in terms of the stress transfer function $C(x_1)$.

On defining

$$r = \frac{G}{2F}, \quad s = \sqrt{\frac{H}{F}}, \tag{198}$$

the most general solution of the differential equation (191) satisfying the symmetry condition $C(x_1) \equiv C(-x_1)$ is given by

- if $s > r$:

$$C(x_1) = P\cosh\frac{px_1}{b}\cos\frac{qx_1}{b} + Q\sinh\frac{px_1}{b}\sin\frac{qx_1}{b}, \tag{199}$$

- if $s = r$:

$$C(x_1) = P\cosh\frac{px_1}{b} + Q\frac{x_1}{b}\sinh\frac{px_1}{b}, \tag{200}$$

- if $s < r$:

$$C(x_1) = P\cosh\frac{px_1}{b}\cosh\frac{qx_1}{b} + Q\sinh\frac{px_1}{b}\sinh\frac{qx_1}{b}, \tag{201}$$

where

$$p = \sqrt{\frac{1}{2}(r+s)}, \quad q = \sqrt{\frac{1}{2}|r-s|} \tag{202}$$

4.6 Solution for Ply Cracks

Consider now a uniform array of ply cracks, having density $\rho = 1/(2L)$, in the 90° ply of the cross-ply laminate as shown in Fig. 6(b). The tractions on the ply crack surfaces must be zero so that from (179) and (178)

$$\sigma_{13}^{90}(L, x_3) = \frac{b}{a}C'(L)x_3 = 0, \quad \text{implying} \quad C'(L) = 0, \tag{203}$$

$$\sigma_{11}^{90}(L, x_3) = -\frac{b}{a}C(L) + \hat{\sigma}_{11}^{90} = 0, \quad \text{implying} \quad C(L) = \frac{a}{b}\hat{\sigma}_{11}^{90} \tag{204}$$

On applying these conditions and on writing $P = A + B, Q = A - B$, the parameters A and B must be selected so that

$$A = -\frac{a}{b}\frac{(p-q)\tanh\frac{(p-q)L}{b}}{\cosh\frac{(p+q)L}{b}}\Lambda\hat{\sigma}_{11}^{90}, \quad B = \frac{a}{b}\frac{(p+q)\tanh\frac{(p+q)L}{b}}{\cosh\frac{(p-q)L}{b}}\Lambda\hat{\sigma}_{11}^{90}, \tag{205}$$

where

$$\frac{1}{\Lambda} = (q+p)\tanh\frac{(p+q)L}{b} + (q-p)\tanh\frac{(p-q)L}{b} \tag{206}$$

The only boundary condition for a damaged laminate that has not been satisfied is given by $(153)_1$. It is clear from (183), (184) and (193) or (194) that it is not possible for this boundary condition to be satisfied by the approximate solution derived. The boundary condition $(153)_1$ is now replaced by the following averaged condition

$$\bar{u}_1^0 = \pm L\varepsilon_A, \quad \text{on} \quad x_1 = \pm L \tag{207}$$

It can then be shown using (196) and (203) that

$$\varepsilon_A = \frac{a\,\Phi\hat{\sigma}_{11}^{90}}{L\tilde{E}_A^0} + \hat{\varepsilon}_A \tag{208}$$

where

$$\Phi = \frac{4\Lambda pq}{p^2 - q^2}\tanh\frac{(p+q)L}{b}\tanh\frac{(p-q)L}{b} \tag{209}$$

On using (157) it can be shown using (180) and (181) that the effective applied transverse stress σ_T defined by $(167)_2$ is given by

$$\sigma_T = \frac{ab}{Lh}\left(\nu_A^0\frac{E_T^0}{E_A^0} - \nu_A^{90}\right)\Phi\hat{\sigma}_{11}^{90} + \hat{\sigma}_T \tag{210}$$

It is assumed that the values of $\sigma_A = \hat{\sigma}_A$, $\varepsilon_T = \hat{\varepsilon}_T$, σ_t and ΔT are known so that the stress-strain relations (161) for an undamaged laminate may be written

$$\begin{aligned} \hat{\varepsilon}_A &= \frac{1}{E_A^{(L)}}\sigma_A - \frac{\nu_A^{(L)}}{E_A^{(L)}}\hat{\sigma}_T - \frac{\nu_a^{(L)}}{E_A^{(L)}}\sigma_t + \alpha_A^{(L)}\,\Delta T\,, \\ \varepsilon_T &= -\frac{\nu_A^{(L)}}{E_A^{(L)}}\sigma_A + \frac{1}{E_T^{(L)}}\hat{\sigma}_T - \frac{\nu_t^{(L)}}{E_T^{(L)}}\sigma_t + \alpha_T^{(L)}\,\Delta T \end{aligned} \tag{211}$$

On using (208) and (210) to eliminate $\hat{\varepsilon}_A$ and $\hat{\sigma}_T$ respectively, it can be shown following extensive algebraic manipulation that

$$\varepsilon_A = \frac{1}{E_A}\,\sigma_A - \frac{\nu_A}{E_A}\,\sigma_T - \frac{\nu_a}{E_A}\,\sigma_t + \alpha_A\,\Delta T, \tag{212}$$

$$\varepsilon_T = -\,\frac{\nu_A}{E_A}\,\sigma_A + \frac{1}{E_T}\,\sigma_T - \frac{\nu_t}{E_T}\,\sigma_t + \alpha_T\,\Delta T, \tag{213}$$

where the thermoelastic constants of the damaged laminate are defined by

$$\frac{1}{E_A} = \frac{1}{E_A^{(L)}} + \left(1 - \nu_A^{(L)}\nu_A^{90}\right)^2 \frac{a}{L}\frac{\tilde{E}_T^{90}}{E_A^{(L)}\tilde{E}_A^{0}}\frac{\Phi}{\xi}, \tag{214}$$

$$\frac{1}{E_T} = \frac{1}{E_T^{(L)}}\left[1 + \left(\nu_A^0\frac{E_T^0}{E_A^0} - \nu_A^{90}\right)^2 \frac{b^2}{h^2}\frac{a}{L}\frac{\tilde{E}_A^0\tilde{E}_T^{90}}{\tilde{E}_A^{(L)}E_T^{(L)}}\frac{\Phi}{\xi}\right], \tag{215}$$

implying $E_T = \xi E_T^{(L)}$,

$$\frac{\nu_A}{E_A} = \frac{\nu_A^{(L)}}{E_A^{(L)}} + \left(1 - \nu_A^{(L)}\nu_A^{90}\right)\left(\nu_A^0\frac{E_T^0}{E_A^0} - \nu_A^{90}\right)\frac{b}{h}\frac{a}{L}\frac{\tilde{E}_T^{90}}{E_A^{(L)}E_T^{(L)}}\frac{\Phi}{\xi}, \tag{216}$$

$$\begin{aligned} \frac{\nu_a}{E_A} &= \frac{\nu_a^{(L)}}{E_A^{(L)}} \\ &+ \left[\frac{\nu_a^{(L)}}{E_A^{(L)}} - \frac{\nu_t^{90}}{E_T^{90}} + \nu_A^{90}\left(\frac{\nu_t^{(L)}}{E_T^{(L)}} - \frac{\nu_a^{90}}{E_A^{90}}\right)\right]\left(1 - \nu_A^{(L)}\nu_A^{90}\right)\frac{a}{L}\frac{\tilde{E}_A^{(L)}}{E_A^{(L)}}\frac{\tilde{E}_T^{90}}{\tilde{E}_A^{0}}\frac{\Phi}{\xi}, \end{aligned} \tag{217}$$

$$\begin{aligned} \frac{\nu_t}{E_T} &= \frac{\nu_t^{(L)}}{E_T^{(L)}} \\ &- \left[\frac{\nu_a^{(L)}}{E_A^{(L)}} - \frac{\nu_t^{90}}{E_T^{90}} + \nu_A^{90}\left(\frac{\nu_t^{(L)}}{E_T^{(L)}} - \frac{\nu_a^{90}}{E_A^{90}}\right)\right]\left(\nu_A^0\frac{E_T^0}{E_A^0} - \nu_A^{90}\right)\frac{b}{h}\frac{a}{L}\frac{\tilde{E}_T^{90}}{E_T^{(L)}}\frac{\Phi}{\xi}, \end{aligned} \tag{218}$$

$$\alpha_A = \alpha_A^{(L)} + \left(1 - \nu_A^{(L)}\nu_A^{90}\right)\left(\alpha_A^{(L)} + \nu_A^{90}\alpha_T^{(L)} - \tilde{\alpha}_T^{90}\right)\frac{a}{L}\frac{\tilde{E}_A^{(L)}}{E_A^{(L)}}\frac{\tilde{E}_T^{90}}{\tilde{E}_A^{0}}\frac{\Phi}{\xi}, \quad (219)$$

$$\alpha_T = \alpha_T^{(L)} - \left(\alpha_A^{(L)} + \nu_A^{90}\alpha_T^{(L)} - \tilde{\alpha}_T^{90}\right)\left(\nu_A^{0}\frac{E_T^0}{E_A^0} - \nu_A^{90}\right)\frac{b}{h}\frac{a}{L}\frac{\tilde{E}_T^{90}}{E_T^{(L)}}\frac{\Phi}{\xi}, \quad (220)$$

where

$$\xi = 1 - \left(\nu_A^{0}\frac{E_T^0}{E_A^0} - \nu_A^{90}\right)^2\frac{b^2}{h^2}\frac{a}{L}\frac{\tilde{E}_A^0\tilde{E}_T^{90}}{\tilde{E}_A^{(L)}E_T^{(L)}}\Phi. \quad (221)$$

The results (212)-(220) show that the stress-strain relations of a damaged laminate are exactly of the same form as those for an undamaged laminate. The formation of damage affects only the values of the thermoelastic constants, and not the form of the stress-strain relations.

4.7 Through-Thickness Properties of Damaged Laminates

On applying (168) to the ply crack problem being considered

$$\varepsilon_t = \frac{1}{Lh}\int_0^L u_3^0(x_1, h)\mathrm{d}x_1 \quad (222)$$

It can be shown again following extensive algebraic manipulation that through-thickness stress-strain relation is obtained

$$\varepsilon_t = -\frac{\nu_a}{E_A}\sigma_A - \frac{\nu_t}{E_T}\sigma_T + \frac{1}{E_t}\sigma_t + \alpha_t\Delta T, \quad (223)$$

where

$$\frac{\nu_a}{E_A} = \frac{\nu_a^{(L)}}{E_A^{(L)}} + \Omega\left(1 - \nu_A^{90}\nu_A^{(L)}\right)\frac{b}{h}\frac{a}{L}\frac{\tilde{E}_T^{90}}{E_A^{(L)}}\frac{\Phi}{\xi}, \quad (224)$$

$$\frac{\nu_t}{E_T} = \frac{\nu_t^{(L)}}{E_T^{(L)}} - \Omega\left(\nu_A^{0}\frac{E_T^0}{E_A^0} - \nu_A^{90}\right)\frac{b^2}{h^2}\frac{a}{L}\frac{\tilde{E}_A^0\tilde{E}_T^{90}}{\tilde{E}_A^{(L)}E_T^{(L)}}\frac{\Phi}{\xi}, \quad (225)$$

$$\frac{1}{E_t} = \frac{1}{E_t^{(L)}} + \Omega\left\{\frac{\nu_a^{(L)}}{E_A^{(L)}} - \frac{\nu_t^{90}}{E_T^{90}} + \nu_A^{90}\left(\frac{\nu_t^{(L)}}{E_T^{(L)}} - \frac{\nu_a^{90}}{E_A^{90}}\right)\right\}\frac{b}{h}\frac{a}{L}\tilde{E}_T^{90}\frac{\Phi}{\xi}, \quad (226)$$

$$\alpha_t = \alpha_t^{(L)} - \Omega\left(\alpha_A^{(L)} + \nu_A^{90}\alpha_T^{(L)} - \tilde{\alpha}_T^{90}\right)\frac{b}{h}\frac{a}{L}\tilde{E}_T^{90}\frac{\Phi}{\xi}, \quad (227)$$

and where

$$\Omega = \frac{\tilde{\nu}_a^0}{\tilde{E}_A^0} - \frac{\tilde{\nu}_t^{90}}{\tilde{E}_T^{90}} - \frac{\nu_t^{(L)}}{E_T^{(L)}}\left(\nu_A^{0}\frac{E_T^0}{E_A^0} - \nu_A^{90}\right) \quad (228)$$

The relations (217) and (218) are equivalent to the results (224) and (225) because it can be shown that

$$\frac{\nu_a^{(L)}}{E_A^{(L)}} - \frac{\nu_t^{90}}{E_T^{90}} + \nu_A^{90}\left(\frac{\nu_t^{(L)}}{E_T^{(L)}} - \frac{\nu_a^{90}}{E_A^{90}}\right) = \Omega\frac{b}{h}\frac{\tilde{E}_A^0}{\tilde{E}_A^{(L)}} \tag{229}$$

4.8 Example Predictions

The key results of this section are expressions for the various effective thermoelastic properties of a simple cross-ply laminate having an array of uniformly spaced ply cracks. For the example the following ply properties, typical of a transverse isotropic carbon fibre reinforced composite, are used:

$$\begin{array}{llllll}
E_A & = 140.77\,\mathrm{GPa}, & E_T & = 8.85\,\mathrm{GPa}, & E_t & = 8.85\,\mathrm{GPa}, \\
\nu_A & = 0.28, & \nu_a & = 0.28, & \nu_t & = 0.43, \\
\mu_A & = 4.59\,\mathrm{GPa}, & \mu_a & = 4.59\,\mathrm{GPa}, & \mu_t & = 3.09441\,\mathrm{GPa}, \\
\alpha_A & = 0.245\mathrm{x}10^{-6}\mathrm{K}^{-1}, & \alpha_T & = 45.6\mathrm{x}10^{-6}\mathrm{K}^{-1}, & \alpha_t & = 45.6\mathrm{x}10^{-6}\mathrm{K}^{-1}
\end{array}$$

When these ply properties are used in conjunction with the formulae (214)-(220) and (224)-(227), for a set of ply crack densities in the range 0 – 4 cracks/mm, the results shown in Fig. 7 are obtained. The results shown assume the following identifications:

$$\begin{array}{llllll}
E_A & \equiv \mathrm{EA}, & E_T & \equiv \mathrm{ET}, & E_t & \equiv \mathrm{Est}, \\
\nu_A & \equiv \mathrm{nuA}, & \nu_a & \equiv \mathrm{nusa}, & \nu_t & \equiv \mathrm{nust}, \\
\mu_A & \equiv \mathrm{muA}, & \mu_a & \equiv \mathrm{musa}, & \mu_t & \equiv \mathrm{must}, \\
\alpha_A & \equiv \mathrm{alA}, & \alpha_T & \equiv \mathrm{alT}, & \alpha_t & \equiv \mathrm{alst}
\end{array}$$

It is noted that for ply crack densities exceeding 2/mm, the effective properties no longer depend on the crack density. Also, it is seen that the effective in-plane transverse modulus E_T is hardly affected by ply cracking, and that the effective axial thermal expansion coefficient is affected a great deal by ply cracking.

Similar situations arise for ply cracking in the 90° plies of general symmetric laminates for a variety of laminate configurations considered in the WWFE III International Exercise (Kaddour et al., 2013a) concerned with the assessment of damage models for composite laminates. Results analogous to those derived here for cross-ply laminates have been derived for general symmetric laminates (McCartney, 2013a,b) and assessed/discussed by the organisers of the Exercise (Kaddour et al., 2013b). However, the author recommends that a great deal of caution is applied when considering the comparison models as the information presented by the organisers

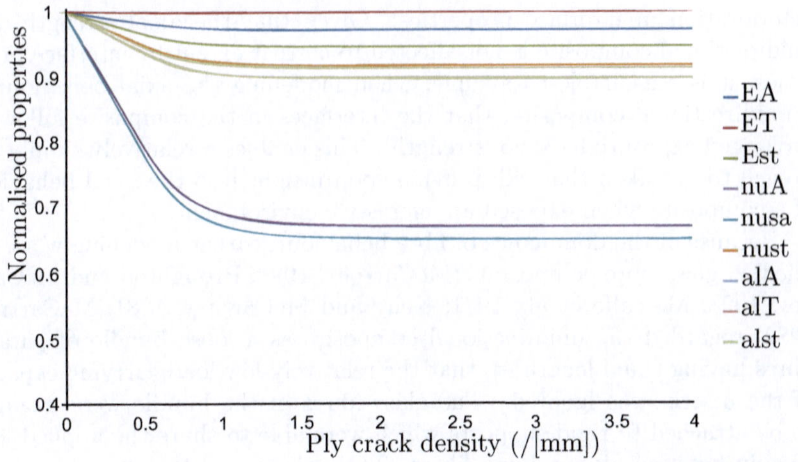

Figure 7. Predictions of the normalised effective properties of a simple cross-ply laminate as a function of the density of a uniform distribution of ply cracks in the 90° ply

(Kaddour et al., 2013b) relating to the generalised plane strain model developed by the author is wholly misleading, and conclusions are not justified by the information presented to the exercise by the author (see McCartney (2013b)).

5 Model of Composite Degradation Due to Environmental Damage

The axial strength of unidirectional fibre reinforced composites is controlled by the strength of the fibres. In cross-ply laminates the axial strength of the laminate is controlled to a large degree by the strength of the fibres in the 0° plies. Fibre strength is statistical in nature due to the presence of defects both on fibre surfaces, and in their interior. The effect of interface properties on axial strength are of secondary importance, and modelling their effect on axial strength requires the use of sophisticated stress transfer models and Monte Carlo simulation techniques. For unprotected glass fibres, it is well known that the environmental exposure of the composite leads to time dependent reductions in fibre strength. The strength reduction of the fibres results because of the progressive growth of fibre defects caused by stress corrosion cracking at a microscopic level.

Environmental exposure, provided that it is saturated, can lead to a deterioration in interface properties. Given that the axial strength of a unidirectional composite is not affected to a great extent by interface properties, it is reasonable to assume, when modelling the axial behaviour of a unidirectional composite, that the interfaces in the composite following prolonged exposure have no 'strength'. This enables a relatively simply approach to be taken that will provide good insight into the axial behaviour of a composite when exposed an aggressive environment.

Because of the dominance of fibre behaviour, earlier modelling work applied to glass fibre composites (McCartney, 1998; Broughton and McCartney, 1998; Metcalfe et al., 1971; Kelly and McCartney, 1981; McCartney, 1982) regarded the unidirectional composite as a loose bundle of parallel fibres having equal length, so that the relatively low load carrying capacity of the matrix was ignored. The glass fibres in the bundle were assumed to be attached to rigid supports which were able to share the applied load equally between all surviving fibres. The objective of this report is to extend the existing loose bundle model so that the load carrying capacity of the matrix is taken into account when considering the axial behaviour of glass fibre reinforced composites subject to environmental exposure.

5.1 Model Geometry

Consider a unidirectional fibre reinforced composite having a fibre volume fraction V_f and matrix volume fraction V_m such that $V_f + V_m = 1$. The composite has been wholly immersed in an aggressive environment for a sufficient time for the composite to be fully saturated. The application of axial load to the composite leads to the environmental growth of defects in the fibres; a phenomenon well known to afflict glass fibres. The interfaces between the fibres and matrix are regarded as being significantly weakened by the environment to the extent that it can be assumed that the fibres and matrix behave independently in regions of the composite that are well away from the uniaxial loading mechanism where clamping effects become important. This assumption means that the composite can be modelled as a parallel bar model, as shown in Fig. 8.

The fibres in the composite are regarded as acting as a loose bundle forming one bar of the model. The matrix material in the composite is considered as being gathered together to form the other bar of the model which is regarded as homogeneous, i.e. the bar is solid. When a fibre fails the load it carried is shared between the surviving fibres in the bundle and the matrix in such a way that all surviving fibres and matrix experience the same axial strain increment. The fibres are assumed to have the same length so that each surviving fibre has the same stress throughout the progressive

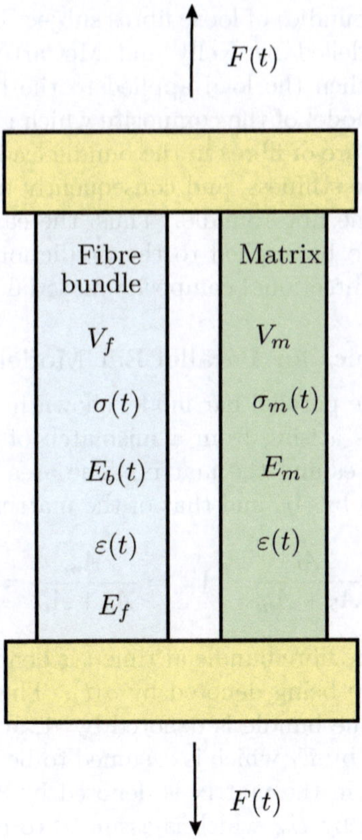

Figure 8. Schematic diagram of the parallel bar model of a unidirectional composite for predicting effects of environmental exposure on axial composite properties

failure process. The composite is subject to a fixed applied load F for all times $t > 0$, where $t = 0$ corresponds to the time when the fixed load F is first applied. Environmental defect growth in the fibres leads to progressive fibre failure until the bundle collapses. It is assumed that bundle collapse corresponds to the catastrophic failure of the composite, i.e. the matrix strength is insufficient to maintain the load when all the fibres have failed. The objective is to develop the parallel bar model of a composite so that it can predict the dependence of composite life t_f on the fixed applied load F, and the dependence of the residual strength $F^*(t)$ of the composite on elapsed time t from the instant of first loading.

The behaviour of bundles of loose fibres subject to environmental degradation has been modelled by Kelly and McCartney (1981); McCartney (1982) for the case when the load applied to the bundle is fixed in time. For the parallel bar model of the composite which is subject to a fixed load F, the progressive failure of fibres in the bundle leads to a time dependence of the effective bundle stiffness, and consequently to a time dependence of the load applied to the fibre bundle. Thus, the earlier modelling requires modification if it is to be applied to the prediction of the behaviour of a uniaxially loaded unidirectional composite material having weak interfaces.

5.2 Basic Mechanics for Parallel Bar Model of a Composite

The analysis of the parallel bar model shown in Fig. 8 will neglect any axial thermal stresses arising from a mismatch of the thermal expansion coefficients of the fibres and the matrix. The area fraction of all fibres in the bundle is denoted by A_b, and that of the matrix is A_m. It follows that

$$V_f = \frac{A_b}{A_b + A_m}, \quad V_m = \frac{A_m}{A_b + A_m} = 1 - V_f \tag{230}$$

The load applied to the fibre bundle at time t is denoted by $F_b(t)$, the stress in each surviving fibre being denoted by $\sigma(t)$. The cross-sectional area of each of the fibres in the bundle is denoted by A, and the axial modulus of each fibres is denoted by E_f which is assumed to be time independent. The axial stress at time t in the matrix is denoted by $\sigma_m(t)$. The modulus of the matrix is denoted by E_m which is assumed to be independent of time. A time dependence could be included to account for visco-elastic effects, or for time-dependence arising from matrix ageing.

The axial strain in all surviving fibres of the bundle and the matrix has the same time dependent value that is denoted by $\varepsilon(t)$. As thermal expansion mismatch effects are neglected it follows that

$$\varepsilon(t) = \frac{\sigma(t)}{E_f} = \frac{\sigma_m(t)}{E_m} \tag{231}$$

The balance of forces in the parallel bar model leads to the equilibrium relation

$$F_b(t) + A_m \sigma_m(t) = F \tag{232}$$

The number of surviving fibres in the bundle at time t is denoted by $N(t)$ so that the load applied to the bundle at time t may be written

$$F_b(t) = N(t) A \sigma(t). \tag{233}$$

On substituting (233) into (232) followed by the elimination of $\sigma_m(t)$ using (231) it is easily shown that the number of fibres surviving at time t is related to the fibre stress $\sigma(t)$ through the following relation that quantitatively characterises the load sharing that occurs when fibres in the composite fail

$$\left[\frac{N(t)}{N_0} + \alpha\right]\sigma(t) = \frac{F}{A_b}, \quad \text{where} \quad \alpha = \frac{V_m E_m}{V_f E_f}, \tag{234}$$

and where $N_0 = N(0)$ and $A_b = N_0 A$. This is the generalisation to a composite material of the relation used in the modelling a loose bundle of fibres subject to environmental degradation (Kelly and McCartney, 1981; McCartney, 1982) which is recovered from (234) on letting $\alpha \to 0$.

It is useful to relate the number of fibres surviving in the bundle at time t to the effective axial modulus of the bundle $E_b(t)$. The effective stress applied to the bundle is defined by

$$\sigma_b(t) = \frac{F_b(t)}{A_b}, \tag{235}$$

and since the axial strain of the bundle and the individual fibres has the value $\sigma(t)$

$$\varepsilon(t) = \frac{\sigma(t)}{E_f} = \frac{\sigma_b(t)}{E_b(t)} = \frac{F_b(t)}{A_b E_b(t)} = \frac{N(t)}{N_0}\frac{\sigma(t)}{E_b(t)}, \tag{236}$$

where use has been made of (233) and (235). Clearly the effective axial modulus of the fibre bundle is given by

$$E_b(t) = \frac{N(t)}{N_0} E_{\mathrm{f}} \tag{237}$$

The effective stress σ_{app} applied to the composite is defined by

$$\sigma_{\mathrm{app}} = \frac{F}{A_b + A_m}, \tag{238}$$

and it can be shown from (230) and (234), together with the fact that $A_b = N_0 A$, that

$$\sigma_{\mathrm{app}}(t) = E_c(t)\,\varepsilon(t), \quad \text{where} \quad E_c(t) = V_{\mathrm{f}} E_b(t) + V_{\mathrm{m}} E_m, \tag{239}$$

where $E_c(t)$ is the effective axial modulus of the composite defined by the rule of mixtures, as to be expected.

5.3 Accounting for Defect Growth

The objective here is to show how the analysis of Kelly and McCartney (1981), developed for loose bundles exposed to an aggressive environment, must be modified for application to a unidirectional composite having weak interfaces. The analysis is based on the assumption that the strength of individual fibres is determined by surface defects whose effective size and distribution along the fibre surface is statistically distributed.

Fibre failure is assumed to be governed by a Griffith type of failure criterion having the form

$$K^2 = y^2\sigma^2 a = K_{\mathrm{Ic}}^2, \tag{240}$$

where K is an effective stress intensity factor for a fibre defect of effective size a subject to a fibre stress s, K_{Ic} is the effective fracture toughness of the fibre material, and where y is a dimensionless parameter designed to account for defect geometry. The aggressive environment leads to defect growth when the fibre is under load. Such defect growth is assumed to be governed by a growth law of the form

$$\frac{\mathrm{d}a}{\mathrm{d}t} = CK^n, \tag{241}$$

where C and n are material constants.

When a constant load is applied to a unidirectional composite, exposed to an aggressive environment to the point of saturation, the fibre defects grow in size according to the growth law (241) eventually leading to fibre failure when the failure criterion (240) is satisfied. Thus fibres progressively fail and the load carried by failed fibres is, for the parallel bar model under discussion, transferred to the surviving fibres and matrix using the load sharing rule (234). The stress in each fibre of the system is thus time dependent. It is useful to present here the relationship that determines the initial defect size $X_0(t)$ that requires a time t to grow to the critical size $a_c(t)$, at which the fibre stress is $s(t)$, under the influence of a time dependent fibre stress history $\sigma(\tau)$; $0 < \tau < t$. The critical defect size at time t is predicted by (240) to be

$$a_c(t) = \left[\frac{K_{Ic}}{y\sigma(t)}\right]^2, \tag{242}$$

and it can be shown on integrating (241) between the limits $X_0(t)$ and $a_c(t)$ that

$$X_0(t) = \frac{K_{\mathrm{Ic}}^2}{y^2}\left[\sigma^{n-2}(t) + (n-2)\lambda\int_0^t \sigma^n(\tau)\mathrm{d}\tau\right]^{\frac{2}{2-n}}, \tag{243}$$

where

$$\lambda = \frac{1}{2} C K_{\rm Ic}^{n-2} y^2 \tag{244}$$

On using (240) it follows from (243) that the initial strength $\sigma_i(t)$ of the fibres, that fail at time t when their stress is $\sigma(t)$, is given by

$$\sigma_i(t) = \left[\sigma^{n-2}(t) + (n-2)\lambda \int_0^t \sigma^n(\tau)\mathrm{d}\tau \right]^{\frac{1}{n-2}} \tag{245}$$

The cross-sectional area of the sample of unidirectional composite is assumed to be large enough for there to be a very large number of fibres. It can then be assumed that the bundle of fibres used in the parallel bar model contains every possible fibre strength that can arise in the statistical distribution. It is assumed that the strength distribution of the fibres is given by the two parameter Weibull distribution (Weibull, 1951) so that, for a large bundle of N_0 fibres, the expected number of fibres N surviving when the stress in each fibre is σ is given by

$$N = N_0 \exp\left[-\left(\frac{\sigma}{\sigma_0}\right)^m \right] \tag{246}$$

where σ_0 is a scaling parameter that will depend on the length of the composite.

5.4 Prediction of Static Strength

It is useful to investigate the prediction of the static strength of a unidirectional composite assuming that the parallel bar model is valid. When using (246) in conjunction with (234) it is easily shown that the total load carried by the composite, when the stress in surviving fibres has the value σ is given by

$$\hat{F} = \hat{\sigma}\left[\alpha + \mathrm{e}^{-\hat{\sigma}^m}\right], \tag{247}$$

where $\hat{F}$ and $\hat{\sigma}$ are a dimensionless normalised load and stress respectively defined by

$$\hat{F} = \frac{F}{N_0 \sigma_0 A}, \quad \hat{\sigma} = \frac{\sigma}{\sigma_0} \tag{248}$$

The static strength of the composite is the maximum value of the load that can be carried by the composite. The maximum load occurs when $\hat{F}$ has a local maximum when plotted as a function of $\hat{\sigma}$. The maximum fibre stress $\sigma_{\rm max}$ satisfies the transcendental equation

$$m\,\hat{\sigma}_{\rm max}^m = 1 + \alpha\,\mathrm{e}^{\hat{\sigma}_{\rm max}^m} \tag{249}$$

The corresponding static strength for the composite is then obtained using

$$\frac{F_{\rm max}}{N_0 A\,\sigma_0} = \hat{F}_{\rm max} = \hat{\sigma}_{\rm max}\left[\alpha + {\rm e}^{-\hat{\sigma}^m_{\rm max}}\right] = \alpha\frac{m\hat{\sigma}^{m+1}_{\rm max}}{m\,\hat{\sigma}^m_{\rm max} - 1}, \tag{250}$$

which is consistent with the known result for a loose bundle Kelly and McCartney (1981) when the limit $\alpha \to 0$ is taken.

The equation (249) governing the maximum fibre stress does not always have a solution as is easily seen by examining the form of the LHS and RHS of (249). On letting $x = \hat{\sigma}^m_{\rm max}$ the critical conditions defining the limit of solutions to (249) may be written

$$mx = 1 + \alpha{\rm e}^x, \quad m = \alpha{\rm e}^x \tag{251}$$

These conditions correspond to the touching of the curves $y = mx - 1$ and $y = \alpha{\rm e}^x$. It is easily seen that the critical condition occurs when

$$x = \ln\frac{m}{\alpha} = \frac{1+m}{m} \tag{252}$$

It is concluded that the equation (249) has a solution only if

$$\alpha < m{\rm e}^{-\left(1+\frac{1}{m}\right)} \tag{253}$$

If this condition is not satisfied then it is deduced that the fibres progressively fail until there is just one surviving fibre which will then fail, i.e. the bundle does not suddenly collapse. The value of the Weibull modulus m for fibres of interest is usually such that the condition (253) is satisfied so that bundle collapse is always expected in practice.

5.5 Prediction of Progressive Damage

At time t the fibres that survive in the composite are those whose initial strengths were greater than $\sigma_i(t)$ defined by (245). It then follows from (246) that the expected number of surviving fibres $N(t)$ at time t is given by

$$\frac{N(t)}{N_0} = \hat{N}(t) = \exp\left[-\left(\frac{\sigma_i(t)}{\sigma_0}\right)^m\right] \tag{254}$$

On substituting (245) in (254)

$$\left(\ln\frac{1}{\hat{N}(t)}\right)^{\frac{n-2}{m}} = \hat{\sigma}^{n-2}(t) + (n-2)\eta\int\limits_0^t \hat{\sigma}^n(\tau){\rm d}\tau, \tag{255}$$

where

$$\eta = \lambda\sigma_0^2 = \frac{1}{2}CK_{\mathrm{Ic}}^{n-2}y^2\sigma_0^2, \tag{256}$$

and where use has been made of the definitions (248), which when applied to the load sharing rule (234) lead to

$$\hat{N}(t) = \frac{\hat{F}}{\hat{\sigma}(t)} - \alpha \tag{257}$$

On substituting (257) into (255)

$$\left(\ln\frac{\hat{\sigma}(t)}{\hat{F} - \alpha\hat{\sigma}(t)}\right)^{\frac{n-2}{m}} = \hat{\sigma}^{n-2}(t) + (n-2)\eta\int\limits_0^t \hat{\sigma}^n(\tau)\mathrm{d}\tau. \tag{258}$$

On differentiating (258) with respect to t, the following differential equation governing the time dependence of the normalised fibre stress $\hat{\sigma}(t)$ is obtained

$$\left[\frac{1}{m}\frac{\hat{F}}{\hat{F} - \alpha\hat{\sigma}(t)}\left(\ln\frac{\hat{\sigma}(t)}{\hat{F} - \alpha\hat{\sigma}(t)}\right)^{\frac{n-m-2}{m}} - \hat{\sigma}^{n-2}(t)\right]\frac{d\hat{\sigma}(t)}{d(\eta t)} = \hat{\sigma}^{n+1}(t) \tag{259}$$

This differential equation is solved by standard numerical techniques subject to the initial condition

$$\hat{\sigma}(0) = s_0, \tag{260}$$

where s_0 is the solution of the transcendental equation

$$\hat{F} = s_0\left[\alpha + \mathrm{e}^{-s_0^m}\right], \tag{261}$$

corresponding to (247), that must be solved numerically.

5.6 Predicting the Failure Stress and Time to Failure

The structure of the differential equation is such that $d\hat{\sigma}/dt \to \infty$ when $\hat{\sigma}(t) \to \hat{\sigma}_f$ where

$$\frac{1}{m}\frac{\hat{F}}{\hat{F} - \alpha\hat{\sigma}_f}\left(\ln\frac{\hat{\sigma}_f}{\hat{F} - \alpha\hat{\sigma}_f}\right)^{\frac{n-m-2}{m}} = \hat{\sigma}_f^{n-2}. \tag{262}$$

The stress $\hat{\sigma}_f$ in the surviving fibres when the composite fails can thus be determined using numerical methods without having to solve the differential equation (259). It should be noted that when $\hat{F} = \hat{F}_{\max}$ the solution of (262) is given by $\hat{\sigma}_f = \hat{\sigma}_{\max}$ where $\hat{\sigma}_{\max}$ and $\hat{F}_{\max}$ are given by (249) and (250) respectively. The time to failure for the composite is denoted by t_f.

The number of surviving fibres just before composite failure is obtained using (254) and is given by

$$\frac{N(t_f)}{N_0} = \hat{N}_f = \mathrm{e}^{-\hat{\sigma}_i^m}, \quad \hat{\sigma}_i = \frac{\sigma_i(t_f)}{\sigma_0} \tag{263}$$

The transcendental equation (262), that usually must be solved numerically, involves the dimensionless loading parameter $\hat{F}$ in a complicated way. It is useful to unravel the dependence on this parameter by using the load sharing rule (257) to express (262) in terms of N_f as follows

$$\hat{F}^{n-2} = \frac{1}{m} \frac{\left[\hat{N}_f + \alpha\right]^{n-1}}{\hat{N}_f} \left(\ln \frac{1}{\hat{N}_f}\right)^{\frac{n-m-2}{m}} \tag{264}$$

Having solved (264) to find $\hat{N}_f$ using numerical methods, the normalised failure stress is obtained, on making use of (234), from the relation

$$\hat{\sigma}_f = \frac{\hat{F}}{\hat{N}_f + \alpha} \tag{265}$$

The time to failure t_f can be predicted only by solving the differential equation (259) in the normalised stress range $s_0 \le \hat{\sigma}(t) \le \hat{\sigma}_f$.

5.7 Predicting Residual Strength

A key requirement concerning the effects of environment on composite degradation is the prediction of the time dependence of the residual strength of a composite. This has already been considered for the case of a loose bundle of fibres (McCartney, 1982). The objective now is to extend the analysis, and simplify it so far as is possible, so that the residual strength of a unidirectional composite with weak interfaces can be predicted. After an elapsed time t from the application of a fixed load F, the load is instantaneously increased until the composite fails catastrophically. Just before the load is suddenly increased the stress in the surviving fibres has the value $\sigma(t)$ and at any stage during the subsequent instantaneous load increase the value of the stress in the fibres is denoted by s. When the fibre stress has the value s the critical defect size has the following value specified by (240)

$$a_c^* = \left(\frac{K_{\mathrm{Ic}}}{y\, s}\right)^2 \tag{266}$$

It is necessary to calculate the original size X^* of the critical defect using the relations (240) and (241). It is easily shown that

$$(X^*)^{\frac{2-n}{2}} = (a_c^*)^{\frac{2-n}{2}} + \frac{1}{2} C\,(n-2) y^n \int_0^t \sigma^n(\tau) \mathrm{d}\tau \tag{267}$$

On using (240) the initial strength of the fibres that are critical at time t when the fibre stress has the value s is denoted by s_i and is given, on using (267), by the relation

$$\hat{s}_i^{n-2} = \hat{s}^{n-2} + (n-2)\eta \int_0^t \hat{\sigma}^n(\tau) \mathrm{d}\tau, \tag{268}$$

where use has been made of (256) and where

$$\hat{s}_i = \frac{s_i}{\sigma_0}, \quad \hat{s} = \frac{s}{\sigma_0} \tag{269}$$

On using (258) the relation (268) may be written in the form

$$\hat{s}_i^{n-2} = \hat{s}^{n-2} - k(t), \tag{270}$$

where

$$k(t) = \hat{\sigma}^{n-2}(t) - \left[\ln \frac{\hat{\sigma}(t)}{\hat{F} - \alpha \hat{\sigma}(t)} \right]^{\frac{n-2}{m}} \tag{271}$$

The load applied to the composite F_s, when the fibre stress has the value s, is obtained from (248) and (257) so that

$$\frac{F_s}{N_0 \sigma_0 A} = \hat{F}_s = \hat{s} \left[\alpha + \hat{N}_s \right], \tag{272}$$

where $\hat{F}_s$ is the normalised applied load and where $\hat{N}_s$ is the normalised number of surviving fibres when the load on the composite is such that the fibre stress has the value s. It follows from (246) that

$$\hat{N}_s = \mathrm{e}^{-\hat{s}_i^m} \tag{273}$$

On substituting in (272) the following expression is derived for the normalised load applied to the composite during a residual strength test

$$\hat{F}_s = \hat{s} \left[\alpha + \mathrm{e}^{-\hat{s}_i^m} \right]. \tag{274}$$

The residual strength $S(t)$ of the composite at time t is the maximum value of F_s when s is varied, or alternatively the maximum value of $\hat{F}_s$ when $\hat{s}$ is varied. Noting that $k(t)$ is independent of $\hat{s}$, the maximum value of $\hat{F}_s$ occurs when $\hat{s}_i = x(t)$ which satisfies the transcendental equation

$$1 + \alpha e^{x^m(t)} = m x^m(t) \left(1 + \frac{k(t)}{x^{n-2}(t)}\right) \tag{275}$$

On using (270) the stress $\sigma_{\max}(t)$ in the surviving fibres just before the composite fails during a residual strength test is obtained from

$$\frac{\sigma_{\max}(t)}{\sigma_0} = \left[x^{n-2}(t) + k(t)\right]^{\frac{1}{n-2}} = \hat{\sigma}_{\max}(t). \tag{276}$$

It then follows from (270) and (274) that the residual strength of the composite $S(t)$ is obtained using

$$\frac{S(t)}{\sigma_0} = \hat{\sigma}_{\max}(t) \left[\alpha + e^{-x^m(t)}\right] = \hat{S}(t). \tag{277}$$

When $t = 0$ it can be shown using (258) that $k(0) = 0$ in which case the transcendental equation (275) reduces to the form (249) which needs to be solved when calculating the static strength of the composite.

5.8 Example Prediction

In order to assess the properties of the model some example predictions have been made to illustrate the principal characteristics. There are four parameters that need to be specified in order to obtain predictions:

1. The Weibull exponent characterising the strength distribution of the fibres before environmental exposure. This parameter, which often has values in the range 4 – 8, appears in the relation (246) defining the expected number of fibre failures for a given fibre stress in a static loading test. The value $m = 8$ will be used here. The value of m is usually obtained from single fibre strength tests.
2. The exponent n appearing in the defect growth law (241). This parameter, which usually has values in the range 3 – 30, is often obtained from stress corrosion cracking tests carried out using monolithic glass testpieces rather than fibres. The value $n = 20$ is used here.
3. The ratio α defined by (234) which takes approximate account of the properties of the fibre and matrix, and also of the fibre volume fraction. For many glass fibre composites of interest, the value of α lies in the range 0 - 0.1. It will be assumed here that $\alpha = 0.025$.

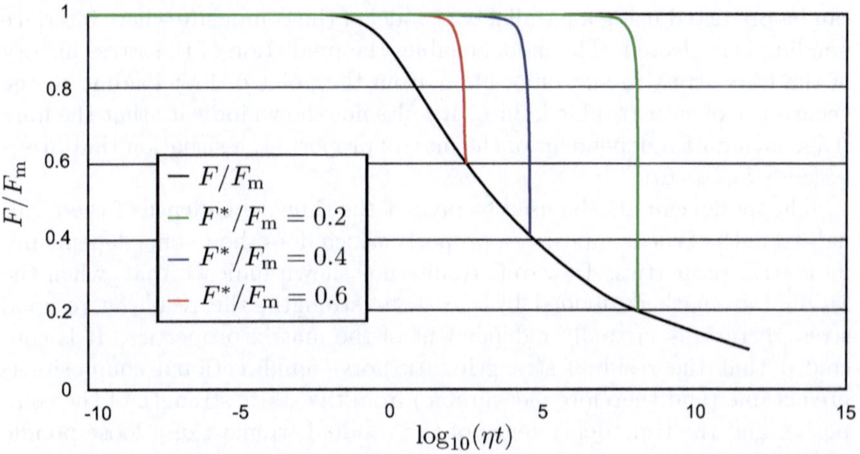

Figure 9. Schematic diagram of the parallel bar model of a unidirectional composite for predicting effects of environmental exposure on axial composite properties

4. The level of loading applied axially to the composite where the model assumes that the ratio F/F_m is given where F is the axial load applied to the composite and F_m is the static strength, i.e. the strength of the composite before environmental exposure. The value of F/F_m always lies in the range 0 – 1.

The Euler-Richardson solution technique (Churchhouse, 1981) is used to solve the ordinary differential equation (259) where the normalised dimensionless time ηt may be regarded as an unknown function of $\hat{\sigma}$. In other words, the differential equation can be used directly to determine an increment in the value of ηt for any given increment in $\hat{\sigma}$. The initial condition is specified by (260) and (261) and the range $s_0 \leq \hat{\sigma} \leq \hat{\sigma}_f$ is subdivided into 100 equal intervals when solving the differential equation. The upper limit $\hat{\sigma}_f$ is determined by the relations (264) and (265). Figure 9 shows the result of solving the differential equation (259) to find the normalised time ηt_f for various values of the loading ratio F/F_m. The normalising parameter η is defined by (256). It is seen that as $F/F_m \to 1$ the lifetime tends to zero. Figure 9 also shows predictions of the normalised residual strength $\hat{S}$ defined by (277), as a function of the normalised time ηt.

The principal conclusion to be drawn from the results presented is that the time dependence of the axial properties of a unidirectional fibre reinforced glass composite subject to environmental exposure under fixed load can be predicted using a parallel bar model of the composite where interface bonding is neglected. The model enables the prediction of the stress history of the fibre stress in surviving fibres from the point of first loading to the occurrence of catastrophic failure. Results not shown indicate that the fibre stress is almost independent of the matrix properties, a situation that arises because $E_{\mathrm{m}} \ll E_{\mathrm{f}}$.

The model can also be used to predict the time dependence of the residual strength of the composite, a property which does show some dependence on matrix properties. However, results not shown indicate that, when the residual strength is divided by the static strength, the resulting residual strength ratio is virtually independent of the matrix properties. It is concluded that the residual strength ratio for a unidirectional composite is predictable (and therefore measurable) from the static strength of the composite, and the time dependence of the residual strength of a loose bundle of fibres.

6 Closing Remarks

A varied set of topics concerning the behaviour of composite materials has been considered in this paper. They concern the estimation of the undamaged properties of plies in terms of fibre and matrix properties, the estimation of the undamaged properties of general symmetric laminates, the consideration of an elegant method of considering cracks in anisotropic materials using orthogonal polynomials, a detailed treatment of ply cracking in a simple cross-ply laminate, and the modelling of the effects of environmental exposure on the lifetime and residual strength of unidirectional composites. Much of the work presented here has not been published before. For the analyses dealing with composite damage, example predictions have been given to help readers understand the capabilities of the various damage models.

It is hoped that readers of this paper will be convinced that analytical modelling, which has been undertaken in some quite complex situations, enables much deeper insight into the modelling of composite material systems than numerical solution methods permit, and provides opportunities for convenient design methods based on relatively compact formulae rather than on data tables and graphs that have to be generated when using numerical methods such as finite element analysis.

Bibliography

W.R. Broughton and L.N. McCartney. Predictive models for assessing long-term performance of polymer matrix composites. Technical Report CMMT(A)95, NPL, Teddington, April 1998.

R.F. Churchhouse. Numerical methods. In W. Ledermann, editor, *Handbook of Applicable Mathematics*, volume 3, pages 319–321, Chichester, 1981. Wiley.

G.M.L. Gladwell and A.H. England. *J. Mech. Appl. Math.*, 30:175, 1977.

Z. Hashin. Analysis of composite materials - a survey. *Trans. ASME. J. Appl. Mech.*, 50:481–505, 1983.

A.S. Kaddour, M.J. Hinton, P.A. Smith, and S. Li. The background to the third world-wide failure exercise. *J. Comp. Mater.*, 47(20-21):2417–2426, 2013a.

A.S. Kaddour, M.J. Hinton, P.A. Smith, and S. Li. A comparison between the predictive capability of matrix cracking, damage and failure criteria for fibre reinforced laminates: part a of the third world-wide failure exercise. *J. Comp. Mater.*, 47(20-21):2749–2779, 2013b.

A. Kelly and L.N. McCartney. Failure by stress corrosion of bundles of fibres. *Proceedings of the Royal Society of London*, A374:475–489, 1981.

J.C. Maxwell. *A Treatise on Electricity and Magnetism*, volume 1. Clarendon Press, Oxford, 1st edition, 1873.

L.N. McCartney. Time dependent strength of large bundles of fibres loaded in corrosive environments. *Fibre Science & Technology*, 16:95–109, 1982.

L.N. McCartney. Theory of stress transfer in a 0-90-0 cross-ply laminate containing a parallel array of transverse cracks. *J. Mech. Phys. Solids*, 40:27–68, 1992.

L.N. McCartney. Model of composite degradation due to environmental damage. Technical Report CMMT(A)124, NPL, Teddington, September 1998.

L.N. McCartney. Maxwell's far-field methodology predicting elastic properties of multiphase composites reinforced with aligned transversely isotropic spheroids. *Phil. Mag.*, 90:4175–4207, 2010.

L.N. McCartney. Derivations of energy-based modelling for ply cracking in general symmetric laminates. *J. Comp. Mater.*, 47(20-21):2641–2673, 2013a.

L.N. McCartney. Energy methods for modelling damage in laminates. *J. Comp. Mater.*, 47(20-21):2613–2640, 2013b.

L.N. McCartney and T.A.E. Gorley. Complex variable method of calculating stress intensity factors for cracks in plates. In A.R. Luxmoore et al., editor, *Proc. 4th Int. Conf. on Numerical Methods in Fracture Mechanics*, Swansea, 1987. Pineridge Press.

L.N. McCartney and A. Kelly. Maxwell's far-field methodology applied to the prediction of properties of multi-phase isotropic particulate composites. *Proc. Roy. Soc.*, A464:423–446, 2008.

A.G. Metcalfe, M.E. Gulden, and G.K. Schmitz. *Glass Technology*, 12: 15–23, 1971.

D.P. Rooke and D.J. Cartwright. *Compendium of Stress Intensity*. Hillingdon Press, Uxbridge, 1st edition, 1976.

G.C. Sih and H. Liebowitz. Mathematical theories of brittle fracture. In H. Liebowitz, editor, *Fracture*, volume II, page 67, San Diego, 1968. Academic Press.

W. Weibull. A statistical distribution function of wide applicability. *Journal of Applied Mechanics*, 19:293–297, 1951.

Analysis of Failure in Composite Structures

Ramesh Talreja *

* Department of Aerospace Engineering & Department of Materials Science and Engineering, College Station, Texas 77843, USA

Abstract Fiber reinforced composite materials provide a high level of structural safety against failure by tolerating damage (failure at microstructure level) while still sustaining significant loads. Using full load-bearing capacity of composite structures requires, however, reliable analysis of failure at micro and macro levels. The current failure theories, which are for homogenized composites, are unable to meet this requirement. This exposition reviews some of the commonly used theories to examine reasons for this limitation. Physical mechanisms underlying failure in composite materials are then discussed, motivating development of failure analysis with multi-scale approaches. Such approaches can also include effect of defects on failure. A comprehensive scheme is put forward for future development of physically based failure analysis of composite structures.

1 Introduction

Composite materials not only offer high specific stiffness and strength properties, they also possess ability to contain cracks while retaining significant load-bearing capacity. This aspect makes it possible to design structures with high levels of safety, particularly when inspection in service environment is not feasible or cost-effective. In fact, the ability of a composite structure to tolerate cracks can be controlled by devising various combinations of fiber architecture parameters, e.g. thickness and stacking sequence of fiber-reinforced layers with straight or woven fibers, fiber volume fraction, and hybrid reinforcement (e.g. mix of glass and carbon fibers). Such inherent advantages of composite materials are only possible if mechanics based analyses are available to relate the fiber architecture parameters to the composite performance. Such is not the case today. The current industry practice is to use ad-hoc strength theories that are unable to address the variety of failure mechanisms that initiate locally from the microstructure designed into a composite material. The inherent problem with these theories is that they are formulated on homogenized composite solids, whereby

H. Altenbach, T. Sadowski (Eds.), *Failure and Damage Analysis of Advanced Materials*,
CISM International Centre for Mechanical Sciences
DOI 10.1007/978-3-7091-1835-1_5

specific knowledge of the microstructure is lost. Although through certain micromechanics analyses it is possible to reflect the microstructure in directional and symmetry properties in an average sense, the details of microstructure that play roles in initiating and developing failure mechanisms are not present in these analyses to determine the governing conditions for failure. An alternative and appropriate way is to conduct failure analysis at the local level and carry its outcome to description of failure conditions at the macro level. This chapter will be aimed at the future development of this alternative approach.

The outline of the chapter is as follows. The conventional failure theories for composite materials will be reviewed first. Focusing on four main theories, the nature of the assumptions made in developing such theories will be scrutinized. The inherent limitations in each theory will be brought out in view of our current understanding of the failure mechanisms and modes in composite materials. Subsequent sections of the chapter will then discuss how the present situation can be remedied. Three specific remedies will be proposed and in each case the ways to develop the remedies will be outlined. The final section will propose a comprehensive scheme that will integrate all three remedies into a rational failure analysis in the future.

2 Conventional Failure Theories for Composite Materials

2.1 Tsai-Hill Failure Theory

Azzi and Tsai (1965) put forth this theory for unidirectional (UD) fiber-reinforced composites. Their starting point was Hill's proposed criterion for yielding in metals of orthotropic symmetry (Hill, 1948). This criterion was proposed to account for directionally dependent yield stress in metals that have been drawn by rolling in one direction. By examining Fig. 1, which sketches a plate with its rolling direction aligned with x-axis, it is clear that the three coordinate planes x-y, y-z, and x-z are each planes of symmetry, thus rendering the rolled plate orthotropic. Although Hill (1948) treated yielding as well as the plastic deformation behavior of an orthotropic solid from initiation of yielding onwards, Azzi and Tsai (1965) restricted themselves only to Hill's proposed criterion for initiation of yielding. This criterion was written as a generalization of the Huber-von Mises criterion for yielding of isotropic solids, which in the x-y-z coordinate system can be expressed as

$$(\sigma_y - \sigma_z)^2 + (\sigma_z - \sigma_x)^2 + (\sigma_x - \sigma_y)^2 + 6(\tau_{yz}^2 + \tau_{xz}^2 + \tau_{xy}^2) = 2Y_0^2 \qquad (1)$$

where Y_0 is the yield stress and σ and τ denote normal and shear stresses,

Figure 1. Illustration of planes of symmetry in a metal plate drawn in the rolling direction aligned with the x-axis

respectively.

Hill generalized this expression for orthotropic solids by replacing $1/2Y_0^2$ for the isotropic case by six multiplying constants, one for each of the terms on the l.h.s of Eq. (1), thus obtaining

$$F(\sigma_y-\sigma_z)^2+G(\sigma_z-\sigma_x)^2+H(\sigma_x-\sigma_y)^2+2L\tau_{yz}^2+2M\tau_{xz}^2+2N\tau_{xy}^2=1 \quad (2)$$

The constants F, G and$H$ can be related to the physical yield stresses X, Y and Z, in the x-, y- and z-directions, respectively, and the constants L, M and N can similarly be related to the yield stresses R, S and T, in shear in the planes y-z, x-z and x-y, respectively. These relations, given by Hill (1948) are as follows

$$\begin{aligned} \frac{1}{X^2}&=G+H, \quad \frac{1}{Y^2}=H+F, \quad \frac{1}{Z^2}=F+G,\\ \frac{1}{R^2}&=2L, \quad\quad \frac{1}{S^2}=2M, \quad\quad \frac{1}{T^2}=2N \end{aligned} \quad (3)$$

It is to be noted that the yield stress values are assumed to be the same in tension and compression, i.e. the Bauschinger effect is assumed not to exist.

At this point, it is worth noting that the yield criterion for orthotropic solids, Eq. (2), cannot be derived from the distortional energy density, as pointed out by Hill (1948), due to the shear strains resulting from hydrostatic pressure. Thus, the generalization of the Huber-von Mises isotropic yield criterion to orthotropic solids, Eq. (2), is not valid in the energy sense. This observation has fundamental implication: While the isotropic

Huber-von Mises yield criterion is supported by an energy concept, this is not the case for Hill's orthotropic yield criterion. In other words, the orthotropic criterion is a stress-based formulation derived as a mathematical generalization of the isotropic case.

In spite of the loss of the Hill criterion's connection to a single critical value of the distortional energy density, the criterion still represents a single mechanism of yielding, although its criticality is now different for loading in the three principal directions and in the three principal planes. We shall return to the implication of this behavior after we describe the Tsai-Hill criterion.

Azzi and Tsai (1965) adapted the Hill criterion, Eq. (2), to "strength" of a UD composite by making the following assumptions.

1. The yield stresses in the Hill criterion can be replaced by critical stresses representing load-bearing capacities (strengths), such that X, Y and Z are the normal stresses to failure in the x-, y- and z-directions, respectively, and R, S and T are the shear stresses to failure in the y-z, x-z and x-y planes, respectively.
2. In the cross-sectional plane the UD composite can be assumed to be isotropic, giving $Y = Z$.

Assuming further that the UD composites are often used as thin layers (plies) in laminates, thereby sufficing to consider in-plane stresses in the x-y plane, Azzi and Tsai (1965) reduced the Hill criterion, Eq. (2), to their proposed failure criterion as

$$\frac{\sigma_1}{X} - \frac{\sigma_1 \sigma_2}{X} + \frac{\sigma_2}{Y} + \frac{\tau_{12}}{T} = 1 \tag{4}$$

where the subscripts x and y of the stress components in the Hill criterion, Eq. (2), are changed to 1 and 2, respectively.

Let us examine assumption 1 stated above and its repercussions. Figure 2 gives a graphical representation of this assumption and what it implies. In each of the three basic loading modes, tension (or compression) along fibers, tension (or compression) across fibers, and shear in the plane of a UD ply, it is implied that the composite ply behaves in a linear elastic manner, terminating the deformation in an abrupt loss of the load-bearing capacity at critical states, $\sigma_1 = X$, $\sigma_2 = Y$ and $\tau_{12} = T$. While in the Hill criterion for orthotropic solids, the underlying mechanism for all six critical states was yielding, in the version of the criterion proposed by Azzi and Tsai (1965) the three critical states cannot be assumed to have the same underlying mechanism. Indeed the mechanisms in the three critical states are widely different, as will be discussed below. This implies that in contrast to the Hill criterion for orthotropic solids, its adaptation by Azzi and Tsai (1965)

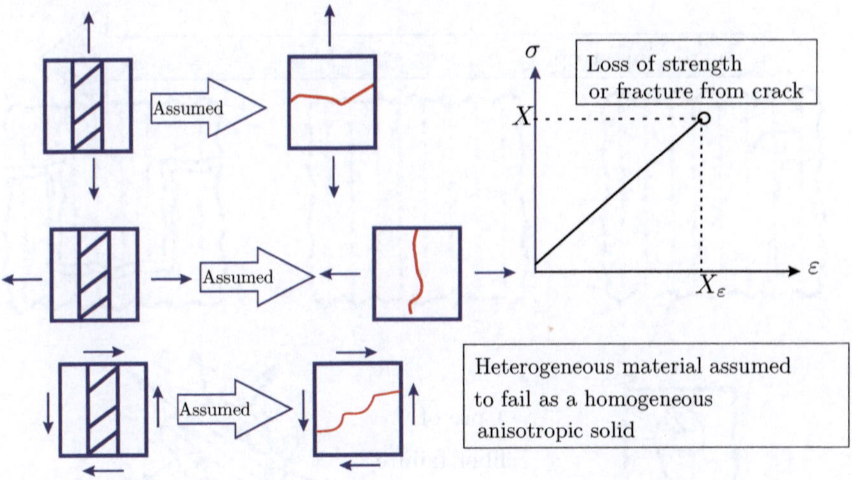

Figure 2. Graphical representation of the assumptions underlying the Tsai-Hill criterion

cannot be reduced to the isotropic case. In other words, the critical yield stress values in the principal directions in the Hill criterion can be equated to obtain the Huber-von Mises criterion, while doing so for the Azzi-Tsai version would not be acceptable on physical grounds because of the different underlying failure mechanisms.

Before we get into the specifics of the failure mechanisms in UD composites, it is noted that the failure modes in tension and compression in the principal directions are different, and therefore, the critical values of the stresses in tension and compression are also different. Since the Bauschinger effect was neglected in the Hill criterion, it was also not present in the Azzi-Tsai version. Assuming a way is found to correct for this, the fact that the mechanisms are different in the two principal directions is sufficient to question the validity of the Azzi-Tsai version from the point of view of its reduction to the isotropic case.

We shall next examine the implications of the Azzi-Tsai version in terms of its representation of the interactive effects of combined stresses in initiating failure in UD composites. For this, we shall consider two stresses, namely tensile axial stress σ_1 and tensile transverse stress σ_2.

Figure 3 gives a schematic depiction of the failure process in UD composites subjected to axial tension. As the force is increased from zero, a few fibers fail initially at their weakest points. These points are randomly

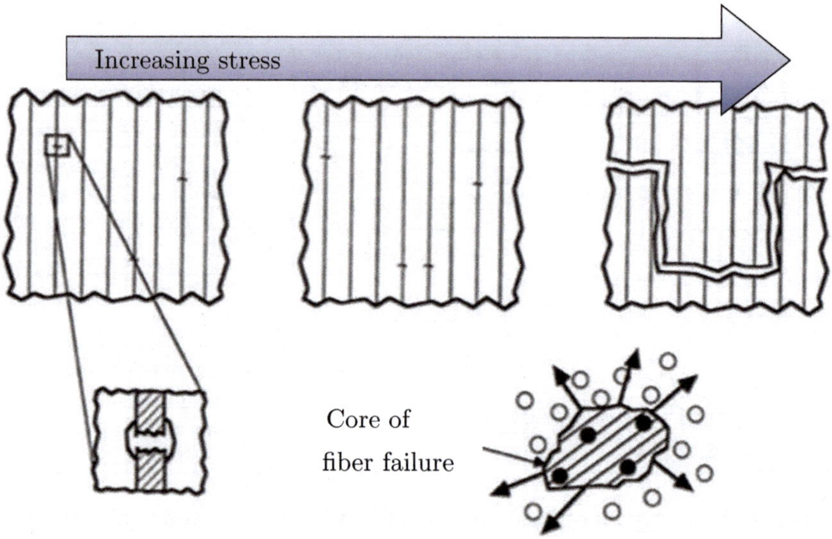

Figure 3. The stochastic process of fiber failures leading to failure of a UD composite in axial tension

distributed in the volume of the composite. The fibers also debond from the matrix locally at the broken fiber ends. As the load is increased further, the previously broken fibers redistribute stresses in the regions surrounding the broken fiber ends, influencing the failure of neighboring fibers. Since the weak points along the length of a fiber are randomly distributed, the process of fiber failures near the stress concentration regions of the previously broken fibers progresses stochastically. Final failure results when a cluster of broken fibers forms a crack that grows unstably. This event is of random nature and as a consequence the value of the applied stress to failure, $\sigma_1 = X$, is not deterministic and can only be described by statistical methods.

Consider next applying a transverse stress σ_2 to a UD composite. To understand initiation of failure we must examine the local stresses developed in the matrix. It can generally be said that these stresses will be triaxial and non-uniformly distributed. Since the fibers in a cross-section are randomly distributed in a practical composite, the initiation of failure will occur when a certain failure condition is satisfied at a stressed point. It is conceivable that the initiation of failure will occur at a point on the fiber/matrix interface since this surface is a potential weak plane. There is experimental

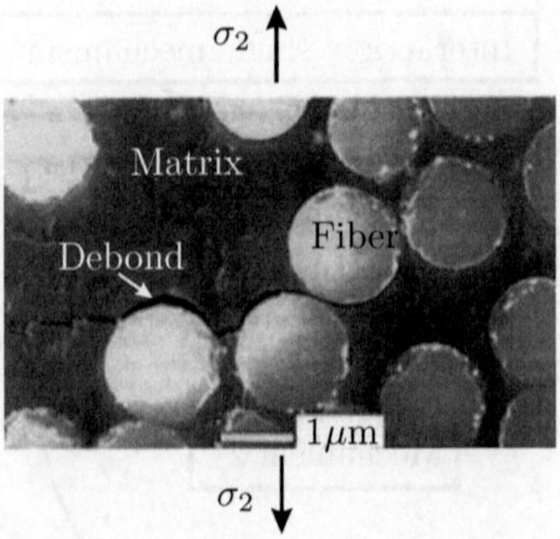

Figure 4. Fiber/matrix debonding in a UD composite subjected to transverse tensile stress (Wood and Bradley, 1997)

evidence in several polymer matrix composites that fiber/matrix debonding occurs; see Fig. 4 as an example taken from Wood and Bradley (1997). Assuming the debonding mechanism to be the first failure mechanism, final failure of the composite will result when a sufficiently large crack from merger of the debond cracks has formed and has grown unstably. While other mechanisms may also be possible for the failure from a transverse stress, for the sake of our purpose to discuss the interaction of σ_1 and σ_2 it will suffice to consider the fiber/matrix debonding.

Consider now the case of a single mechanism underlying failure, such as yielding of a rolled metal sheet, for which Hill (1948) proposed the failure criterion, Eq. (2). Let two tensile stresses, $\sigma_x = \sigma_1$ and $\sigma_y = \sigma_2$ be applied. Eq. (2) then reduces to

$$H\sigma_1^2 - 2H\sigma_1\sigma_2 + (F+H)\sigma_2^2 = 1 \tag{5}$$

This is a quadratic equation illustrated in Fig. 5 by the solid curve. The interaction of the two applied stresses in causing yielding, the underlying failure mechanism, is represented by this equation in a manner consistent with the generalization of isotropic yielding assumed by Hill (1948).

Replacing the yield stresses X and Y, which are related to the constants F and H given by Eq. (3), (with $Y = Z$, as assumed by Azzi and Tsai (1965), by failure stresses corresponding to two different mechanisms renders the quadratic interaction curve in Fig. 5 invalid. If X corresponds to the fiber failure mechanism described above (Fig. 3), and if Y stands for failure from debonding (Fig. 4), then the interaction described by the Azzi-Tsai Eq. (4) is meaningless. Only one of the two failure mechanisms can be critical, not

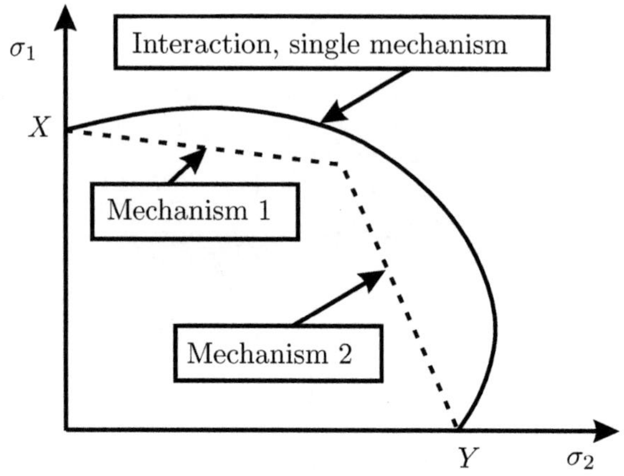

Figure 5. Interaction of two stresses for failure caused by one and two mechanisms

both at the same time. Thus, assuming $\sigma_1 = X$ is critical, the presence of σ_2 could conceivably change (reduce) the failure stress X, until σ_2 itself is large enough to cause failure by debonding, when the criticality switches to this mechanism. The critical mechanism given by $\sigma_2 = Y$ then could similarly be influenced by the presence of σ_1. Figure 5 illustrates the effects on the failure stresses of the two mechanisms, named here mechanism 1 and mechanism 2, by assumed linear reductions of the failure stresses caused by combined application of σ_1 and σ_2. Noteworthy is the discontinuity at the intersection of the relations describing failure stresses of the two mechanisms.

2.2 Tsai-Wu Failure Theory

Tsai and Wu (1971) took their starting point in strength criteria formulation for anisotropic solids put forward by Goldenblat and Kopnov (1965), who proposed that such criteria should be invariant w.r.t. coordinate transformation, and formulated a very general scalar function of stress components as

$$(F_{ij}\sigma_{ij})^{\alpha} + (F_{ijkl}\sigma_{ij}\sigma_{kl})^{\beta} + (F_{ijklmn}\sigma_{ij}\sigma_{kl}\sigma_{mn})^{\gamma} + ... = 1 \tag{6}$$

where F_{ij}, etc. are tensor-valued strength coefficients and α, β, γ are material constants.

For UD composites Tsai and Wu (1971) reduced Eq. (6) to a simpler form given by

$$F_p\sigma_p + F_{pq}\sigma_p\sigma_q = 1 \tag{7}$$

where $p = q = 1$, 2 and 6. Here up to quadratic terms in Eq. (6) are retained and $\alpha = \beta = 1$ is assumed. Also, assuming composite plies are thin, only in-plane stresses are considered, for which the compact Voigt notation for stresses is used, i.e. $\sigma_1 = \sigma_{11}$, $\sigma_2 = \sigma_{22}$, and $\sigma_6 = \sigma_{12} = \sigma_{21}$.

Equation (6) represents a quadric surface in the $(\sigma_1, \sigma_2, \sigma_6)$ coordinate system. Generally, it can describe 17 possible surfaces. Keeping to surfaces of real-valued roots along any radial stress path, only ellipsoids and elliptical paraboloids are possible. The latter are not acceptable for finite strength in all stress states. Finally, for real ellipsoids the following conditions must be satisfied by the strength coefficients in Eq. (7)

$$\begin{aligned} F_{11}F_{22} - F_{12}^2 > 0 \\ F_{21}F_{66} - F_{26}^2 > 0 \\ F_{11}F_{66} - F_{16}^2 > 0 \end{aligned} \tag{8}$$

The geometrical interpretation of the strength coefficients is as follows. F_1, F_2 and F_6 give the center coordinates of the ellipsoid, F_{11}, F_{22} and F_{66} determine the size of the ellipsoid, and F_{12}, F_{16} and F_{26} provide the inclinations of the ellipsoid w.r.t. the planes $\sigma_1\sigma_2$, $\sigma_1\sigma_6$ and $\sigma_2\sigma_6$, respectively.

Aligning the coordinate axes x_1 and x_2 with directions parallel and normal to the fibers in the plane of the composite, respectively, it can be seen that the orthotropic symmetry implies same strength in positive and negative shear stress σ_6. This results in the strength coefficients

$$F_6 = F_{16} = F_{26} = 0$$

Equation (7) now reduces to

$$F_1\sigma_1 + F_2\sigma_2 + 2F_{12}\sigma_1\sigma_2 + F_{22}\sigma_2^2 + F_{66}\sigma_6^2 = 1 \tag{9}$$

The strength coefficients F_1 and F_2 are nonzero if the strength values in tension and compression are different in the axial and transverse directions. These coefficients and the corresponding coefficients F_{11} and F_{22} can be expressed in terms of the strengths as

$$\begin{aligned} F_1 &= \frac{X' - X}{XX'}, \\ F_{11} &= \frac{1}{XX'}, \\ F_2 &= \frac{Y' - Y}{YY'}, \\ F_{22} &= \frac{1}{YY'} \end{aligned} \tag{10}$$

where the primed quantities are compression strength values.

As noted above, the shear strength does not depend on the sign of the shear stress σ_6 and therefore the coefficient F_{66} is simply $1/T^2$, where T, as before, is the in-plane shear strength.

The remaining strength coefficient F_{12} cannot be determined uniquely. Its geometrical interpretation, as noted above, is the amount of inclination the ellipsoid has w.r.t. the $\sigma_1\sigma_2$ coordinate plane. One possible way to determine its value is by setting the shear stress $\sigma_6 = 0$ and keeping a constant normal stress ratio $\sigma_2/\sigma_1 = B$. Then, from Eqs. (9) and (10), one obtains

$$F_{12} = \frac{1}{BP^2}\left[1 - P\left(\frac{1}{X} - \frac{1}{X'} + \frac{B}{Y} - \frac{B}{Y'}\right) - P^2\left(\frac{1}{XX'} + \frac{B^2}{YY'}\right)\right] \tag{11}$$

where P is the strength under the biaxiality ratio B. Obviously, the dependency of the strength coefficient F_{12} on B makes it non-unique. Performing a biaxial test on a UD composite is also not an easy task. Tsai and Wu (1971) proposed various practical ways of testing to estimate the value of the strength coefficient.

The non-unique value of F_{12} is, however, restricted by the convex surface requirement given by Eq. (8). Defining a dimensionless form of the coefficient as

$$F_{12}^* = \frac{F_{12}}{\sqrt{F_{11}F_{22}}} \tag{12}$$

from the first of the set of Eqs. (8) one gets

$$-1 < F_{12}^* < 1 \tag{13}$$

It should be noted that any value of this dimensionless coefficient within the range (-1, 1) assures only that the ellipsoid describing the strength envelope is a real, closed surface. A wide range of ellipsoids will thus be admissible.

As an example, if the value of the dimensionless coefficient is -0.5, then the resulting ellipsoid represents a mathematical generalization of the Huber-von Mises yield criterion proposed by Hill (1948) when reduced to in-plane stresses.

As a summarizing remark on the Tsai-Wu failure theory it can be stated that it is an effective scheme to describe limiting stress states (strength) of orthotropic solids. However, the strength coefficient F_{12} while allowing flexibility in selection of the ellipsoid also induces ambiguity since there is no physical principle to guide its selection. In fact there is no physics underlying the failure theory other than the implication that a single-valued limiting stress state called strength exists. The representation of the limiting stress state by a single, smooth ellipsoid by the theory implies also that there is a single underlying failure mechanism leading to the limiting stress condition of failure. As discussed above, in UD composites this is not the case.

2.3 Hashin's Failure Theory

Hashin (1980) motivated his failure theory for UD fiber-reinforced composites by observing the inherent difficulties in the Tsai-Wu theory when applied to these materials. Arguing that the limiting stress states represented by the Tsai-Wu ellipsoid cannot be valid when widely different failure modes operate, Hashin (1980) proposed instead to incorporate these modes by a piecewise smooth representation. To capture the attractive feature of invariance w.r.t. coordinate transformation, present in the Tsai-Wu formulation via tensor-valued strength coefficients, Hashin proposed instead to formulate the failure conditions using stress invariants. He chose these invariants for transversely isotropic symmetry arguing that the fiber distribution in the UD composite cross-section is usually random.

Based on the available knowledge then, Hashin (1980) recognized that failure under a uniaxial stress σ_1 occurred by fiber failure, while the stress σ_2 caused failure (cracking) in the matrix. He viewed the role of the shear stress σ_6 in the combined stress application as influencing the fiber and matrix failure modes. For failure in matrix he resorted to the concept of a "failure plane", whose inclination in the matrix was governed by the normal and shear stresses acting on the plane. This notion of an inclined plane of failure comes from failure in soils, originating in the classical work of Coulomb (1776), where it was assumed that failure was the outcome of overcoming cohesion and internal friction in isotropic soils.

Leaving full details to Hashin (1980), the thrust of his theory will be summarized here. To begin, the formulation of limit states is in terms of the stress invariants for transversely isotropic solids with the cross-sectional plane of a UD composite as the plane of isotropy. Using the stress components in the coordinate axis x_1 aligned with the fiber direction, x_2 axis normal to fibers in the plane of a UD composite layer and x_3 axis in the thickness direction of the layer, the stress invariants for transverse isotropic symmetry are

$$\begin{aligned} I_1 &= \sigma_{11}, \\ I_2 &= \sigma_{22} + \sigma_{33}, \\ I_3 &= \sigma_{23}^2 + \sigma_{22}\sigma_{33}, \\ I_4 &= \sigma_{12}^2 + \sigma_{13}^2, \\ I_5 &= 2\sigma_{12}\sigma_{23}\sigma_{13} - \sigma_{22}\sigma_{13}^2 - \sigma_{33}\sigma_{12}^2 \end{aligned} \tag{14}$$

A general quadratic polynomial in stress components constructed from these invariants will have the following form (Hashin, 1980)

$$A_1 I_1 + B_1 I_1^2 + A_2 I_2 + B_2 I_2^2 + C_{12} I_1 I_2 + A_3 I_3 + A_4 I_4 = 1 \tag{15}$$

Considering only in-plane stresses in a UD composite, and using the Voigt notation, Eq. (15) can be reduced to the following form

$$A_1\sigma_1 + B_1\sigma_1^2 + (A_2 - A_3)\sigma_2 + B_2\sigma_2^2 + C_{12}\sigma_1\sigma_2 + A_4\sigma_6^2 = 1 \tag{16}$$

It is noted here that this equation is identical to the Tsai-Wu failure criterion, Eq. (9), when the constants in the two equations are equated as follows

$$F_1 = A_1, F_2 = A_2 - A_3, F_{11} = B_1, F_{22} = B_2, 2F_{12} = C_{12}, F_{66} = A_4 \tag{17}$$

Thus, using a tensor polynomial, as done by Tsai and Wu (1971), or a polynomial with stress invariants, as done by Hashin (1980), leads to the same result for the case of UD composites with in-plane stresses.

Further development of Eq. (16) by Hashin (1980) differs from that of Tsai and Wu (1971) in that Hashin makes a set of assumptions concerning failure modes to adapt Eq. (16) to those failure modes, while Tsai and Wu treat only experimental determination of the strength coefficients.

For the fiber failure mode, Hashin (1980) separates failure in tension from that in compression. For tension failure mode he assumes influence of the shear stress such that the interaction of the axial stress σ_1 and the shear stress σ_6 can be represented by a quadratic curve given by

$$F_{11}\sigma_1^2 + F_{66}\sigma_6^2 = 1 \tag{18}$$

where the Tsai-Wu strength coefficients are used. It is noted that Hashin (1980) neglects the linear term in σ_1 present in Eq. (16). In terms of the composite strength values, the criterion for fiber failure mode, Eq. (18), can be written as

$$\left(\frac{\sigma_1}{X}\right)^2 + \left(\frac{\sigma_6}{T}\right)^2 = 1 \tag{19}$$

Not being sure of this criterion, Hashin (1980) suggested to simply use the maximum stress criterion $\sigma_1 = X$ that carries no influence of the shear stress σ_6.

For fiber failure mode in compression, Hashin (1980) admitted that there was not a clear understanding of the shear stress effect on this failure mode. Again, he suggested using the maximum stress criterion, $\sigma_1 = X'$.

For matrix failure modes, Hashin (1980) assumed a failure plane to be in the matrix without cutting through fibers. Such a plane must be parallel to fibers, but can be generally inclined w.r.t. a reference axial plane. Assuming the angle θ to describe this inclination, the failure function can be expressed as

$$f(\sigma_2, \sigma_6, \theta) = 1 \tag{20}$$

Hashin (1980) suggested that the value of θ that maximizes the function f would give the inclination of the failure plane. However, he did not offer a procedure for determining that value. Instead, he proposed the following matrix failure criteria.

For tensile matrix failure, Hashin (1980) suggested a quadratic interaction between σ_2 and σ_6, as for the tensile fiber failure mode, giving

$$\left(\frac{\sigma_2}{Y}\right)^2 + \left(\frac{\sigma_6}{T}\right)^2 = 1 \tag{21}$$

After elaborate arguments concerning the effect of shear stress on the matrix failure mode in compression, Hashin (1980) formulated the following criterion that requires distinguishing between shear strength in the axial and transverse directions

$$\left(\frac{\sigma_1}{2T'}\right)^2 + \left[\left(\frac{Y'}{2T'}\right)^2 - 1\right]\frac{\sigma_2}{Y'} + \frac{\sigma_6{}^2}{T} = 1 \tag{22}$$

where the prime on Y denotes the strength in compression and that on T indicates the transverse shear strength.

It is noted that no practical method exists to experimentally determine the "transverse" shear strength. Applying a pure shear stress in the plane of a UD composite, e.g. by torsion on a thin tube with fibers running in the circumferential direction, produces failure along fibers.

It would be correct to recognize that Hashin (1980) was the first to introduce physical considerations in developing a failure theory for fiber-reinforced composites. As we shall see later, the inherent difficulty in incorporating knowledge of failure mechanisms into a formulation that operates on homogenized solids still cannot be overcome.

2.4 Puck's Failure Theory

Puck (1992) followed the failure theory framework of Hashin (1980) and proposed an elaborate scheme for implementation of his theory in Puck and Schürmann (1998). A lot of new terminology and symbols have been introduced in Puck's theory, but we will try to keep to the conventional forms for convenience of relating the theory to previous works.

As in Hashin (1980), Puck's theory (Puck, 1992; Puck and Schürmann, 1998) recognizes failure in UD composites to be in fiber failure and matrix failure modes. The latter was renamed by Puck as inter fiber failure mode, but we shall continue to refer to it as matrix failure mode, understanding that at a point in the composite this failure mode is restricted to the matrix. Furthermore, as proposed by Hashin (1980), the matrix failure is assumed to occur on a plane of inclination θ w.r.t. the axial plane x_1x_3. As noted above, Hashin (1980) proposed that this angle would maximize a failure function, Eq. (20). However, he did not propose the form of this function.

Puck's theory is largely concerned with formulation of failure conditions for the assumed potential failure plane whose inclination is given by the angle θ. These conditions are guided by the original idea of Coulomb (1776), and later developed by Mohr (1882); Paul (1961), that failure on a plane occurs when certain resistances, related to its cohesion and internal friction, are overcome. The nature of this failure is brittle, as no inelastic deformation in the failure process is involved. In fact in Coulomb's failure model, failure is seen as slipping of two rigid blocks along the inclined plane.

In Puck's theory the failure conditions for the matrix failure plane are based on the traction vector on the plane. This vector is resolved in directions normal and tangential to the failure plane, denoted, respectively, by σ_n and $\tau_{n\psi}$, which is further resolved in a longitudinal component τ_{nl} and a transverse component τ_{nt} (Fig. 6). Following Coulomb's basic idea, failure on a plane is assumed to occur when "stressing" of a plane either by the action of a single traction component, or by the combined action of the traction components reaches a limit state. The maximum values reached by individual traction components at the limit (failure) states are viewed as the "resistances" of the matrix material (in the presence of fibers) to failure. These resistances are denoted in Puck's papers by R with superscripts $+$ or

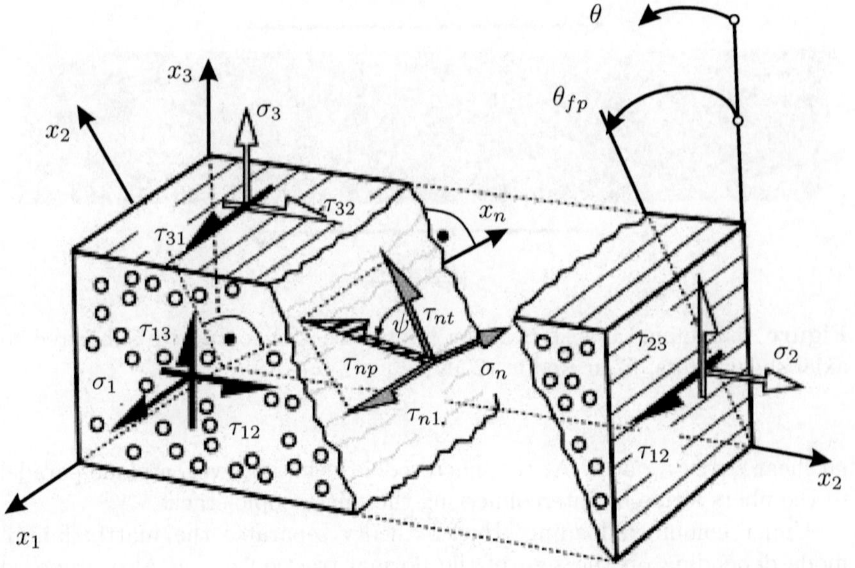

Figure 6. The assumed matrix failure plane inclined at the angle θ_{fp} with tractions σ_n, τ_{nl} and τ_{nt} acting on the plane. Stresses acting on the UD composite are also indicated, from Puck et al. (2002)

- assigned for tension or compression, respectively, and subscripts indicating resistances to the corresponding traction component acting individually. Thus, four resistances ("basic strengths") are attributed to a failure plane. Two of these, namely, tensile and compressive strengths caused by $\sigma_n > 0$ and $\sigma_n < 0$, respectively, are the usual strengths denoted conventionally as Y and Y', respectively, e.g. in Eqs. (21) and (22) above. The remaining two resistances, given by the limit state values of τ_{nl} and τ_{nt}, are the conventional "axial" shear strength and "transverse" shear strength, respectively. These appear as T and T', respectively, in Hashin's Eq. (22) above. As noted in remarks above related to that equation, there is no practical way of experimentally determining T' since applying pure shear in the plane of a UD composite invariably results in failure along fibers, i.e. at the limit state value T. As noted by Puck and Schürmann (1998), and observed experimentally by Redon (2000), Fig. 7, the failure in axial shear is actually along planes inclined to the fiber direction at a microscopic scale. Thus, microscopically, the fracture resistance is in tension normal to the inclined planes that are rotated w.r.t. the inclined failure plane along which the ax-

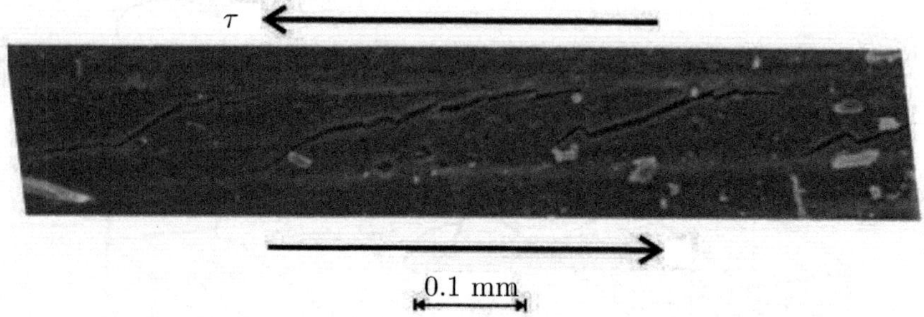

Figure 7. Inclined cracks between fibers in a UD composite subjected to axial shear stress, from Redon (2000)

ial shear stress τ_{nl} acts. At the macroscopic level, however, a plane parallel to the fibers forms by interconnecting the microscopic cracks.

Under combined loading, Puck's theory separates the matrix failure mode depending on the sign of the normal traction σ_n, as also proposed by Hashin (1980). However, while Hashin proposed Eqs. (21) and (22) for tension and compression on the failure plane, respectively, Puck's theory introduces an additional failure mode at high values of compressive stress σ_2. Furthermore, at points corresponding to switchover from one failure mode to another, Puck's theory introduces tangents on the failure envelope as additional parameters. These four parameters (Fig. 8), called "inclination" parameters, are the additional four parameters, which are needed to complete the failure description.

The four inclination parameters will be denoted by simpler symbols here than those given in papers by Puck and associates. Thus, referring to Fig. 8, the p-parameter indicating tangent to the (σ_n, τ_{nl}) curve on the tension side will be denoted p_1 and the one on the compression side will be denoted p_2. Similarly, the parameters corresponding to the (σ_n, τ_{nt}) curve will be named p_3 and p_4. The failure condition proposed in Puck's theory for the combined application of $\sigma_2 > 0$ and σ_6, called mode A, where failure occurs parallel to the x_1x_3 plane, i.e. $\theta = 0$, is

$$\left[\left(\frac{\sigma_6}{T}\right)^2 + \left(1 - p_1\frac{Y}{T}\right)^2 \left(\frac{\sigma_2}{Y}\right)^2\right]^{1/2} + p_1\frac{\sigma_2}{T}^2 = 1 \qquad (23)$$

It can be noted that this failure condition coincides with that of Hashin's, Eq. (21), for $p_1 = 0$.

For combined stresses $\sigma_2 < 0$ and σ_6, called mode B in Puck's theory,

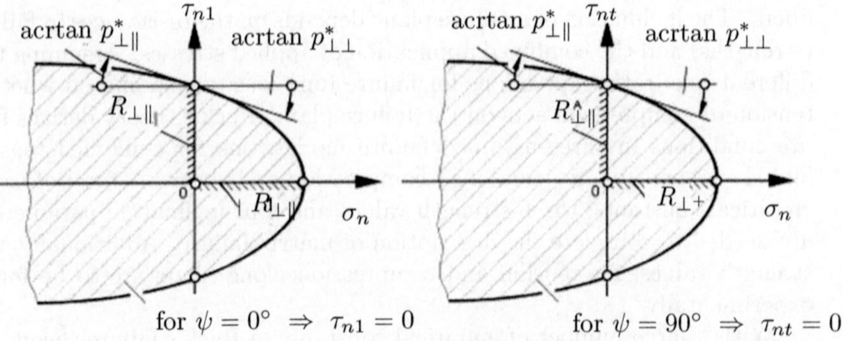

Figure 8. Master failure envelopes indicating the four inclination parameters (Puck et al., 2002)

where failure occurs on planes parallel to the x_1x_3 plane, i.e. $\theta = 0$, as in mode A, the failure condition is given by

$$\left[\left(\frac{\sigma_6}{T}\right)^2 + \left(p_2\frac{\sigma_2}{T}\right)^2\right]^{1/2} + p_2\frac{\sigma_2}{T} = 1 \tag{24}$$

Puck et al. (2002) recommend values of p_2 between 0.2 and 0.3 based on test data for glass/epoxy and carbon/epoxy composites. Squaring both sides of Eq. (24) and neglecting quadratic terms in p_2 gives

$$\left(\frac{\sigma_6}{T}\right)^2 + 2p_2\frac{\sigma_2\sigma_6}{T^2} = 1 \tag{25}$$

This equation expresses approximately the beneficial effect of the compressive stress $\sigma_2 < 0$ on failure caused by the in-plane shear stress σ_6.

Finally, for combined stresses $\sigma_2 < 0$ and σ_6, where the failure plane angle $\theta > 0$, called mode C in Puck's theory, the proposed failure condition is

$$\left[\frac{\sigma_6}{2(1+p_4)T}\right]^2 + \left(\frac{\sigma_2}{Y'}\right)^2 + \frac{\sigma_2}{Y'} = 0 \tag{26}$$

The inclination parameter p_4 appearing in this equation depends on the failure plane angle, which is difficult to determine experimentally (Puck et al., 2002). It is noted that for $\sigma_6 = 0$, Eq. (26) gives the solution $\sigma_2 = -Y'$, the compressive strength transverse to fibers.

In summarizing Puck's theory for UD composite layers, it can be stated that the foundation of the theory lies in the assumption that failure in the matrix (between fibers) occurs on certain planes that are parallel to the fibers. The inclination of a failure plane depends on the resistances to failure (strengths) and the combined application of applied stresses. Assuming two different quadratic expressions for failure functions, depending on whether tension or compression acts on the failure plane, Puck's theory derives failure conditions in three assumed failure modes, one for combined tension and shear, and two for combined compression and shear. A total of seven empirical constants, three strength values and four inclination parameters, are needed to complete the description of matrix failure. Additionally, two strength values, for tension and compression along fibers are to be found experimentally.

With a large number of empirical constants in Puck's failure theory, its ability to describe failure data is better than all previous failure theories. However, some of the seven constants associated with failure in the matrix are difficult to determine, even for a UD composite layer.

3 Limitations of Phenomenological Failure Theories

The four failure theories reviewed above represent the essential features of phenomenological approaches to describing limit states for failure in UD composites. Common to all such theories is the objective of characterizing the limit states by envelopes (surfaces in stress component space) such that stress states at a point in the composite can be kept within the envelopes to avoid failure. The Azzi-Tsai adaptation of Hill's yield criterion for orthotropic layers has limit states that are not applicable to composites, as discussed above. The Tsai-Wu formulation of limit states is completely geometrical with no particular reference to failure in composite materials. It also implicitly assumes a single underlying mechanism in all combinations of stresses prior to the limit states. Hashin's theory, and later Puck's theory, make failure mode distinctions as fiber failure and matrix failure, and further qualify these depending on tension or compression.

An inherent limitation in the phenomenological failure theories comes from their formulation on homogeneous anisotropic solids. Specifically for UD composites, the anisotropy takes the form of orthotropic or transversely isotropic symmetry, if the fiber direction is known. Then the failure can be attributed to fibers or matrix, and accordingly separated in modes. However, due to the homogenized composite description, the failure modes cannot be analyzed. For instance, the statistical nature of the fiber failure mode in tension cannot be taken into account since the statistical properties of fiber strength are not included in the homogenized description of the composite. The effects of fibers on failure initiation and progression in the

matrix and at the fiber/matrix interfaces also cannot be specifically taken into account. These effects are known to depend on the local stress fields in the matrix and at the fiber/matrix interfaces. When the composite has been homogenized, the information needed to determine these stress fields is lost. Any attempt to account for the local effects in the phenomenological theories therefore leads to assumptions, resulting in unknowns, which can only be found indirectly via testing.
The phenomenological theories operate on limit states (strengths) that are end values of stresses at final failure. Thus, initiation of failure and its progression to total loss of load carrying capacity (strength) are not part of the description in these theories. The entire failure path is determined by the local stress fields, which are not available in the phenomenological failure theories.
The limit states in the phenomenological failure theories are single failure events. UD composites of high fiber volume fraction, e.g. those made with glass/epoxy and carbon/epoxy in pre-impregnated form (prepregs), do show abrupt failure that appears as brittle fracture. However, when these UD layers are stacked to form laminates, their failure shows significant progression. In fact a variety of failure events occurs in laminates that cannot be described in terms of the limit state values (strengths) of UD layers. The failure mechanisms in laminates, often described as damage because of their significant progression, have been studied since the late 1970s. The vast literature in this field, known as damage mechanics, deals with the initiation, progression and criticality of damage as well as the effects produced on the thermo-mechanical response of laminates (Talreja, 1994; Talreja and Singh, 2012). From structural applications points of view, the response characteristics of a composite structure can be degraded to unacceptable levels much before the limit states given by conventional strengths are reached. The usefulness of the phenomenological failure theories then becomes questionable.

Another limitation of the phenomenological failure theories lies in their inability to account for manufacturing defects that are inevitable in practical composite structures. In recent years, the composite structural applications have increased in non-aerospace fields such as wind turbine blades and automotive structures, where cost requirements do not allow high levels of quality control of manufacturing processes and limit in-service inspection. The importance of accounting for manufacturing defects in the design phase has therefore become vital. There is ample evidence that manufacturing defects influence failure initiation and progression. One example of the effect of manufacturing induced voids on fiber/matrix debonding is shown in Fig. 9, taken from Wood and Bradley (1997). Such voids were difficult to observe

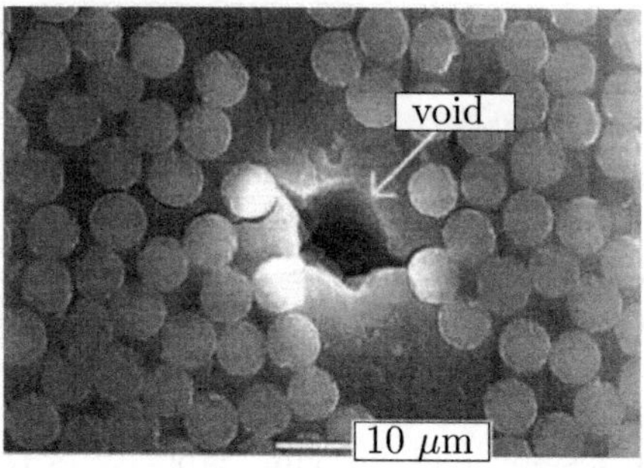

Figure 9. The presence of a void in the matrix initiates fiber/matrix debonding under transverse loading of a UD composite (Wood and Bradley, 1997)

in the past, but can now be characterized by micro computed tomography (Lambert et al., 2012). The detailed information concerning the size, shape and location of voids makes it possible to analyze their effects on failure initiation and progression at microstructure levels considering the local stress fields (Huang and Talreja, 2005; Chowdhury et al., 2008; Ricotta et al., 2008). A failure analysis strategy that can reduce manufacturing cost by evaluating the effects of defects on structural performance is now feasible (Talreja, 2009, 2013).

4 A Comprehensive Failure Assessment Scheme for Composite Laminates

To overcome the limitations of phenomenological failure theories for composite laminates following three remedies are proposed.

- **Remedy #1: Multi-scale analysis of failure**
  Composite materials are *heterogeneous* solids with distinct interfaces that act as *weak planes* under favorable local stress conditions. Under general imposed loading on a composite, failure of these planes triggers subsequent events of the failure process. Although the failure progression can take different paths depending on the fiber architecture and the nature of the imposed loading, the sequence of the events on those paths often involve linkages of the cracks formed at interfaces and in the matrix, and fiber failures in the later stages. The state of criticality in the performance of the composite depends on the designed function, and it can be reached before what is conventionally

described as failure, i.e. separation of the material in parts seen at the macro level. The intensity of the distributed micro level cracks can, for instance, degrade the deformational characteristics, measured as stiffness properties at the macro level. The composite structure in some applications would be considered to have failed to perform when that occurs to an undesirable level. In other situations, the inability to carry imposed loads will be seen as failure to perform the designed function. In any case, determining the criticality conditions associated with failure requires analyses of the first events of failure at the micro level and their subsequent development leading to macro level failure. Thus the failure prediction necessarily involves a multi-scale analysis.

- **Remedy #2: Analysis of constrained failure**
  Composite materials are designed with selected fiber architecture to meet the needs imposed by the service environment. Thus, as an example, thin plies of UD composites with straight fibers are stacked in different orientations to create laminates. Other more complex fiber architectures are generated by using woven fabrics instead of straight fibers, as dictated by cost and performance requirements. In any case, the failure process is significantly altered by the presence of interfaces between layers containing straight fibers or woven fabrics. Until these interfaces fail, i.e. delamination occurs, the failure process within the layers is subjected to what is described as a mutual "constraint" imposed on each other by the layers with differently oriented fibers. The phenomenological failure theories, described above, are for "unconstrained" UD composites. It is common to apply these theories in a so-called "ply-by-ply" failure analysis of laminates. This ignores the failure progression induced by the ply constraints and is therefore a source of significant errors in the resulting failure predictions.
- **Remedy #3: Analysis of manufacturing defects**
  As noted before, current failure theories for composite materials are formulated on homogenized solids with account made of anisotropy induced by fiber orientations. Real composites, however, contain defects resulting from the particular manufacturing process. These defects either initiate failure or affect the failure initiation from weak sites such as interfaces between constituents and between layers. Traditional approach to analysis of defects has been to embed selected defects into the homogenized solid for assessing their effects. This is inadequate for analyzing the interactions between the defects and the composite microstructure.

A comprehensive failure analysis strategy that integrates the three remedies is depicted in Fig. 10. The initial part of the failure analysis is focused

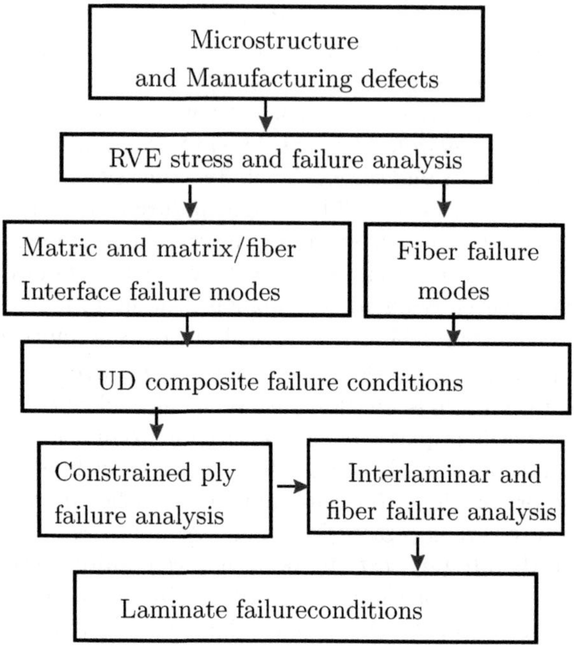

Figure 10. A schematic depiction of a comprehensive failure assessment strategy integrating multi-scale analysis, constrained failure progression and effects of defects

on determining conditions for failure initiation in UD composites. Here one starts by analyzing a representative volume element (RVE) of the UD composite that is constructed based on observations of the microstructure, which invariably has irregularities of fiber distribution as well as defects such as voids in matrix. The local stress fields are calculated and failure conditions are imposed depending on whether the failure mode is due to fiber failure or due to failure in matrix or at the fiber/matrix interfaces. The output of this analysis is expressed in terms of the stress states averaged over the respective RVE. These expressions replace the phenomenological failure criteria and instead represent failure initiation in a UD composite. Further failure analysis concerns the failure progression within UD plies of a given laminate. This failure process consists of multiple ply cracking induced by the constraint on a given ply by its neighboring plies (to which it is bonded). Further progression of damage involved interlaminar cracking (delamination) and fiber failure. For review of these mechanisms and their effects on the laminate mechanical response, see Talreja and Singh (2012).

The final failure assessment of a given laminate is not expected to provide analytical expressions describing failure envelopes. Instead, a generic computational methodology is proposed for failure assessment that will ascertain whether the loading on a composite structure will avoid failure.

5 Conclusion

This chapter has critically examined the current phenomenological failure theories for UD composites for the purpose of assessment of failure in composite structures. Four commonly used theories have been scrutinized. Rather than assessing the ability of any of the theories in curve-fitting test data, their underlying assumptions have been examined. It has been argued that none of the theories can predict the conditions under which failure in a composite laminate will occur. At best these are curve-fitting schemes that require increasing number of empirical constants for improved fit to the data. Although the early such failure theories were formulated when the knowledge about the physical mechanisms of failure initiation and progression in composite materials was meager, today a wealth of such knowledge exists. What's more, the microstructure details of composite materials, including the irregularities and defects induced by manufacturing, can be characterized adequately to conduct representative stress analyses at the local level. Based on these the failure process can be analyzed with computational schemes that allow proper physically based failure analysis. This is the direction in which future work in composite failure should be taken. The phenomenological theories should be gradually phased out until the new failure analysis methodologies are matured.

Bibliography

V.D. Azzi and S. Tsai. Anisotropic strength of composites. *Experimental Mechanics*, 5:283–288, 1965.

K.A. Chowdhury, R. Talreja, and A.A. Benzerga. Effects of manufacturing-induced voids on local failure in polymer-based composites. *Journal of Engineering Materials and Technology*, 130(2):0210101–0210109, 2008.

C.A. Coulomb. Essai sur une application des regles de maximis et minimis a quelques problemes de statique, relatifs a l'architecture. *Memoires de Mathematique de l'academie Royale des Sciences Paris*, 7:343–382, 1776.

I.I. Goldenblat and V.A. Kopnov. Strength criteria for anisotropic materials. *Izvestia Academy Nauk USSR. Mechanika*, (6):77–83, 1965.

Z. Hashin. Failure criteria for unidirectional fiber composites. *Trans ASME. J Applied Mechanics*, 47:329–334, 1980.

R. Hill. A theory of the yielding and plastic flow of anisotropic materials. *Proc Roy Soc A*, 193:281–297, 1948.

H. Huang and R. Talreja. Effects of void geometry on elastic properties of unidirectional fiber reinforced composites. *Comp Sci Tech*, 65:1964–1981, 2005.

J. Lambert, A.R. Chambers, I. Sinclair, and S.M. Spearing. 3d damage characterisation and the role of voids in the fatigue of wind turbine blade materials. *Comp Sci Tech*, 72:337–343, 2012.

O. Mohr. Über die darstellung des spannungszustandes und des deformationszustandes eines körperelementes und über die anwendung derselben in der festigkeitslehre. *Civilingenieur*, XXVIII:113–156, 1882.

B. Paul. A modification of the Coulomb-Mohr theory of fracture. *Trans ASME. J Appl Mech*, 28(2):259–268, 1961.

A. Puck. Ein Bruchkriterium gibt die Richtung an. *Kunststoffe*, 82(7): 607–610, 1992.

A. Puck and H. Schürmann. Failure analysis of frp laminates by means of physically based phenomenological models. *Comp Sci Tech*, 58:1045–1067, 1998.

A. Puck, J. Kopp, and M. Knops. Guidelines for the determination of the parameters in Puck's action plane strength criterion. *Comp Sci Tech*, 62:371–378, 2002.

O. Redon. Fatigue damage development and failure in unidirectional and angle-ply glass fibre/carbon fibre hybrid laminates. Technical Report Risø-R-1168, Risø National Laboratory, Roskilde, Denmark, 2000.

M. Ricotta, M. Quaresimin, and R. Talreja. Mode i strain energy release rate in composite laminates in the presence of voids. *Comp Sci Tech*, 68: 2616–2623, 2008.

R. Talreja, editor. *Damage Mechanics of Composite Materials*. Elsevier, Amsterdam, 1994.

R. Talreja. Defect damage mechanics: broader strategy for performance evaluation of composites. *Plastics, Rubber and Composites*, 38:49–54, 2009.

R. Talreja. Studies on the failure analysis of composite materials with manufacturing defects. *Mechanics of Composite Materials*, 49:35–44, 2013.

R. Talreja and C.V. Singh, editors. *Damage and Failure of Composite Materials*. Cambridge University Press, Cambridge, 2012.

S.W. Tsai and E.M. Wu. A general theory of strength for anisotropic materials. *J. Composite Materials*, 5:58–80, 1971.

C.A. Wood and W.L. Bradley. Determination of the effect of seawater on the interfacial strength of an interlayer e-glass/graphite/epoxy composite by *in situ* observation of transverse cracking in an environmental sem. *Composites Science and Technology*, 57(8):1033–1043, 1997.

GPSR Compliance

The European Union's (EU) General Product Safety Regulation (GPSR) is a set of rules that requires consumer products to be safe and our obligations to ensure this.

If you have any concerns about our products, you can contact us on ProductSafety@springernature.com

In case Publisher is established outside the EU, the EU authorized representative is:

Springer Nature Customer Service Center GmbH
Europaplatz 3
69115 Heidelberg, Germany

Batch number: 10725825

Printed by Printforce, the Netherlands